T0182944

Undergraduate Texts in Physics

Series Editors

Kurt H. Becker, NYU Polytechnic School of Engineering, Brooklyn, NY, USA

Jean-Marc Di Meglio, Matière et Systèmes Complexes, Université Paris Diderot, Bâtiment Condorcet, Paris, France

Sadri D. Hassani, Department of Physics, Loomis Laboratory, University of Illinois at Urbana-Champaign, Urbana, IL, USA

Morten Hjorth-Jensen, Department of Physics, Blindern, University of Oslo, Oslo, Norway

Michael Inglis, Patchogue, NY, USA

Bill Munro, NTT Basic Research Laboratories, Optical Science Laboratories, Atsugi, Kanagawa, Japan

Susan Scott, Department of Quantum Science, Australian National University, Acton, ACT, Australia

Martin Stutzmann, Walter Schottky Institute, Technical University of Munich, Garching, Bayern, Germany

Undergraduate Texts in Physics (UTP) publishes authoritative texts covering topics encountered in a physics undergraduate syllabus. Each title in the series is suitable as an adopted text for undergraduate courses, typically containing practice problems, worked examples, chapter summaries, and suggestions for further reading. UTP titles should provide an exceptionally clear and concise treatment of a subject at undergraduate level, usually based on a successful lecture course. Core and elective subjects are considered for inclusion in UTP.

UTP books will be ideal candidates for course adoption, providing lecturers with a firm basis for development of lecture series, and students with an essential reference for their studies and beyond.

More information about this series at http://www.springer.com/series/15593

Alexander Belyaev • Douglas Ross

The Basics of Nuclear and Particle Physics

 Springer

Alexander Belyaev
Department of Physics and Astronomy
University of Southampton
Southampton, UK

Douglas Ross
Department of Physics and Astronomy
University of Southampton
Southampton, UK

ISSN 2510-411X   ISSN 2510-4128   (electronic)
Undergraduate Texts in Physics
ISBN 978-3-030-80115-1   ISBN 978-3-030-80116-8   (eBook)
https://doi.org/10.1007/978-3-030-80116-8

© The Editor(s) (if applicable) and The Author(s), under exclusive license to Springer Nature Switzerland
AG 2021
This work is subject to copyright. All rights are solely and exclusively licensed by the Publisher, whether
the whole or part of the material is concerned, specifically the rights of translation, reprinting, reuse
of illustrations, recitation, broadcasting, reproduction on microfilms or in any other physical way, and
transmission or information storage and retrieval, electronic adaptation, computer software, or by similar
or dissimilar methodology now known or hereafter developed.
The use of general descriptive names, registered names, trademarks, service marks, etc. in this publication
does not imply, even in the absence of a specific statement, that such names are exempt from the relevant
protective laws and regulations and therefore free for general use.
The publisher, the authors, and the editors are safe to assume that the advice and information in this book
are believed to be true and accurate at the date of publication. Neither the publisher nor the authors or
the editors give a warranty, expressed or implied, with respect to the material contained herein or for any
errors or omissions that may have been made. The publisher remains neutral with regard to jurisdictional
claims in published maps and institutional affiliations.

Cover Image: © Michael Gilbert / Science Photo Library

This Springer imprint is published by the registered company Springer Nature Switzerland AG
The registered company address is: Gewerbestrasse 11, 6330 Cham, Switzerland

This book is dedicated to Elena and Jackie with love and gratitude for their constant support.

Preface

The Scope of This Book

This book is intended to be a first course in nuclear and Particle Physics at an undergraduate level. We discuss the main experimental techniques – both past and present – for investigating the properties of nuclei and particles as well as describing the enormous progress that has been made in theoretical developments which explain this behaviour. We have described, albeit briefly, the seminal experiments from the end of the nineteenth century right up to the date of publication, which have had significant effect on the development of the subject.

Typically, such a course would be taken by undergraduates in their third year. It is designed to cover all the required core material. There is, almost certainly, more material than would normally be required for a core course. For some chapters, the supplementary material, which could be omitted without damaging the flow of the remainder of the book, appears in the last section. The final chapter is almost certainly outside the scope of a core course and has been added to give the reader a taster of current developments in Particle Physics.

The book assumes that the reader has a reasonable grounding in Quantum Physics, Atomic Physics, Electromagnetism and Thermal Physics. However, we do not expect the reader to remember all that they have learnt in the prerequisite courses, and all results that are used are explicitly displayed in the text. The hope is that once the memory has been jogged in this way, the reader will recall the relevant material sufficiently to be able to follow its usage here.

On the other hand, the book has been designed so that it is possible to follow all the material in the book without having studied relativistic quantum mechanics (e.g. the Dirac equation), quantum field theory or group theory. These are topics which are usually taken only at the postgraduate level and therefore outside the scope of a book designed to cover a first course on nuclear and Particle Physics. There is a brief explanation of the qualitative ideas of quantum field theory, which should be accessible to undergraduates with a basic knowledge of Quantum Physics, and it has been necessary on very few occasions to quote results which can only be derived using quantum field theory. Likewise, no knowledge of formal group theory is required, although it is expected that the reader is familiar with the concept of symmetry and what this implies for quantum states.

We do *not* use four-vector notation in special relativity. Although many of the expressions would have been simpler using such notation, not all students will have come across this notation in a first course on special relativity.

Chapter Structures

Needless to say, it is impossible to write a textbook on any branch of physics, without using mathematics and mathematical derivations. In this book, we have limited the mathematical derivations displayed in the main body of the text to those that are essential to be able to follow the development of the theory. Derivations of some of the results appear in appendices to some of the chapters. The reader who does not care to follow such sustained mathematical threads is welcome to skip these appendices entirely, provided they are prepared to accept the relevant results at "face value", without understanding where they come from. The appendices are there for readers who are unwilling to accept a result unless they have seen it derived – or at least seen an outline derivation. Note that some of the results derived refer more generally to results of quantum or Thermal Physics. These appendices are admittedly rather terse, and the reader may prefer to consult other literature in order to find more detailed and comprehensive derivations of the results.

Like all other branches of science, Nuclear Physics and Particle Physics introduce a plethora of technical terms, as well as making use of technical terms from other branches of physics. Usually, the first time a *"technical term"* is used, it is displayed in italics between inverted commas (as shown). This indicates that the term is defined in the extensive glossary at the back of the book. Subsequent instances of the term are typeset in the normal way, and the reader can refer to the glossary to be reminded of its meaning. A similar approach is applied to the frequent use of the acronyms, which are common parlance of nuclear or particle physicists. The first time an acronym is used, the full set of words represented by the acronym is displayed in the text. Subsequently, the acronym is usually used without explanation, but there is a table of acronyms in the back of the book. The same is true (mainly in the first part of the book) for the use of chemical symbols. Names of elements appear in full for the first time they are used. Subsequently, a given isotope is written in the form $^A_Z Ch$, where Z is the atomic number, A the atomic mass number and Ch is the chemical symbol for the element. At the back of the book there is also a table of chemical symbols for all elements used. All this will slow the reader down considerably. This has been done deliberately as an effective way to familiarize the reader with the relevant technical terms, acronyms and chemical symbols, so that references to the glossary or tables at the back of the book are eventually obviated.

Each chapter ends with a bullet-point summary of the salient features of the chapter. These summaries indicate the minimum that the reader should have taken away from each chapter. This is followed by a set of between three and five problems (with the exception of the final chapter). The degree of difficulty of these problems varies widely – from inserting numbers into formulae given in the text to those requiring a thorough understanding of the material in the chapter. All of the readers should be able to answer at least one problem in each chapter, but the reader should not be disheartened if they cannot solve all the problems, particularly after the first reading of the chapter. At the end of the book we also provide solutions for those problems which have numerical answers.

Units

The energies, momenta, masses and distance scales of nuclear and Particle Physics are many orders of magnitude smaller than the standard International System (SI) units. It is therefore usual for nuclear and particle physicists to use nuclear units in which energies are measured in MeV, momenta in MeV/c, masses in MeV/c^2 and distances in fermi. These units and their relation to SI units are explained in a section on units at the beginning of the book. Particle experimentalists tend to work in cgs units (centimetres, grams and seconds as opposed to metres, kilograms and seconds) – so that in Chapter 14 we use the cgs system.

Theoretical particle physicists tend to use "Natural Units" in which the reduced Planck's constant and the speed of light in a vacuum are set equal to unity. This greatly simplifies many of the expressions used in Particle Physics. However, we have chosen *not* to adopt this system as we feel it is likely to cause confusion to readers who are not used to this system of units. The reader should be aware, however, that using natural units, masses are often quoted in MeV, which strictly speaking means MeV/c^2. Likewise, momenta are sometimes quoted in MeV, whereas what is really meant is MeV/c.

Citations

Since the discovery of radioactivity at the end of the nineteenth century and the discovery of the nucleus at the beginning of the twentieth century through to the ongoing development of Particle Physics, which began in the middle of the twentieth century, tens of thousands of physicists have published hundreds of thousands of papers. We needed to be very selective in the people we mention explicitly in the text. These are the names of those who, in our opinion, have first been responsible for the various theoretical and experimental discoveries that we discuss in the book. The majority of the people mentioned were awarded a Nobel Prize for their contribution to the subject. This book is *not* intended to contain a comprehensive history, and there is likely to be controversy over who discovered something or measured something first.

Likewise, we have been selective in the citations to papers. In many cases, an author will have written several papers on the same topic, and we have limited ourselves in the bibliography to one paper per topic. It is, in any case, not our advice that the reader should attempt to understand the original papers, which are written by specialists for the benefit of other specialists. The references are included for completeness and to give due credit to experimentalists and theorists who wrote the original papers. Furthermore, until the middle of the twentieth century, English was not universally accepted as the language for scientific publications, so many of the earlier papers were written in the native language of the author. Even after English became universally accepted in western countries, publications in the former Soviet Union continued to appear in Russian. Translations of all these papers are available, but citations in this bibliography are to the original versions and in some cases are likely to stretch the reader's linguistic skills.

We hope that the reader enjoys this book and finds it both informative and interesting. The reader may find some repetition in cases where a given concept or idea is discussed in different chapters or even different sections of the same chapter. Once again this is deliberate – *repetitio est mater studorium*.[1]

Finally, the reader is welcome to ask either of us any question related to the text. Emails should be sent to:

a.belyaev@soton.ac.uk

or

d.a.ross@soton.ac.uk

We will endeavour to answer any queries promptly.

Southampton, UK   Alexander Belyaev
Southampton, UK   Douglas Ross
May 2021

---

[1] Repetition is the mother of all learning.

Acknowledgements

We are grateful to Chris Sachrajda, Lydia Mikhailovna Shcheglova and Anatoly Nikolaevich Solomin for reading parts of this manuscript and making useful comments and suggestions.

Contents

Units and Constants

Although the SI system of units is accepted as universal for all of physics, nuclear and Particle Physics deal with scales of mass, energy and length, which are extremely small in such a system of units. Therefore, it is convenient to use an associated system of units (nuclear units) whose magnitudes are the order of magnitude of nuclear and particle scales. The conversion of these units to SI units together with powers of 10 and a selection of masses and main physical constants are shown in the tables below.

	Nuclear units	SI unit equivalent
Length	fermi (fm)	10^{-15} m
Area	barn (b)	10^{-28} m^2
Energy	MeV (10^6 eV)	1.6022×10^{-13} J
Momentum	MeV/c	5.3446×10^{-22} kg m/s
Mass	MeV/c^2	1.7827×10^{-30} kg
Electric charge	a.u. of charge	1.6022×10^{-19} C

Powers of 10

10^3	kilo (k)	10^{-3} milli (m)
10^6	Mega (M)	10^{-6} micro (μ)
10^9	Giga (G)	10^{-9} nano (n)
10^{12}	Tera (T)	10^{-12} pico (p)
10^{15}	Penta (P)	10^{-15} femto (f)

Masses of stable sub-atomic particles[a]

Particle	kg	MeV/c^2
Electron	9.1094×10^{-30}	0.51098
Proton	1.6726×10^{-27}	938.27
Neutron	1.6749×10^{-27}	939.57

[a]Masses and other physical constants are given here to five significant figures. In the text it will usually be sufficient to truncate these to three significant figures

Main physical constants

	Nuclear units	SI units
Speed of light, c	2.9979×10^{23} fm s^{-1}	2.9979×10^{8} m s^{-1}
Reduced Planck's constant, $\hbar$	6.5821×10^{-22} MeV s	1.0546×10^{-34} J s
Conversion constant, $\hbar c$	197.32 MeV fm	3.1616×10^{-26} J m
Electron charge, e	1 a.u.	1.6022×10^{-19} C
Permittivity of vacuum, ε_0	5.5267×10^{-2} a.u.^{2} MeV^{-1} fm^{-1}	8.8542×10^{-12} F m^{-1}
Boltzmann constant, k_B	8.6173×10^{-11} MeV K^{-1}	1.3806×10^{-23} J K^{-1}
Fermi coupling constant, G_F	1.1664×10^{-5} GeV^{-2}	4.5437×10^{15} J^{-2}
Fine structure constant, α	1/137.04	1/137.04

The fine structure constant, $\alpha (\equiv e^2/4\pi\varepsilon_0\hbar c)$, is dimensionless and therefore the same in all systems of units

Historical Introduction

Ever since ancient times, people have tried to understand nature and its fundamental building blocks. We know of several "theories" which came from the ancient philosophers. More than 2000 years ago, Empedocles (490–430 BC) suggested that all matter was made up of four elements: water, earth, air and fire. Democritus (460–370 BC) suggested that the universe consists of empty space and an (almost) infinite number of invisible particles which differ from each other in form, position and arrangement. He called them *atoms* (meaning "indivisible" in Greek).

From that time, our understanding of the fundamental building blocks of nature has evolved into a powerful science of elementary Particle Physics, or, Particle Physics in short. The main difference between this science and philosophy is that science verifies its theoretical predictions by experiment, whilst experiment in its turn provides ground for a further theoretical development. Theory and experiment are essentially interacting components of Particle Physics as a science. That is exactly how the Standard Model, which represents our current understanding of fundamental particles and their interactions, has been established.

Nuclear Physics was the predecessor of Particle Physics and is still studied as a separate subject. Both nuclear and Particle Physics have a lot of common elements and a strong historical connection, hence the title of this book.

In Table 1, we present a timeline of Particle Physics in a very brief way. A more detailed history can be found, for example, at http://en.wikipedia.org/wiki/Timeline_of_particle_physics. Rapid development of Nuclear Physics started at the end of the nineteenth century and it is often associated with the discovery of the proton in 1886 by Eugene Goldstein [1] in the form of positively charged rays in the discharge tube, using a perforated cathode. Ten years later, in 1896, the electron was discovered by Joseph J. Thompson, in the form of cathode rays in discharge tube experiments [2]. One of the most important milestones in the early history of Nuclear Physics was the Geiger-Marsden experiments in 1911, at the suggestion of Ernest Rutherford, which provided evidence for atomic structure with a nucleus in the centre and electrons orbiting around it. We start this book by describing this experiment and Rutherford's interpretation of the physics behind it. Since then, many exciting discoveries have been made and amongst them there were several

Table 1 A very brief timeline of nuclear and Particle Physics

1886	Eugene Goldstein discovered a positively charged sub-atomic particle
1897	Joseph J. Thomson discovered the electron
1904	Hantaro Nagaoka postulated a model of the atom in which electrons orbit around a positively charged centre
1909	Robert Millikan measured the charge and mass of the electron
1911	Ernest Rutherford discovered the nucleus of an atom
1913	Niels Bohr introduced his atomic theory
1919	Ernest Rutherford discovered the proton
1920s	Modern atomic theory was developed by Werner Heisenberg, Louis de Broglie and Erwin Schroedinger
1932	James Chadwick discovered the neutron
1936	Discovery of the muon by Carl Anderson and Seth Neddermeyer during a study of cosmic radiation
1947	Cecil Powell, César Lattes and Giuseppe Occhialini discovered the pion
1947	George Rochester and Clifford Butler discovered the kaon (K-meson), the first particle with "strangeness"
1964	Evidence for the first three "flavours" (up, down and strange) discovered
1974	Burton Richter and Samuel Ting discovered the J/ψ particle, demonstrating the existence of a fourth quark flavour (charm)
1977	Upsilon particle discovered at Fermilab, demonstrating the existence of the fifth quark flavour (bottom)
1983	Carlo Rubbia and Simon Van der Meer discovered the W- and Z- bosons at CERN SPS
1995	Sixth quark flavour (top) discovered at Fermilab
2000	Tau neutrino proved to be distinct from other neutrinos at Fermilab
2012	The Higgs boson was discovered at the LHC

essential milestones of our understanding and confirmation of the Standard Model. In 1936 Carl Anderson and Seth Neddermeyer discovered the muon in the study of the cosmic rays [3]. It was noticed that particles curved differently from electrons and other known particles when passed through a magnetic field. This discovery of the second generation of charged leptons was the truly unexpected opening of the world of the elementary particles beyond the first generation. In 1947 Cecil Powell, César Lattes and Giuseppe Occhialini [4](University of Bristol) discovered the the first "meson" – the pion (π-meson) which was predicted by Hideki Yukawa in 1935 as a carrier of the strong force. This discovery together with discovery of the kaon (K-meson) by Clifford Butler and George Rochester in 1947 [5] (University of Manchester) gave a hint about the composite nature of strongly bounded states such as baryons and mesons. Another vitally important discovery was the discovery of the W-[6, 7] and Z-[8, 9] bosons at CERN[1] SPS by Carlo Rubbia and Simon Van der Meer in 1983. This discovery was the confirmation that weak interactions

---

[1]Centre Européen pour la Recherche Nucléaire, the European Organization for Nuclear Research.

Standard Model of Elementary Particles

Fig. 1 A summary of elementary particles of the Standard Model and their interactions. [*Source*: https://commons.wikimedia.org/wiki/File:Standard_Model_of_Elementary_Particles.svg]

were mediated by massive particles, as predicted by the Standard Model. The origin of the mass of these particles – the Higgs mechanism – was not confirmed until the announcement of the discovery of the Higgs boson in July 2012 by the ATLAS and CMS collaborations [10, 11]. This was a very important historical event, as it completed the Standard Model particle content. In Fig. 1 we present the complete list of particles of the Standard Model which will be discussed later in this book.

The discovery of the Higgs boson does not preclude particles and theories beyond the Standard Model: on the contrary, there are several fundamental motivations to go beyond the Standard Model and there are several promising theories beyond the Standard Model, which are consistent with the properties of the discovered Higgs boson. These theories are being tested at various current experiments and will continue to be tested in promising future experiments. Whereas the Geiger-Marsden experiment used a probe, whose momentum, p, was such that the de Broglie wavelength, λ, given by the well-known relation

$$\lambda = \frac{h}{p}, \tag{1}$$

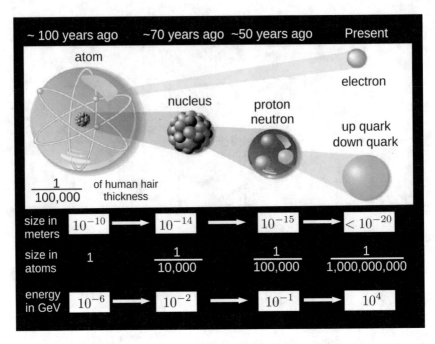

Fig. 2 Timeline of the energy scale accessible in Particle Physics, and the corresponding size resolution. [*Source*: https://www.quora.com/What-size-are-the-particles-of-an-atom-in-relation-to-its-size (edited)]

was much larger than the nucleus, modern scattering experiments use projectiles of much larger momentum and therefore much shorter de Broglie wavelength in order to probe the internal the structure of nuclei. In (1) h is Planck's constant, and the magnitude of the momentum, p, related to its total energy E and the mass, m as $E = \sqrt{(pc)^2 + (mc^2)^2}$. For a massless particle, one can see that the de Broglie wavelength is inversely proportional to its energy. This is also approximately true for a massive particle with sufficiently large energy. One can use this fact to resolve the structure of the probed nucleus when the wavelength of the probe particle is comparable or smaller than the size of the nucleus. In this case, i.e. when the energy of the probe particle is large enough, there will be diffractive scattering off the nucleus, thereby resolving its structure at the scale of the de Broglie wavelength of the probing projectile. However, if the energy of the probe particle is too low, the de Broglie wavelength will be larger than the size of the nucleus and there will be no diffraction and its internal structure will be not be resolved. The principle behind (1) relating the distance probed and the momentum of the probing projectile, is one of the main foundations of Particle Physics. At the present energy, the Large Hadron Collider (LHC), which has now reached 13 TeV (1.3×10^7 MeV) probes the distances as low as 10^{-20} m (10^{-5} fm) (and the structure of the particles at these distances)! The timeline evolution of the energy/distance scale accessible in Particle Physics is presented in Fig. 2.

The well-known rest energy of a particle of mass m

$$E = mc^2 \tag{2}$$

provides the possibility of producing new heavy unknown particles. It opens another way to explore new theories beyond the Standard Model. We expect the timeline of accessible energy Particle Physics to be updated in the near future with new energy frontiers and new discoveries, uncovering new physics beyond the Standard Model.

Part I
Nuclear Structure

Chapter 1
Rutherford Scattering

1.1 The "Plum Pudding" Model

In 1897, Joseph J. Thompson discovered electrons from experiments with cathode rays [2]. He showed that cathode rays were *not* waves as originally thought, but negatively charged particles (originally called "corpuscles"). He estimated that these had a mass of order of one thousandth of the mass of an atom and that they were the same as the negatively charged particles emitted from a β-decay radioactive source.

This discovery led him to postulate a model which consisted of a number of (small) negatively charged electrons embedded in a positively charged "dough", so as to give the atom its overall electric neutrality. This was known as the "Plum Pudding" model [12].

1.2 The Geiger-Marsden Experiments

As we shall see later the "Plum Pudding" model predicts that a charged particle which is moving through such a positively charged "dough" will experience a very weak electric force and will only undergo very small angular deflections. In order to verify this, Hans Geiger and Ernest Marsden, at the behest of Ernest Rutherford, carried out three experiments between 1908 and 1910 in which α-particles from a radioactive source were incident on a very thin foil of gold (gold was selected because it can be beaten very thin – the foil used by Geiger and Marsden had a thickness of 400 nm). The entire apparatus was encased in a tube, which was evacuated in order to minimize energy loss of the α-particles before they scattered off the foil. A schematic sketch of the experimental setup is shown in Fig. 1.1.

In the first experiment, [13] a screen was placed behind the gold foil and scintillations caused by the α-particles landing on the screen, were observed with a travelling microscope. Although most (86%) of the α-particles passed through

© The Author(s), under exclusive license to Springer Nature Switzerland AG 2021
A. Belyaev, D. Ross, *The Basics of Nuclear and Particle Physics*, Undergraduate
Texts in Physics, https://doi.org/10.1007/978-3-030-80116-8_1

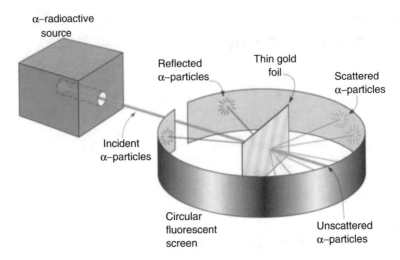

Fig. 1.1 Schematic setup of the Geiger-Marsden experiment [*Source*: https://chemistryonline.guru/wp-content/uploads/2017/07/ruth3.png (edited)]

with a deflection of less than 1°, a substantial angular spread of scintillations was observed.

In the second experiment [14], the screen was placed on the incident side of the gold foil in order to observe reflected α-particles. The screen was protected from direct α-particles by placing an impenetrable lead plate in the direct path of the particles. They nevertheless observed that about one particle in 8000 was reflected by the foil, implying that there had been scattering through an angle of greater than 90° – way above the limit predicted by the "Plum Pudding" model.

In a third experiment [15], a year later, Geiger and Marsden used several different foils of different thickness and made of different materials. In this experiment, they managed to determine the most probable deflection angle. They showed that the most probable angle of scattering:

1. Increased with increasing thickness of the foil,
2. Increased with the atomic mass of the material the foil,
3. Decreased with increasing velocity of the incident α-particles.

1.3 Rutherford's Scattering Formula

Rutherford's surprise at the results of the Geiger-Marsden experiment, particularly the fact that some of the α-particles were scattered though an angle of more than 90°, led him to state during a lecture at Cambridge University:

Fig. 1.2 Head-on collision between an α-particle and a nucleus, (N). The point of closest approach, (P), is a distance D from the nucleus

> It was almost as incredible as if you fired a 15-inch shell at a piece of tissue paper and it came back and hit you. On consideration, I realized that this scattering backward must be the result of a single collision. . .

In 1911, he adopted the model postulated 7 years earlier by the Japanese physicist Hantaro Nagaoka [16]. This model comprised of a small positively charged nucleus at the centre of an atom with electrons orbiting around it. Within this model, Rutherford calculated the probability of scattering of the α-particles through an angle θ [17] under the following assumptions:

- The atom contains a nucleus of charge Ze, where Z is the atomic number of the atom (i.e. the number of electrons in the neutral atom),
- The nucleus can be treated as a point particle,
- The nucleus is sufficiently massive compared with the mass of the incident α-particle that the nuclear recoil may be neglected,
- The laws of classical mechanics and Electromagnetism can be applied and that no other forces are present,
- The collision is elastic.

If the collision between the nucleus and incident particle, with kinetic energy T and electric charge ze[1] were head-on, as shown in Fig. 1.2, the distance of closest approach D is obtained by equating the initial kinetic energy to the Coulomb energy at closest approach, i.e.

$$T = \frac{zZe^2}{4\pi\varepsilon_0 D}, \tag{1.1}$$

so that the distance of closest approach is given by

$$D = \frac{zZe^2}{4\pi\varepsilon_0 T}, \tag{1.2}$$

at which point the α-particle reverses direction.

In general, the collision is not head-on, but is described by a quantity, b, called the *"impact parameter"*. This is the perpendicular distance between the nucleus and the initial line of the incident projectile, as shown in Fig. 1.3.

---

[1] For an incident α-particle $z = 2$.

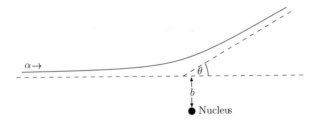

Fig. 1.3 The dashed lines are asymptotic to the incident and final directions of motion of the α-particle. The angle, θ, between them is the scattering angle. The impact parameter, b, is the perpendicular distance between nucleus and the initial line of motion

In the case of head-on collision, $b = 0$ and the α-particle reverses direction, i.e. the scattering angle is 180° (π radians). For non-zero values of the impact parameter the scattering angle will be less, since we know from Coulomb's law that the further away the positively charged α-particle is from the positively charged nucleus, the weaker the electric force between the particles and consequently the smaller the scattering angle. We also expect the scattering angle to increase with increasing product of the electric charges of the nucleus and the α-particle.

The exact relation between impact parameter, b, and scattering angle is given by

$$\tan\left(\frac{\theta}{2}\right) = \frac{D}{2b}. \tag{1.3}$$

The derivation of this formula is rather complicated (it is alleged that Rutherford had a famous dynamics textbook open on his desk when he performed his calculations) and this has been placed in Appendix 1. The reader who is happy to accept the formula (1.3), without seeing its derivation, need not study the appendix. Note, however, that this formula is consistent for head-on collision when $b = 0$ and $\theta = \pi$, and the angle decreases with increasing value of b, as expected.

1.4 Flux and Cross Section

The "*flux*", F, of incident particles is defined as the number of incident particles arriving per unit area per second at the target.

The number of particles, $dN(b)$, with impact parameter between b and $b + db$ is the flux multiplied by the area between two concentric circles of radius b and $b + db$ (see Fig. 1.4)

$$dN(b) = F\,2\pi b\,db. \tag{1.4}$$

Fig. 1.4 Area of an
infinitesimal slice, *db*, of
impact parameter

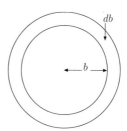

Differentiating (1.3) gives us

$$db = -\frac{D}{4\sin^2(\theta/2)}d\theta \tag{1.5}$$

which allows us to write an expression for the number of α-particles scattered
through an angle between θ and $\theta + d\theta$ after substituting (1.5) and (1.3) into (1.4):

$$dN(\theta) = F\pi\frac{D^2}{4}\frac{\cos(\theta/2)}{\sin^3(\theta/2)}d\theta \tag{1.6}$$

(the minus sign has been dropped as it merely indicates that as b increases, the
scattering angle θ decreases – $dN(\theta)$ must be positive).

The *"differential cross section"*, $d\sigma/d\theta$, with respect to the scattering angle is
the number of scatterings between θ and $\theta + d\theta$ per unit flux, per unit range of
angle, i.e.

$$\frac{d\sigma}{d\theta} = \frac{dN(\theta)}{Fd\theta} = \pi\frac{D^2}{4}\frac{\cos(\theta/2)}{\sin^3(\theta/2)}.$$

It is more usual to quote the differential cross section with respect to a given
interval of solid angle, $d\Omega$. Solid angle is defined such that an area element, dS, of
a sphere of radius r subtends a solid angle (at the centre of the sphere)

$$d\Omega = \frac{dS}{r^2}.$$

The unit of solid angle is the *"steradian"* (sr). Solid angle is related to the scattering
angle θ and the *"azimuthal angle"*, ϕ, by

$$d\Omega = \sin\theta d\theta d\phi. \tag{1.7}$$

The relation between the number of events, the flux, differential solid angle $d\Omega$ and
differential cross section is given by

$$\frac{dN}{d\Omega} = F\frac{d\sigma}{d\Omega},$$

in analogy with the relation for differential cross section with respect to scattering angle, $d\theta$.

The integration over the azimuthal angle just gives a factor of 2π, since the (double) differential cross section is independent of the azimuthal angle, ϕ, so we may write

$$\frac{d\sigma}{d\theta} = \int_0^{2\pi} d\phi \frac{d^2\sigma}{d\theta d\phi} = 2\pi \frac{d^2\sigma}{d\theta d\phi}$$

so that

$$\frac{d^2\sigma}{d\theta d\phi} = \frac{D^2}{8} \frac{\cos(\theta/2)}{\sin^3(\theta/2)}$$

and substitute $d\theta d\phi$ by $d\Omega$ (using the relation (1.7)) to obtain

$$\frac{d\sigma}{d\Omega} = \frac{D^2}{8} \frac{\cos(\theta/2)}{\sin^3(\theta/2)} \frac{1}{2\sin(\theta/2)\cos(\theta/2)} = \frac{D^2}{16\sin^4(\theta/2)}. \tag{1.8}$$

Using (1.2) we have

$$\frac{d\sigma}{d\Omega} = \left(\frac{Zze^2}{4\pi\varepsilon_0 T}\right)^2 \frac{1}{16\sin^4(\theta/2)} = \left(\frac{Zz\alpha\hbar c}{T}\right)^2 \frac{1}{16\sin^4(\theta/2)}. \tag{1.9}$$

In the last step we have expressed the square of the electron charge in terms of the dimensionless fine structure constant, α ($\approx 1/137$):

$$\frac{e^2}{4\pi\varepsilon_0} = \alpha\hbar c.$$

Throughout this book, it is (usually) convenient to use this notation.

It is sometimes more convenient to express this differential cross section in terms of the momentum, p, of the incident particle rather than its kinetic energy, T,

$$\frac{d\sigma}{d\Omega} = \left(\frac{Zz\alpha\hbar c m_\alpha}{p^2}\right)^2 \frac{1}{4\sin^4(\theta/2)}. \tag{1.10}$$

Differential cross sections have the dimension of an area. These are usually quoted in terms of "barns" (10^{-28} m^2), so that, for example, 1 millibarn (mb) is an area of 10^{-31} m^2. A cross section of 1 fm^2 corresponds to 10 mb. If a gold ($Z=79$) target is bombarded with α-particles with kinetic energy 5 MeV, the differential cross section at a scattering angle of 45° is (inserting numerical values into (1.9))

$$\left(\frac{79 \times 2 \times 197.3\,[\text{MeV fm}]}{137 \times 2 \times 5\,[\text{MeV}]}\right)^2 \frac{1}{4\sin^4(22.5^\circ)} = 6035\,\text{fm}^2/\text{sr} \equiv 60.35\,\text{b/sr}.$$

In terms of fermi (fm) (10^{-15} m), the distance of closest approach is given by

$$D = 1.44\frac{zZ}{T}\,\text{fm},$$

where the kinetic energy T is given in MeV. Gold has atomic number 79 and the energy of the α-particles from polonium or radium decay is around 5 MeV so the distance of closest approach is around 45 fm.

The total number of particles scattered into a given solid angle is the differential cross section multiplied by the flux, multiplied by the number of nuclei in the foil – or more precisely in the part of the foil that is "illuminated" by the incident α-particles. We assume that the foil is sufficiently thin so that multiple scatterings are very unlikely and we can treat the nuclei as though they all lie in a single plane. The mass of a nucleus with atomic mass number A is given to a very good approximation by Am_p, where m_p is the mass of the proton. The total number of nuclei per unit area of foil is given by

$$\rho\delta\frac{1}{Am_p},$$

where ρ is the density and δ is the thickness of the foil. This means that the fraction, $dN(\theta)/N$, of α particles scattered into a small interval of solid angle $d\Omega$ is given by

$$\frac{dN(\theta)}{N} = \rho\delta\frac{1}{Am_p}\frac{d\sigma}{d\Omega}d\Omega. \tag{1.11}$$

Therefore, if we place a detector with an acceptance area dS at a distance r from the foil and at an angle θ to the direction of the incident α-particles, the fraction of incident α-particles which enter the detector is given by replacing $d\Omega$ by dS/r^2 in (1.11). Thus finally, if N α-particles are fired at the gold foil, the number of scintillations expected per unit area at scattering angle θ is given by

$$\frac{dN(\theta)}{dS} = N\frac{\rho\delta}{Am_p r^2}\left(\frac{zZ\alpha\hbar c}{m_\alpha v^2}\right)^2\frac{1}{4\sin^4(\theta/2)}, \tag{1.12}$$

where we have replaced the (non-relativistic) momentum of the α-particle, p, by $m_\alpha v$.

We notice that the differential cross section diverges as the scattering angle goes to zero. However, (1.3) tells us that small angle scattering corresponds to a large impact parameter. The formula derived by Rutherford assumes scattering off an isolated nucleus. If the impact parameter becomes of the order of the atomic radius,

R, of gold (which is the order of magnitude of the separation between gold nuclei in the foil), then this assumption becomes invalid and the formula is only reliable for values of the impact parameter $b \ll R$, or (using (1.3) in the small angle approximation), $\theta \gg D/R$. The atomic radius of gold is 1.44×10^5 fm, so with a typical value of distance of closest approach, D of 45 fm, we see that the Rutherford scattering formula is good down to angles of around 3×10^{-4} radians – but the apparent divergent in the limit $\theta \to 0$ is avoided.

A further breakdown of the Rutherford formula for the scattering cross section can occur when the assumption of the point-like nature of the nucleus becomes invalid. This can occur for higher energy incident projectiles for which the distance of closest approach (which is inversely proportional to the kinetic energy of the incident particle) becomes comparable with the nuclear radius (of the order of a few fm). The deviation from the Rutherford formula would be larger for smaller values of the impact parameter, where the incident α-particle gets closer to the centre of the nucleus. A small impact parameter corresponds to large scattering angle, so the deviation from the Rutherford prediction for large energy incident particles is enhanced at large scattering angle.

1.5   Inconsistency of the "Plum Pudding" Model

Let us consider what would be expected if the "Plum Pudding" model were indeed correct.

We know from Gauss' law that at a distance r from the centre of the atom, the electric field is determined by the charge enclosed in a sphere of radius r surrounding the centre of the atom.

The volume of a sphere of radius r is proportional to r^3. Therefore for r smaller than the radius, R, of the atom, the electric charge enclosed with a sphere of radius r is a fraction r^3/R^3 of the total electric charge (assuming a uniform distribution of electric charge throughout the "dough"), so that the magnitude of the electric field at a distance r from the centre of the atom is given by

$$\left(\frac{r^3}{R^3} \right) \frac{Ze}{4\pi \varepsilon_0 r^2}, \ (r \leq R) .$$

This is a maximum for $r = R$. This means that the scattering angle cannot be larger than the scattering angle corresponding to impact parameter $b = R$. For values of impact parameter $b < R$, the scattering angle *decreases* as b decreases.

We have seen above that for α-particles with typical kinetic energy of 5 MeV, this corresponds to a maximum scattering angle of around 3×10^{-4} radians ($\approx 0.017°$). Such an angle would have been far too small to be observed in any of the Geiger-Marsden experiments and they certainly would not have observed any scattering exceeding $90°$.

In fact, the scattering from the "Plum Pudding" model is expected to be even smaller as the above estimate neglects any attractive force between the α-particle and the electrons (the "plums") embedded in the "dough".

1.6 Confirmation of Rutherford Scattering Cross Section

In 1913, Geiger and Marsden [18] performed a far more accurate experiment to check the details of Rutherford's formula (1.12). They checked the dependence of the rate on the scattering angle and found consistency with the prediction

$$N(\theta) \propto \frac{1}{\sin^4(\theta/2)}.$$

Their results, shown in Fig. 1.5, agree remarkably well.

By using foils of different thickness, they showed that the number of particles scattered through a given angle was proportional to the thickness of the foil, and by using foils made from different metals (tin, silver, copper and aluminium) they were able to show that this number was proportional to the square of the atomic number, Z, of the material of the foil.

They were able to slow down the incident α-particles, by placing thin sheets of mica immediately in front of the radioactive source. From this they were able to verify that the number of scattered particles was inversely proportional to the fourth power of their velocity, as indicated in (1.12).

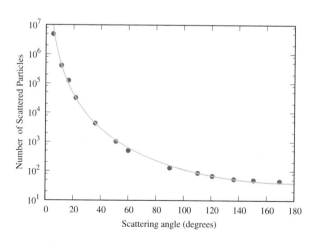

Fig. 1.5 Comparison of results from the Geiger-Marsden experiment (1913) and Rutherford's scattering formula (blue line)

1.7   What About Quantum Effects?

One might ask whether it was correct to assume that classical physics was applicable for the description of Rutherford scattering, which probes sub-atomic scales where we might expect quantum effects to be significant. Of course, at the time of Rutherford's calculation, Quantum Physics was unknown, but nowadays we know that the incident α-particle has an associated de Broglie wave, and that, in general, a wave scattering from a regular configuration of gold atoms will produce a diffraction pattern. The angular scale of such diffraction patterns is of the order of the de Broglie wavelength divided by the mean separation of the gold atoms in the foil.

The de Broglie wavelength, λ, is given by

$$\lambda = \frac{h}{m_\alpha v} = \frac{h}{\sqrt{2m_\alpha T}},$$

(h is Planck's constant), the mass of an α-particle is 6.6×10^{-27} kg, and for α-particles with kinetic energy 5 MeV (8×10^{-13} J) this gives a wavelength

$$\lambda \approx 6 \times 10^{-15}\,\text{m}.$$

In contrast, the separation of the gold atoms is around 170 nm.

This means that the effect of diffraction from the gold atoms is negligible. On the other hand, the size of the nucleus itself is indeed of the order of the de Broglie wavelength of the incident particles, so that for projectiles with somewhat smaller wavelengths, diffraction patterns can be observed from diffraction off single nuclei and these patterns can yield useful information about the structure of nuclei. This is the subject of Chap. 2.

Summary

- Thomson proposed the "Plum Pudding" model in which negatively charged electrons were embedded in a positively charged "dough" which extends throughout the atom.
- A charged particle passing through such a "Plum Pudding" would only experience a small electric field and suffer only a tiny deflection.
- The experiments of Geiger and Marsden, in which α-particles were fired through a thin metal foil, produced much larger deflections and even a small number of scintillations with scattering angles exceeding $90°$.
- Rutherford assumed a model of the atom with a point-like positively charged nucleus at the centre. Using classical Electromagnetism and Classical Mechanics, he calculated the differential scattering cross section as a function of scattering angle. He found a rapid decrease with increasing scattering angle, but neverthe-

less large scattering angles were possible, in contrast to the predictions of the "Plum Pudding model".

- An apparent divergence of the predicted scattering cross section at zero scattering angle is an indication that if the impact parameter becomes of the order of the inter-atomic separation, one can no longer consider scattering off a single nucleus.

- After Rutherford's calculation, Geiger and Marsden performed a further experiment in which they verified not only the angular dependence of Rutherford's formula, but also the dependence of the scattering cross section on the atomic number of the metal used in the foil, and on the velocity of the incident α-particles.

- One is justified in neglecting quantum effects, which would lead to a diffraction pattern from interference of the de Broglie waves off adjacent atoms in the foil, because the de Broglie wavelength of the α-particles is several orders of magnitude smaller than the inter-atomic separation.

Problems

Problem 1.1 α-particles with kinetic energy 5 MeV are incident on a gold ($^{197}_{79}$Au) foil of thickness 400 nm at a rate of 10^8 particles per second. A detector of area 10^{-3} m^2 is placed at an angle of 30° to the direction of the incident α-particles at a distance of 2 m from the foil.
Calculate how many α-particles per second reach the detector. [The density of gold is 1.93×10^4 kg m^{-3}.]

Problem 1.2 If the nuclear radius of gold is 7 fm, above what threshold energy of incident α-particles would you expect to see deviations from the Rutherford scattering formula?
For incident particles with energies just above that threshold would you expect to see these deviations at small or large scattering angles? State your reason.

Problem 1.3 In the original Geiger-Marsden experiment the scattering angle of the observed α-particles varied from 1.5° to 150°. What fraction of these particles were scattered through an angle greater than 90°?

Appendix 1 – Relation Between Scattering Angle and Impact Parameter

The relation between impact parameter, b, and scattering angle, θ, is derived using Newton's second law of motion, Coulomb's law for the force between the α-particle and the nucleus, and conservation of angular momentum.

The initial and final momenta, p_i, p_f, have equal magnitude, p, (elastic scattering with no nuclear recoil is assumed). If we take p_i to be along the z-axis and the scattering to be in the $x - z$ plane, then in Cartesian coordinates these two vectors are given by

$$p_i = p\,(0, 0, 1)$$

$$p_f = p\,(\sin\theta, 0, \cos\theta)$$

and the momentum transfer is given by

$$q \equiv p_f - p_i = p\,(\sin\theta, 0, (\cos\theta - 1)).$$

Using Pythagoras' theorem and some trigonometric manipulation, the momentum transfer, q, has a magnitude

$$q = 2p\sin\left(\frac{\theta}{2}\right). \tag{1.13}$$

The direction of the vector q is along the line joining the nucleus to the point of closest approach of the α-particle. It bisects the vectors p_i and p_f, making an angle $(\pi - \theta)/2$ with each, as can be seen from Fig. 1.6.

The position vector, $\mathbf{r}$, from the nucleus and the α-particle is given in terms of two-dimensional polar coordinates (r, ϕ) with the nucleus as the origin. The angle ϕ is set such that $\phi = 0$ is at the point of closest approach, where $\mathbf{r}$ lies along the vector q.

From Newton's second law of motion, the rate of change of momentum in the direction of q is the component (in that direction) of the force acting on the α-particle due to the electric charge of the nucleus. By Coulomb's law the magnitude of the force is

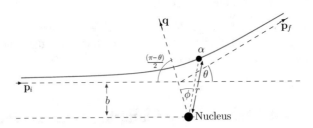

Fig. 1.6 The position of the α-particle is described in terms of the distance, r between the nucleus and the α-particle and ϕ, the angle at the nucleus between the direction of the α-particle and the direction of the momentum transfer, $q = p_f - p_i$

$$F = \frac{zZe^2}{4\pi\varepsilon_0 r^2},$$

where Ze is the electric charge of the nucleus, and ze is the electric charge of the incident particle (for an α-particle $z = 2$). Using the expression (1.2), relating kinetic energy, T, and the closest approach for head-on collision, D, one finds

$$F = \frac{TD}{r^2}. \tag{1.14}$$

At time t, the component of this force in the direction of q is

$$F_q(t) = \frac{TD}{r^2} \cos\phi(t). \tag{1.15}$$

This is equal to rate of change of momentum, so that the total change of momentum is given by

$$q = \int_{-\infty}^{\infty} \frac{TD}{r^2} \cos\phi(t)\, dt. \tag{1.16}$$

We can replace integration over time by integration over the angle ϕ using the relation

$$\frac{d\phi}{dt} dt = d\phi. \tag{1.17}$$

The time derivative of ϕ can be obtained from the conservation of angular momentum, L, where

$$L = m_\alpha r^2 \frac{d\phi}{dt}. \tag{1.18}$$

The initial angular momentum is given by

$$L = bp.$$

Inserting this into (1.18) yields

$$\frac{d\phi}{dt} = \frac{bp}{m_\alpha r^2}.$$

We see from Fig. 1.6 that the limits on the angle ϕ at very early and very late times are given by

$$\lim_{t\to\pm\infty} \phi(t) = \pm\frac{(\pi - \theta)}{2}$$

so that (1.16) becomes

$$q = \int_{-(\pi-\theta)/2}^{(\pi-\theta)/2} \frac{T\,D\,m_\alpha\,r^2}{r^2\,b\,p} \cos\phi\,d\phi = \int_{-(\pi-\theta)/2}^{(\pi-\theta)/2} \frac{D\,p}{2b} \cos\phi\,d\phi, \qquad (1.19)$$

where we have used $T = p^2/2m_\alpha$. Note that (the time-dependent) factor r^2 has cancelled, so that the integral over ϕ may be performed easily, yielding

$$q = \frac{Dp}{b} \sin\left(\frac{1}{2}(\pi - \theta)\right). \qquad (1.20)$$

Now using (1.13) we get

$$2p \sin\left(\frac{\theta}{2}\right) = \frac{Dp}{b} \sin\left(\frac{1}{2}(\pi - \theta)\right) \qquad (1.21)$$

from which it follows that

$$\tan\left(\frac{\theta}{2}\right) = \frac{D}{2b}. \qquad (1.22)$$

Chapter 2
Nuclear Size and Shape

2.1 What Is "Size"?

When we talk of the size of a macroscopic object, we all understand what is meant. With sub-microscopic objects such as atoms or nuclei, we need to be a bit more careful since the classical picture of electrons moving in fixed orbits around the nucleus is incompatible with Quantum Physics.

Due to Heisenberg's uncertainty principle, we cannot know exactly where an electron is in an atom. We can, however, determine its wavefunction $\Psi(r)$ (provided we can solve the Schroedinger equation for this wavefunction). The interpretation of the wavefunction is that the probability density, $P(r)$, (probability per unit volume) at position r is given by

$$P(r) = |\Psi(r)|^2, \tag{2.1}$$

and the size of the atom is the distance from the nucleus beyond which there is only a small probability of finding the electron.[1]

More precisely, the radius, R, of an atom is defined as the expectation value of the radial coordinate, r, of the argument, r of the wavefunction, $\Psi(r)$,

$$R \equiv \int r \, |\Psi(r)|^2 \, d^3r. \tag{2.2}$$

Unfortunately, we can only solve the Schroedinger equation exactly for the hydrogen atom. Nevertheless, this serves as a good example. In its ground state, the wavefunction of a hydrogen atom is given by

---

[1] The wavefunction, Ψ has space-dimension $-\frac{3}{2}$, so that $P(r)$ has the dimension of inverse volume.

© The Author(s), under exclusive license to Springer Nature Switzerland AG 2021
A. Belyaev, D. Ross, *The Basics of Nuclear and Particle Physics*, Undergraduate
Texts in Physics, https://doi.org/10.1007/978-3-030-80116-8_2

$$\Psi_{000}(r) \;=\; \frac{1}{\sqrt{\pi a_0^3}} e^{-r/a_0} \;,$$

where a_0 is the Bohr radius and takes the value 5.292×10^{-11} m (52920 fm).

The expectation value of the radial coordinate, R, can be calculated from (2.2) and takes the value

$$R \;=\; \frac{3}{2} a_0.$$

Using the expression for the probability density, (2.1), and integrating appropriately, we find that the probability of finding the electron within the atomic radius, R, is 58%, so it has a little less than an even chance of being found outside what we have defined as the atomic radius. On the other hand, the probability of finding the electron within twice the atomic radius is 94%, i.e. it is very unlikely to be found more than two atomic radii from the nucleus.

The nucleus consists of two types of *"nucleons"* – positively charged protons and uncharged neutrons. The nuclear radius is defined in analogy with the atomic radius as the expectation value of the radial component using the wavefunction of either type of nucleon. There is also a quantity called the *"charge radius"*, which is the expectation value of the radial coordinate using the wavefunction of the protons only. This is a measure of the distance from the centre over which there is a substantial probability of finding a proton – this is therefore a measure of the range over which the electric charge density is not too small.

The nuclear radius will depend on the atomic mass number, A, of the nucleus, which is the total number of nucleons (protons or neutrons). In the first approximation, we would expect the volume of the nucleus to be proportional to the total number of nucleons, i.e.

$$R \;=\; r_0 A^{1/3}. \tag{2.3}$$

For the charge radius, this works fairly well with the constant r_0 set to 1.22 fm. However, this is not totally correct as we would expect some dependence of the charge radius on the electromagnetic force between the protons, and we therefore expect the formula (2.3) to have a correction which depends on the difference between the number of protons and the number of neutrons in the nucleus. A better fit is obtained from the formula

$$R \;=\; r_A \left(1 + \beta \frac{Z}{A} \right) A^{1/3} \tag{2.4}$$

with (see e.g. [19]) $r_A = 0.718$ fm and $\beta = 0.556$. Alternative formulae which give an even more accurate estimate for the charge radii of nuclei have been developed, but these involve a larger number of free parameters, which need to be fitted. On the other hand, the formula (2.4) works well for the heavier nuclei with $A > 40$ (for

$A = 20$ there is an error of approximately 10%, and the agreement gets worse for even lighter nuclei).

The simpler expression (2.3) still works quite well and will be used as a valid approximation throughout this book.

Nuclear radii are of the order of a few fm. We have seen above that atomic radii are of the order of 50,000 fm. To get some understanding of the meaning of the ratio of such scales, one should note that if the atom were the size of a football, the nucleus would only be the size of a single grain of sand!

2.2   The Limitations of Rutherford Scattering

As explained in the previous chapter, one of the axioms that was postulated in order to derive the Rutherford scattering formula was that the nucleus could be considered to be a point particle. The finite size of a nucleus will lead to deviations from this formula when the incident projectile particle has sufficient energy to be able to penetrate the repulsive electric field down to a distance which is of the order of the nuclear size.

It is difficult to produce α-particles with sufficient energy to probe the charge distribution of the nucleus, so we use high energy electrons instead.

For electrons the projectile charge number, z, is set to unity in the Rutherford scattering formula. There is one further change which is due to the fact that these electrons are moving relativistically with a velocity v close to the speed of light, c. This correction depends on the spin of the incident particle. In the case of a singly charged, spin-$\frac{1}{2}$ projectile of mass m (such as an electron) the correction factor (first calculated by Nevill Mott [20]) is

$$\left(1 + \frac{p^2}{m^2 c^2} \cos^2\left(\frac{\theta}{2}\right)\right),$$

so that the (Mott) differential cross section is

$$\frac{d\sigma}{d\Omega}\bigg|_{\text{Mott}} = \left(\frac{Z\alpha\hbar c}{p^2}\right)^2 \frac{1}{4 \sin^4(\theta/2)} \left[m^2 + \frac{p^2}{c^2} \cos^2\left(\frac{\theta}{2}\right)\right]. \qquad (2.5)$$

In the non-relativistic limit, it is the first term in the square bracket that dominates and we recover the Rutherford scattering formula, (1.10) (with $z = 1$), whereas in the ultra-relativistic case it is the second term that dominates and the differential cross section becomes insensitive to the mass of the incident particle.

2.3   Diffraction

We account for the charge distribution of the nucleus by writing the differential cross section as

$$\frac{d\sigma}{d\Omega} = \frac{d\sigma}{d\Omega}\bigg|_{\text{Mott}} |F(q^2)|^2. \tag{2.6}$$

The correction factor $F(q^2)$ is called the *"electric form factor"* and $\mathbf{q}$ is the momentum transferred by the electron in the scattering, i.e. the difference between the final electron momentum, $\boldsymbol{p}_f$, and the initial momentum, $\boldsymbol{p}_i$, both of which have the same magnitude, p, but differ in direction by the scattering angle, θ. This electric form factor, $F(q^2)$, is in general a complex quantity, and it is the square modulus of this complex quantity which enters into the expression for the differential cross section.

From Fig. 2.1, and a little trigonometry, the magnitude of the momentum transferred (also shown in Appendix 1) is given by

$$q = 2p \sin\left(\frac{\theta}{2}\right). \tag{2.7}$$

To understand the structure of the electric form factor, we need to consider quantum effects. Recall that in Quantum Physics the electron behaves as a wave with de Broglie wavelength

$$\lambda = h/p.$$

More precisely, we can relate the "wave-vector", $\mathbf{k}$, to the momentum $\mathbf{p}$,

$$p = \hbar k, \tag{2.8}$$

where the magnitude, k, of the wave-vector, $\boldsymbol{k}$, is $2\pi/\lambda$ and its direction is the direction of the wave-motion.

We know from considering waves in optics that when a wavefront is incident on an object whose dimensions are of the order of the wavelength, the scattered wave displays maxima and minima. Precisely the same thing happens to de Broglie waves. When the de Broglie wavelength of the incident particle is of the order of the nuclear radius we get a diffraction pattern, in which the differential cross section

Fig. 2.1 Diagram showing the relation between momentum transfer and the scattering angle

Fig. 2.2 Diffraction from an impenetrable sphere

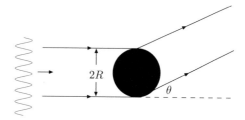

Fig. 2.3 Wave diffracted from the centre O, and wave diffracted from point P, whose polar coordinates are (r, ϑ, ϕ), have path difference $OB - AP$. $\mathbf{q}$ is the momentum transferred by the incident electron and is determined by the scattering angle, θ. Its direction is taken to be the polar axis. (Take care not to confuse the polar angle, ϑ, of the point P, with the scattering angle, θ)

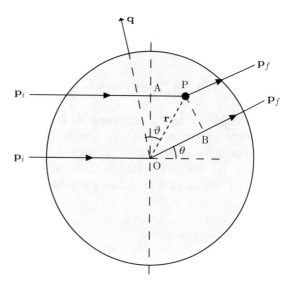

displays maxima and minima in directions when the waves from different parts of the nucleus are in phase or out of phase.

As a simple example, suppose that the nucleus is a solid sphere of radius R, as shown in Fig. 2.2, with an infinite potential inside the sphere and zero potential outside, so that the electron cannot penetrate the sphere.

The wave that passes below the nucleus travels a distance $2R \sin \theta$ further than the wave that passes above the nucleus. If this difference is equal to $\lambda/2$, $3\lambda/2 \cdots$, then we get destructive interference. At these angles the differential cross section vanishes.

The real case is a little more complicated than that. Let us compare the phase of the wave that passes through the centre of the nucleus, taken to be the origin, O, with that of the wave that passes though a point, P, whose vector from the origin is $\mathbf{r}$. This is shown in Fig. 2.3. At the wavefront OA, which passes through the origin, the two waves are in phase. But at the diffracted wavefront, PB, the wave that passes through the origin has travelled a distance, OB, which is the component of the vector $\mathbf{r}$ in the direction of the outgoing momentum, $\boldsymbol{p}_f$, and can be written as $\hat{\boldsymbol{p}}_f \cdot \boldsymbol{r}$, where $\hat{\boldsymbol{p}}_f$ is the unit vector in the direction of $\boldsymbol{p}_f$. Between the origin and the point B, the wave has undergone a phase advance of

$$\Delta\Phi_1 = \frac{2\pi}{\lambda}\hat{p}_f \cdot r = \frac{p_f \cdot r}{\hbar}, \tag{2.9}$$

where in the last step we have used the modified de Broglie relation (2.8). On the other hand, the wave that passes through the point P, has travelled the distance AP, which is the component of r in the direction of the incoming momentum, p_i, and can be written as $\hat{p}_i \cdot r$, and its phase advance is

$$\Delta\Phi_2 = \frac{2\pi}{\lambda}\hat{p}_i \cdot r = \frac{p_i \cdot r}{\hbar}. \tag{2.10}$$

From (2.9) and (2.10) the phase difference is

$$\Delta\Phi = \frac{(p_f - p_i) \cdot r}{\hbar} = \frac{q \cdot r}{\hbar}. \tag{2.11}$$

The amplitude of the component of the diffracted wave from the point P is proportional to the charge density $\rho_p(r)$ at that point (the subscript p indicates that since this is the electric charge density – it is actually the density of protons in the nucleus). The total diffracted wave (at scattering angle θ) is the sum of the waves diffracted from all such points, each with its relative phase factor

$$e^{iq \cdot r/\hbar}.$$

The weighted sum (integral) over all such points generates the electric form factor,

$$F(q^2) = \frac{1}{Ze} \int d^3r \, \rho_p(r) e^{iq \cdot r/\hbar}, \tag{2.12}$$

which is the Fourier transform of the charge distribution. Note that for $q = 0$, the electric form factor is unity – $F(0) = 1$.

The vector, r, has spherical polar coordinates (r, ϑ, ϕ) and the volume measure may be written as

$$d^3r = r^2 \sin\vartheta \, dr d\vartheta d\phi.$$

Furthermore, if we take the polar axis to be in the direction of the momentum transfer, q, then we have

$$q \cdot r = qr \cos\vartheta.$$

For the case of a spherically symmetric charge distribution, $\rho_p(r)$, the angular integration may be done easily, and this leads to

$$F(q^2) = \frac{4\pi\hbar}{Zeq} \int r \, \rho_p(r) \sin\left(\frac{qr}{\hbar}\right) dr. \tag{2.13}$$

Thus we see that a study of the diffractive scattering of electrons from a nucleus can give us information about the charge distribution inside the nucleus.

We have actually already considered two special cases of the charge distribution.

1. The case of a point-like nucleus assumed for the Rutherford (or Mott) scattering. In this case the charge density is written as

$$\rho_p(r) = Ze\delta^3(r),$$

and the electric form factor is $F(q^2) = 1$, i.e. there is no correction to the Mott scattering formula (as expected).

2. The case of the impenetrable sphere of radius R can be treated by assuming a spherically symmetric charge density

$$\rho_p(r) = \frac{Ze}{4\pi R^2}\delta(r - R),$$

giving rise to an electric form factor

$$F(q^2) = \frac{\hbar}{qR}\sin(qR/\hbar).$$

The differential cross section is proportional to $|F(q^2)|^2$ which has maxima approximately situated at

$$qR \approx \left(n + \frac{1}{2}\right)\pi\hbar$$

and minima (actually zeroes) for

$$qR = n\pi\hbar, \quad (n > 0).$$

As a more realistic example, we consider the charge distribution of a uniformly charged sphere is a constant for $r < R$ and zero outside

$$\rho_p(r) = \frac{3Ze}{4\pi R^3}, \quad r < R$$
$$= 0 \quad r > R.$$

The integral in the Fourier transform (2.13) can be done analytically using integration by parts to give

Fig. 2.4 Diffraction pattern from a uniformly charged sphere

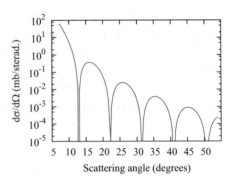

Fig. 2.5 Scattering cross section data of ^{40}Ca bombarded with electrons with energy 1040 MeV

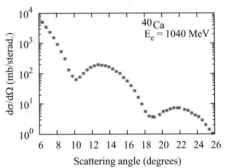

$$F(q^2) = 3 \left(\frac{\hbar}{qR} \right)^3 \left(\sin\left(\frac{qR}{\hbar} \right) - \frac{qR}{\hbar} \cos\left(\frac{qR}{\hbar} \right) \right).$$

Feeding this back into (2.6) for the diffractive differential cross section, we get a differential cross section shown in Fig. 2.4. This is not quite what is observed in experiment which is more like the example of scattering of electrons of energy 1040 MeV against a $^{40}_{20}$Ca (calcium) nucleus[2] shown in Fig. 2.5 [21]. We see that although there are oscillations in the differential cross section, it never actually vanishes. The reason for the discrepancy is that the uniformly charged sphere model for the charge distribution is unrealistic. The charge distribution rapidly becomes small as r exceeds a few fm, but certainly does not suddenly drop to zero for $r > R$.

Nevertheless, a reasonable estimate (to within about 30%) of the nuclear radius R can be obtained from the position of the first minimum of the diffraction pattern. We assume that this first minimum occurs when

$$\frac{qR}{\hbar} \approx \pi.$$

---

[2]The notation ^A_Z{Ch} is used to denote a nucleus of the chemical element, whose atomic number is Z, chemical symbol is {Ch}, and atomic mass number is A.

Using (2.7) to relate the momentum transfer to the scattering angle, we therefore see that for an incident particle with momentum p, and the first minimum at a scattering angle θ_1, the nuclear radius is given (approximately) by

$$R \approx \frac{\pi \hbar}{2p \sin (\theta_1/2)} . \tag{2.14}$$

In the case of electron scattering off calcium, we see from Fig. 2.5 that the first minimum occurs at $10°$. The momentum of the (ultra-relativistic) electron is 1040 MeV/c. Inserting these values and $\hbar c = 197$ MeV fm, we find the radius, R, of the $^{40}_{20}$Ca nucleus is

$$R \approx 3.4 \, \text{fm}.$$

2.4   The Saxon–Woods Distribution

A more realistic model for the charge distribution is the Saxon–Woods distribution [22] for which

$$\rho_p(r) = \rho_0 f_{R,\delta}(r) ,$$

where R is the nuclear radius and the function $f_{a,b}(r)$ is called a *"Saxon–Woods potential"*, with parameters (a, b). Such a potential is given by

$$f_{a,b}(r) = \frac{1}{1 + e^{(r-a)/b}}, \tag{2.15}$$

The overall normalization, ρ_0 is chosen such that total charge is Ze, i.e.

$$Ze = 4\pi \rho_0 \int r^2 \, dr \, \frac{1}{1 + \exp((r - R)/\delta)}.$$

The Saxon–Woods distribution is shown in Fig. 2.6. We take R to be the nuclear charge radius and δ is the "surface depth" – it measures the range in r over which the charge distribution is substantially reduced from its value at $r = R$. It is a parameter, which needs to be fit to data for every nucleus.

This leads to a differential cross section which is shown in Fig. 2.7, where we have taken the values $R = 3.4$ fm and $\delta = 0.58$ fm. We see that this predicted differential cross section has dips but no zeros and is much more similar in shape to the experimental results.

In fact, the Saxon–Woods model fits data from most nuclei rather well for nuclei with atomic mass number $A > 40$, with the charge radius given by (2.4) for atomic number, Z, and atomic mass number, A, and the parameter δ in the range 0.4–0.5 fm.

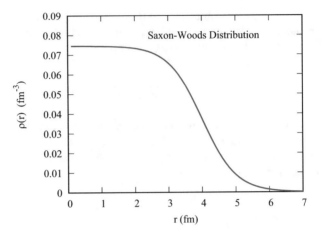

Fig. 2.6 Saxon–Woods charge distribution

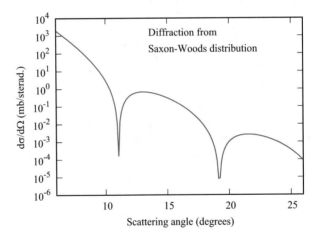

Fig. 2.7 Scattering cross section prediction using Saxon–Woods potential

The Saxon–Woods model has been further refined, e.g. by multiplying the charge density by a polynomial in r. This leads to an enhancement of the quality of the fit to data, but at the expense of introducing more parameters.

2.5   Electric Quadrupole Moments

So far, we have assumed that the charge distribution is spherically symmetric. If that were the case we would have

$$< x^2 > = < y^2 > = < z^2 > = \frac{1}{3} < r^2 >,$$

where

$$< x^2 > = \frac{1}{Ze} \int x^2 \rho(\mathbf{r}) d^3\mathbf{r},$$

etc.

However, for many nuclei this is not the case. They usually still have an axis of symmetry, which we set to be the z-axis, so that they are symmetric in the $x - y$ plane but asymmetric in the $x - z$ or $y - z$ planes. In polar coordinates (r, ϑ, ϕ), this means that the charge distribution is a function of the polar angle, ϑ, but for nuclei which still maintain one axis of symmetry (the z-axis), the charge distribution is independent of the azimuthal angle, ϕ.

We can still determine a charge radius

$$R = \int r \rho(\mathbf{r}) d^3\mathbf{r},$$

even if $\rho(\mathbf{r})$ is not a function of the radial component, r, alone. However, if we look at the expectation values of the squares of individual component of $\mathbf{r}$

$$\langle x^2 \rangle = \int x^2 \rho(\mathbf{r}) d^3\mathbf{r}, \quad \langle y^2 \rangle = \int y^2 \rho(\mathbf{r}) d^3\mathbf{r}, \quad \langle z^2 \rangle = \int z^2 \rho(\mathbf{r}) d^3\mathbf{r},$$

we find that these are not equal. Nuclei which nevertheless have an axis of symmetry have the same values of $\langle x^2 \rangle$ and $\langle y^2 \rangle$, but a different value for $\langle z^2 \rangle$.

Nuclei with a charge distribution that is not spherically symmetric possess an *"electric quadrupole moment"* defined (with respect to the z-axis) as

$$Q = 3\langle z^2 \rangle - \langle r^2 \rangle = \int (3z^2 - r^2) \rho(\mathbf{r}) d^3\mathbf{r}, \tag{2.16}$$

where

$$\langle r^2 \rangle = \langle x^2 \rangle + \langle y^2 \rangle + \langle z^2 \rangle.$$

For a charge distribution which has one axis of symmetry (the z-axis), we may also write the electric quadrupole moment as

$$Q = 2 \left(\langle z^2 \rangle - \langle x^2 \rangle \right). \tag{2.17}$$

The electric quadrupole moment is a measure of the deviation of the electric charge distribution of the nucleus from spherical symmetry.

For example, the charge distribution (in terms of spherical polar coordinates (r, ϑ, ϕ)) could be of the form

$$\rho(r) = a(r) + b(r) |\cos \vartheta| .$$

In polar coordinates $z = r \cos \vartheta$ and therefore

$$\langle z^2 \rangle = \int_0^\infty r^2 dr \int_0^\pi \sin \vartheta \, d\vartheta \int_0^{2\pi} d\phi \, r^2 \cos^2 \vartheta \, (a(r) + b(r) |\cos \vartheta|) .$$

Performing the angular integrals over ϕ and ϑ, we get

$$\langle z^2 \rangle = 4\pi \int dr r^4 \left(\frac{1}{3} a(r) + \frac{1}{4} b(r) \right) .$$

On the other hand, $x = r \sin \vartheta \cos \phi$, so that

$$\langle x^2 \rangle = \int_0^\infty r^2 dr \int_0^\pi \sin \vartheta \, d\vartheta \int_0^{2\pi} d\phi \, r^2 \sin^2 \vartheta \cos^2 \phi \, [a(r) + b(r) |\cos \vartheta|]$$

$$= 4\pi \int dr \, r^4 \left[\frac{1}{3} a(r) + \frac{1}{8} b(r) \right] .$$

The electric quadrupole moment is therefore

$$Q = \pi \int dr r^4 b(r).$$

Q/e has dimensions of area and is therefore usually quoted in millibarns (mb).

Nuclei that possess an electric quadrupole moment but are nevertheless symmetric about the z-axis have a shape which is an oblate spheroid for $Q < 0$ (i.e. $\langle z^2 \rangle < \langle x^2 \rangle$) or a prolate spheroid for $Q > 0$ (i.e. $\langle z^2 \rangle > \langle x^2 \rangle$). Examples of these shapes are shown in Fig. 2.8.

For nuclei whose charge distribution is not spherically symmetric, the electric form factor, given by (2.12), turns out to be complex.

Fig. 2.8 The charge distribution shown on the left is an oblate ellipsoid and represents a nucleus with a negative Q. The charge distribution shown on the right is prolate ellipsoid and represents a nucleus with a positive Q

On the other hand, the electric dipole moment, which is a vector defined by

$$d = \int r \rho_p(r) d^3 r,$$

is almost zero. The reason for this is that to a very good approximation, the wavefunction of a proton in a nucleus is a *"parity"* eigenfunction, i.e.

$$\Psi(r) = \pm \Psi(-r)$$

which implies

$$\rho(r) = |\Psi(r)|^2 = |\Psi(-r)|^2 = \rho(-r),$$

so that the electric dipole moment vanishes by symmetric integration.

2.6 Strong Force Distribution

The protons and neutrons inside a nucleus are held together by a strong nuclear force. This has to be strong enough to overcome the Coulomb repulsion between the protons, but unlike the Coulomb force, it extends only over a short range of a few fm.

Electrons are used to probe the charge distribution of the target nuclei, because they interact with the electric field, but not with the strong forces. Likewise, scattering of neutrons from a nucleus can be used to probe the strong force distribution, but not the electric charge distribution since neutrons are uncharged but interact strongly.

As in the case of electron scattering, the cross section of neutron scattering displays a diffraction pattern if the de Broglie wavelength of the neutrons is of the order of the nuclear size. In such a case the wave from different parts of the nucleus interfere to produce diffraction maxima and minima at different scattering angles. This can be seen in Fig. 2.9 in which neutrons with kinetic energy 14 MeV are scattered from a Ni (nickel) nucleus. The de Broglie wavelength of neutrons with kinetic energy 14 MeV is approximately 1.2 fm so that the first minimum of the differential cross section, at a scattering angle of 42°, implies an effective radius of the strong force distribution of a few fm. This is similar to the charge radius of the nucleus. We would expect the total nucleon distribution to have the same range as the proton (charge) distribution. However, whereas the Coulomb potential from the nucleus is long-range, being attenuated as the inverse of the distance from the centre of the nucleus, the strong force is rapidly attenuated and becomes negligible after a few fm from the nucleus.

The differential cross section can be expressed in terms of a *"scattering amplitude"*, $f(\theta, \phi)$ (usually a function of the scattering angle, θ, only, but could

Fig. 2.9 Differential
scattering cross section of
neutrons with kinetic energy
14 MeV off a nickel target

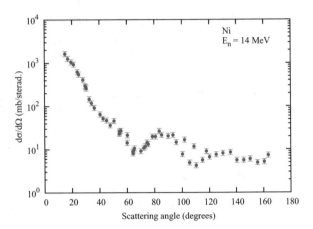

in some cases depend on the azimuthal angle, ϕ, of the outgoing particle), via the
relation

$$\frac{d\sigma}{d\Omega} = |f(\theta, \phi)|^2. \tag{2.18}$$

For an incident particle moving in some general potential $V(r)$, the scattering
amplitude to first order in the potential[3] by the *"Born approximation"* [23]

$$f(\theta, \phi) = \frac{m}{2\pi\hbar^2} \int d^3r \, V(r) e^{i q \cdot r/\hbar}, \tag{2.19}$$

i.e. it is proportional to the Fourier transform of the potential as a function of the
momentum transfer, q, which is related to the scattering angle, θ by (2.7). A brief
derivation of this result can be found in Appendix 2. Again, the reader who is happy
to accept (2.19) at face value need not read the appendix. The result can also be
obtained using Fermi's golden rule [24], but use of the Born approximation is more
rigorous, and not particularly difficult to follow.

Whereas the Coulomb potential due to electromagnetic interactions is well-
understood and can be applied at all energy scales, the only theory we have
of the *"strong interactions"* is Quantum Chromodynamics (QCD), (described in
Chap. 20), but this cannot be applied at the relatively low energies of neutron-
nucleus scattering.

Instead we describe neutron scattering using the Optical Model in which there
is an effective potential, which can be inserted into (2.19) in order to obtain the
scattering amplitude and hence the scattering differential cross section.

---

[3]Whereas such an approximation is clearly valid for an electromagnetic potential which is weak, it
is by no means clear that it is a good approximation for strong nuclear forces. Nevertheless, it has
been used quite successfully to describe neutron scattering data.

The details of the optical model are beyond the scope of this book, but the essential idea is that the effective potential is based on the Saxon–Woods potential, and this, in general, leads to a diffraction pattern. The effective potential is more complicated than the simple Saxon–Woods charge distribution used to describe electron–nucleus scattering. The potential is taken to be complex, with the imaginary part describing the inelastic scattering cross section in which the incident neutron is absorbed by the target nucleus or in which the nucleus splits into smaller nuclei – a process called *"fission"* (which will be discussed in Chap. 9). The real and imaginary parts are both taken to be Saxon–Woods potential (2.15) each with its own coefficient and parameters, a, b.

Furthermore, a spin–orbit coupling term is added to account for the strong interaction between the orbital angular momentum **L** and the spin σ of the neutron (in analogy with the spin–orbit coupling in Atomic Physics). This term is proportional to the radial derivative of a Saxon–Woods potential, and again there is, in general, a real and an imaginary component.

The total effective potential is therefore of the form

$$V(r) \;=\; V f_{a_1,b_1}(r) + i W f_{a_2,b_2}(r) + \left(V_{so}\, g_{a_3,b_3}(r) + i W_{so}\, g_{a_4,b_4}(r)\right) \boldsymbol{L}\cdot\boldsymbol{\sigma}, \qquad (2.20)$$

where

$$g_{a,b}(r) \;=\; \frac{1}{r}\frac{d}{dr} f_{a,b}(r).$$

Despite the fact that the imaginary part of the spin–orbit term can usually be neglected ($W_{so} \approx 0$), the model still has 9 free parameters, and in addition the coefficients, V, W and V_{SO} have a linear dependence on the energy of the incident neutron. These parameters have to be fit to data. However, notwithstanding this uncomfortably large number of free parameters, the model describes neutron scattering well, as can be seen from Fig. 2.10, which shows the fit to data for neutron scattering using 11 MeV neutrons scattering off several different target nuclei.

Summary

- The "radius" of a nucleus is the quantum expectation value of the radial component (distance from the centre of the nucleus), with the probability density defined in terms of the wavefunction for all the nucleons. The "charge radius" is defined similarly but using the wavefunction of the protons only.
- The Mott scattering cross section formula is a correction to the Rutherford scattering cross section formula which accounts for the relativistic behaviour of a spin-$\frac{1}{2}$ incident particle.
- In a scattering experiment in which the incident particle has a sufficiently large momentum that its de Broglie wavelength is of the order of the nuclear radius, a

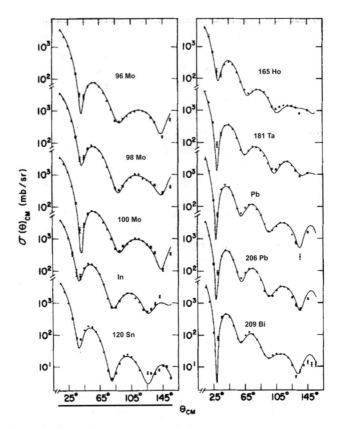

Fig. 2.10 The best fit of the Optical Model to the neutron scattering cross section with neutron energy 11 MeV, off various nuclear target. [Reprinted from [25] by permission from Elsevier]

diffraction pattern is observed with maxima (minima) corresponding to scattering angles at which the waves diffracted from different parts of the nucleus are respectively in (out of) phase.

- The effect of diffraction from a charged incident particle (usually an electron) is to multiply the Mott scattering cross section by the square modulus of the electric form factor of the nucleus. This electric form factor is proportional to the (three-dimensional) Fourier transform of the electric charge distribution.
- Nuclei whose electric charge distribution is not spherically symmetric possess an electric quadrupole moment, which is a measure of the asymmetry of the charge distribution. It is defined as the difference between the expectation value of $3z^2$ and the expectation value of r^2.
- The strong force of a nucleus can be probed using neutron scattering. This also yields a diffraction pattern, but this pattern is controlled by the strong interaction between the nucleus and the neutron. The strong force potential is not understood quantitatively. Nevertheless neutron scattering data are well-described using the

optical model in which the scattering amplitude is determined from an effective potential, which is complex and possesses an additional term describing the spin–orbit interaction. Each of these components is taken to be a Saxon–Woods potential with different parameters and coefficients.

Problems

The following definite integrals are required in order to be able to solve these problems:

$$\int_0^\infty dr\ \sin(br)e^{-ar} = \frac{b}{(a^2+b^2)}, \quad \int_0^\infty dr\ r\sin(br)e^{-ar} = \frac{2ab}{(a^2+b^2)^2}$$

$$\int_0^\infty dr\ r^n e^{-ar} = \frac{n!}{a^{(n+1)}}, \quad \int_0^\pi d\vartheta\ \cos^{2n}\vartheta\ \sin\vartheta = \frac{2}{2n+1},$$

$$\int_0^{2\pi} d\phi\ \cos^2\phi = \pi.$$

Problem 2.1 In an experiment in which electrons with energy 420 MeV are scattered off an oxygen nucleus, the first minimum is observed at a scattering angle of $43°$ followed by a maximum at a scattering angle of $51°$. At what scattering angles would you expect to find the corresponding minimum and maximum when the electron energy is reduced to 360 MeV?

Problem 2.2 Estimate the scattering angle of the first diffraction minimum for the scattering of electrons with energy 225 MeV off a $^{58}_{28}\text{Ni}$ nucleus.
[Take the nuclear radius to be $1.22A^{1/3}$ fm.]

Problem 2.3 Consider a nucleus with atomic number Z and a spherically symmetric electric charge distribution

$$\rho(r) = Ze\frac{a^3}{8\pi}e^{-ar}.$$

Find an expression for the electric form factor from such a distribution and hence write down an expression for the angular dependence of the differential cross section for scattering of an ultra-relativistic ($v \approx c$) incident particle with momentum p.

Problem 2.4 Calculate the scattering amplitude for a particle with mass m and momentum p, moving in a Yukawa potential

$$V(r) = V_0\frac{e^{-\mu r}}{r}.$$

By taking the limit $\mu \to 0$ and an appropriate expression for the coefficient V_0. Verify that this reproduces the Rutherford scattering formula for the differential cross section of an incident particle with momentum p in a Coulomb potential due to a nucleus with electric charge Ze.

Problem 2.5 Consider a charge distribution for a nucleus of atomic number Z,

$$\rho(r) = \rho_0 \left(r + \epsilon r^2 \cos^2 \vartheta\right) e^{-ar}.$$

(a) Find an expression for ρ_0 in terms of Z, ϵ and a.
(b) Find an expression for the charge radius, R_c in terms of Z, ϵ and a.
(c) Find an expression for the electric quadrupole moment Q in terms of Z, ϵ and a.

Appendix 2 – The Born Approximation

A particle of mass m and initial momentum of magnitude $\hbar k$ moving in a localized potential $V(r)$ obeys the (time independent) Schroedinger equation

$$\frac{\hbar^2}{2m} \left(\nabla^2 + k^2\right) \Psi(r) = V(r)\Psi(r). \tag{2.21}$$

The total energy is $\hbar^2 k^2/2m$ and at sufficiently large r, where the potential is negligible, this total energy is just the kinetic energy of a particle with mass m and momentum of magnitude $\hbar k$.

We seek a solution to this equation which in the limit $V(r) \to 0$ is the wavefunction for a free particle moving in the direction $\hat{k}_i$, i.e.

$$\Psi(r) \xrightarrow{V \to 0} e^{ik_i \cdot r}. \tag{2.22}$$

Such a wavefunction obeys the implicit (transcendental) equation

$$\Psi(r) = e^{ik_i \cdot r} - \frac{2m}{\hbar^2} \int d^3 r' \frac{e^{ik|r-r'|}}{4\pi |r - r'|} V(r') \Psi(r'). \tag{2.23}$$

This can be seen using

$$\left(\nabla^2 + k^2\right) \frac{e^{ik|r-r'|}}{4\pi |r - r'|} = -\delta^3(r - r'). \tag{2.24}$$

The Born approximation consists of solving this transcendental equation iteratively up to leading order in the potential, V.

$$\Psi(r) \approx e^{ik_i \cdot r} - \frac{m}{2\pi \hbar^2} \int d^3 r' \frac{e^{ik|r-r'|}}{|r-r'|} V(r') e^{ik_i \cdot r'}. \tag{2.25}$$

Now we will take the limit $r \gg R$, where R is the distance beyond which the potential becomes negligible. In this limit, the integrand in (2.25) only has support for $r' \ll r$, so that we have

$$|r - r'| \approx r - r' \cdot \hat{r} + \mathcal{O}\left(\frac{r'^2}{r}\right),$$

($\hat{r}$ is the unit vector in the direction $\mathbf{r}$.) The final wave-vector (final momentum divided by $\hbar$) is the component of the momentum in the direction $\mathbf{r}$

$$k_f = k\hat{r},$$

The denominator $|r - r'|$ in (2.25) is approximated by r so that we can take e^{ikr}/r outside the integral, leaving

$$\Psi(r) \approx e^{ik_i \cdot r} - \frac{e^{ikr}}{r} \frac{m}{2\pi \hbar^2} \int d^3 r' \, V(r') e^{i(k_i - k_f) \cdot r'}, \tag{2.26}$$

which we may write as

$$\Psi(r) \approx e^{ik_i \cdot r} + \frac{e^{ikr}}{r} f(\vartheta, \phi), \tag{2.27}$$

where the scattering amplitude, $f(\vartheta, \phi)$, is given by

$$f(\theta, \phi) = -\frac{m}{2\pi \hbar^2} \int d^3 r' \, V(r') e^{i(k_i - k_f) \cdot r'}. \tag{2.28}$$

The particle flux (number of particles emitted per unit time into unit area) is

$$F = \frac{-i\hbar}{2m} \left(\Psi^*(r) \nabla \Psi(r) - \Psi(r) \nabla \Psi^*(r) \right). \tag{2.29}$$

The outgoing flux in the direction (ϑ, ϕ) is therefore

$$F_{\vartheta,\phi} = \frac{\hbar k}{m r^2} |f(\vartheta, \phi)|^2. \tag{2.30}$$

At a distance r from the scattering centre, an element of area dA subtends a solid angle $d\Omega = dA/r^2$ so that the rate of particles emitted into an element, $d\Omega$, of solid angle in the direction (ϑ, ϕ) is then

$$N_{\vartheta,\phi} d\Omega = F_{\vartheta,\phi} r^2 d\Omega = \frac{\hbar k}{m} |f(\vartheta, \phi)|^2. \tag{2.31}$$

To find the differential cross section we divide this by the incoming flux, which is

$$F_i = \frac{\hbar k}{m},$$ (2.32)

so that the differential cross section is

$$\frac{d\sigma}{d\Omega} = |f(\vartheta, \phi)|^2,$$ (2.33)

with $f(\vartheta, \phi)$ given in the Born approximation by (2.19).

Chapter 3
Nuclear Masses and the Semi-Empirical Mass Formula

3.1 Some Nuclear Nomenclature

We start this chapter with some definitions of important terms, commonly used in Nuclear Physics. We remind the reader that all these technical terms are defined in the glossary so there is no need to remember them – just consult the glossary whenever necessary:

- *"Nucleon"*: A proton or neutron.
- *"Atomic Number"*, *Z*: The number of protons in a nuclide. This determines the electric charge, Ze, of the nucleus. It is equal to the number of electrons in the corresponding (neutral) atom and therefore determines the chemical properties of that atom.
- *"Atomic Mass number"*, *A*: The number of nucleons in a nucleus.
  We also find it convenient to use the notation $N \equiv A - Z$ to indicate the number of neutrons.
- *"Nuclide"* ($\mathcal{N}$): A nucleus with a specified value of atomic mass number, A, and atomic number, Z. This is usually written as $^A_Z\{\text{Ch}\}$ where Ch is the chemical symbol for an atom with Z electrons. Thus, for example, $^{60}_{28}\text{Ni}$ means nickel with 28 protons and a further 32 neutrons.
- *"Isobar"*: This word is derived from the Greek words *iso* meaning the "same" and *baros* meaning "weight". It means a nucleus with a given atomic mass number, *A*, but a different atomic number, *Z*. Such nuclei have approximately *but not exactly* the same mass, but the atoms have different atomic and chemical properties.
- *"Isotope"*: A nucleus with a given atomic number, *Z*, but different atomic mass numbers, *A*, i.e. different number of neutrons, *N*. Isotopes have very similar atomic and chemical behaviour but may have very different nuclear properties.

© The Author(s), under exclusive license to Springer Nature Switzerland AG 2021
A. Belyaev, D. Ross, *The Basics of Nuclear and Particle Physics*, Undergraduate Texts in Physics, https://doi.org/10.1007/978-3-030-80116-8_3

- *"Isotone"*: A nucleus with a given number of neutrons, N, but a different number of protons. (This concept is not very useful and so this term does not occur very often.)
- *"Ion"*: An atom with one or more electrons missing (positive ion) or one or more extra electrons (negative ion).
- *"Mirror Nuclides"*: Two nuclide in which the number of protons in one nuclide is equal to the number of neutrons in the other and vice versa.

3.2   Measuring Nuclear Masses

The masses of nuclei can now be measured with very high accuracy. For some nuclides the error is less than one part in a hundred million. There are three main methods used for measuring the masses of nuclei with such a high precision:

1. **Penning trap** The Penning trap was named after Franz Penning, but actually originally built in 1961 by Hans Georg Dehmelt (unpublished). Dehmelt was inspired by the vacuum gauge built by Penning, which measures low pressures from the current in an electric discharge in a gas.

   A Penning trap is a cylindrical device in which ions with charge q and mass m are trapped in the radial direction by means of a uniform axial magnetic field, B, and in the axial direction by a non-uniform quadrupole electric field. The trapped ions have three modes of vibration, as shown in Fig. 3.1. They orbit in the radial plane with an angular frequency ω_- called the *"magnetron angular frequency"*, they oscillate in that plane with a larger angular frequency, ω_+, called the *"modified cyclotron angular frequency"*, and they also oscillate in the axial direction with *"axial angular frequency"* ω_z. The cyclotron angular frequency, which is the sum of the modified cyclotron angular frequency and the

Fig. 3.1 The three modes of oscillation of ions in a Penning trap [*Source*: https://www.med. physik.uni-muenchen.de/ research/nuclear-science/ nuclear-masses/mlltrap/ pictures/a_kap2_eigen1.jpg (edited)]

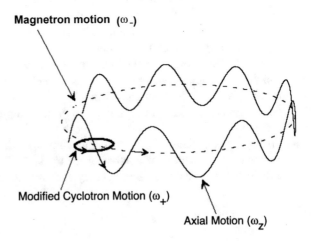

Magnetron motion (ω_-)

Modified Cyclotron Motion (ω_+)

Axial Motion (ω_z)

magnetron angular frequency, is related to the mass-to-charge ratio of the ions and the magnitude of the applied magnetic field by

$$\omega_c \equiv \omega_+ + \omega_- = \frac{qB}{m}. \tag{3.1}$$

This frequency is determined by applying an azimuthal quadrupole radio-frequency (RF) oscillating electric field. When the frequency of this RF field is exactly equal to the cyclotron frequency, a resonance is achieved and the energy gain of the ions reaches a maximum. To detect this maximum, the ions are ejected from the trap and their time of flight is measured by the external ion detector. The resonance corresponds to minimum time of flight. The system has to be calibrated using ions of some well-known stable nuclei. The best mass determining precision $(\delta m/m)$ is $\simeq 10^{-9}$, for light ions, and $\simeq 10^{-7}$ for heavier ones.

Penning traps are used in many laboratories worldwide, including CERN to store antimatter such as antiprotons.

2. **Storage Ring Mass Spectrometer**

This is a storage ring for ions such as the Experimental Storage Ring (ESR) at the GSI Helmholzzentrum in Darmstadt, Germany.[1] The frequency, f, of rotation around the ring depends on the mass-to-charge ratio of the ions, such that the difference, Δf, between rotation frequencies between two nuclides with similar mass (and the same electric charge) is given by

$$\frac{\Delta f}{f} \approx -\frac{\Delta m}{m}. \tag{3.2}$$

This device is useful for determining the masses of certain nuclides using the known masses of standard reference nuclides. The above relation has (relativistic) corrections due to the spread of velocities of the ions in the storage ring. Errors caused by such corrections are suppressed by one of two techniques:

a. **The Schottky mass spectrometer:** This uses electron cooling of the ions. This is the process in which the ions scatter off electrons due to the Coulomb interactions between them. They slow down (cool) through the exchange of momentum and reach equilibrium when all the particles have the same momentum.

b. **The isochronous mass spectrometer:** This is based on a time-of-flight measurement and exploits the fact that the trajectories of the slower ions are slightly shorter than those of the faster ions, which compensates for the error caused by the velocity discrepancy.

---

[1] There are two other such storage ring facilities – HRFL-CSR at the Chinese Academy of Sciences in Lanzhou, and the Rare RI Ring at the Riken Nishina Centre in Saitama, Japan.

Measurements using this type of spectrometer can be conducted very quickly, so that this is a very suitable device for measuring the masses of unstable nuclides, with lifetimes down to only a few microseconds.

3. **Multiple Reflection Time-of-Flight Mass Spectrometer**

   This is a device in which the ions are reflected between two electrostatic mirrors and traverse the path between the two mirrors over 1000 times. This increases the time of flight, T, which can then be measured very precisely. Accuracies of better than one part in ten million have been achieved using this device. The time of flight, T, depends on the ratio of the mass, m, of the ions and their electric charge, q

   $$T = \alpha \sqrt{\frac{m}{q}} + \beta. \tag{3.3}$$

The constants α and β are determined by calibrating the spectrometer with two standard nuclides whose masses are known to a sufficiently high accuracy. Further details of the working of this device can be found in [26].

3.3 Binding Energy

Initially, one might have thought that the mass of a nuclide was simply the total mass of the nucleons $Zm_p + Nm_n$. In reality, the mass of a given nuclide is slightly less than this, by a quantity known as the *"mass defect"*, $\Delta m_\mathcal{N}$. When nucleons bind to form a nucleus, there is a binding energy, $B(A, Z)$, which is the energy that would be required to pull the nucleons apart. The binding energy depends on both the atomic mass number and the atomic number.

From Einstein's Special Theory of Relativity (and the one physics equation that is familiar to almost everybody), the mass of the nuclide is therefore less than the sum of the masses of the nucleons by an amount equal to this binding energy divided by the square of the speed of light.

$$\Delta m_\mathcal{N} = \frac{B(A, Z)}{c^2}, \tag{3.4}$$

so that the expression for the mass, $m_\mathcal{N}$, of a given nuclide is

$$m_\mathcal{N} = Zm_p + Nm_n - \frac{B(A, Z)}{c^2}. \tag{3.5}$$

The binding energy is due to the strong short-range nuclear force that binds the nucleons together, as well as the long-range electromagnetic (Coulomb) repulsion between the positively charged protons. However, the strong nuclear force is far less understood than the electromagnetic force, so that the binding energy due to the

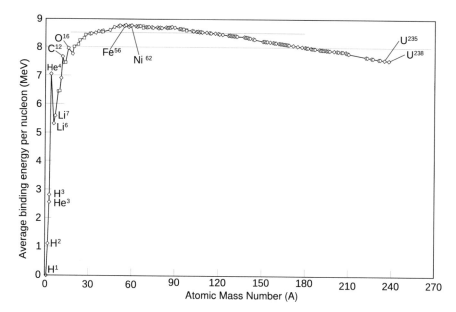

Fig. 3.2 The binding energy per nucleon plotted against atomic mass number. The most stable nuclei are those with atomic mass number of around 62, such as nickel [*Source*: https://upload.wikimedia.org/wikipedia/commons/thumb/5/53/Binding_energy_curve_-_common_isotopes.svg/500px-Binding_energy_curve_-_common_isotopes.svg.png (edited)]

strong force cannot, even in principle, be calculated analytically. All we know about it is that it is very short range, so that it acts only between neighbouring nucleons, whereas the long-range electromagnetic interactions extend over the entire nucleus (and beyond).

Binding energies per nucleon increase sharply with increasing atomic mass number, A, peaking at $^{62}_{28}\text{Ni}$ (nickel)[2] and then decreasing slowly for the more massive nuclei (see Fig. 3.2). This means that the nuclei of $^{62}_{28}\text{Ni}$ and nuclei with similar mass are particularly stable, whereas the more massive nuclei are less stable and can decay into lighter nuclei by fission that will be discussed in detail in Chap. 9.

3.4 Semi-Empirical Mass Formula

For all but the very lightest of nuclei, the binding energy is well reproduced by a semi-empirical formula based on the idea, originally postulated by George Gamow

---

[2]Because of the small mass difference between protons and neutrons, it is actually the isotope $^{56}_{26}\text{Fe}$ (iron), which has the largest binding energy per unit mass, but $^{62}_{28}\text{Ni}$ has the largest binding energy per nucleon.

in 1930 [27], in which the nucleus can be thought of as a liquid drop composed of nucleons (the *"Liquid Drop Model"*). The volume of the liquid drop is proportional to the number of nucleons, A. The semi-empirical mass formula for the binding energy of a given nuclide, as a function of atomic number, Z, and atomic mass number, A, was derived independently by Carl von Weizsäcker [28] in 1935 and Hans Bethe [29] a year later. The formula is also known as the Bethe–Weizsäcker mass formula.

It is "semi-empirical" in that it contains a number of terms whose dependence on Z and A is derived from physics, but which have coefficients that are unknown and have been fit to experimental data on nuclear binding energies. There are five terms, some of which come from a purely classical description of a liquid drop comprising a number of nucleons, and others that are purely quantum effects:

1. **Volume term:** Each nucleon has a binding energy that binds it to the nucleus. The interactions are short range, so to a good approximation the nucleons only interact with their nearest neighbours. Therefore we get a term proportional to the volume of the liquid drop. This in turn is proportional to the number of nucleons, A, with a coefficient, a_V:

$$a_V A.$$

2. **Surface term:** By assuming that each nucleon interacts with its nearest neighbours, and that therefore we get a term proportional to A, the binding energy is overestimated because the nucleons at the surface of the liquid drop only interact with other nucleons inside the nucleus, so that their binding energy is reduced. This leads to a reduction of the binding energy proportional to the surface area of the drop. For a liquid drop comprising of A nucleons the surface area is proportional to $A^{2/3}$ (treating the drop as spherical). The surface term is therefore of the form:

$$-a_S A^{2/3}.$$

3. **Coulomb term:** The binding energy arising from the strong force is somewhat reduced owing to the Coulomb repulsion between the protons. Since the electromagnetic interactions are long range, we expect each proton to interact with all other protons and therefore the Coulomb term is proportional to the square of the nuclear charge, Ze. Furthermore, by Coulomb's law the repulsive potential is expected to be inversely proportional to the nuclear radius.
A uniformly charged sphere of radius R and charge Ze has an electrostatic potential

$$\frac{3Z^2 e^2}{20\pi \varepsilon_0 R}.$$

Using the fact that the nuclear radius is (to a very good approximation) proportional to $A^{1/3}$, treating the nucleus as a uniformly charged sphere, the Coulomb term is of the form

$$-a_C \frac{Z^2}{A^{1/3}}.$$

4. **Asymmetry term:** This is a quantum effect arising from the Pauli exclusion principle that only allows two protons or two neutrons (with opposite spin) in each energy state. If a nucleus contains the same number of protons and neutrons, then the protons and neutrons fill the available energy levels to the maximum energy level (the *"Fermi energy"*). If, on the other hand, one of the protons is replaced by a neutron (or vice versa), then that neutron would be required by the exclusion principle to occupy a higher energy state, since all the ones below it are already occupied. If more protons are replaced by neutrons, these neutrons will have to occupy even higher energy levels. This is illustrated in Fig. 3.3.

Note that the binding energy gives a *negative* contribution to the total energy of the nucleus, so that the nuclide with an equal number, Z, of protons and neutrons, N, has a larger binding energy.

The ground-state energy, E_0, for n identical non-interacting spin-$\frac{1}{2}$ particles of mass m confined to a volume, V, is given by[3]

$$E_0 = \pi^{4/3} \frac{\hbar^2}{10m} \frac{(3n)^{5/3}}{V^{2/3}}. \tag{3.6}$$

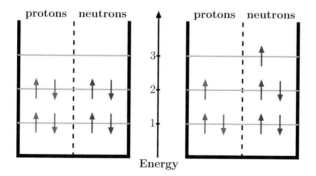

Energy

Fig. 3.3 The left-hand diagram shows a nuclide with equal numbers (4) of protons and neutrons. In the ground state of the nucleus, these fill the lowest 2 available energy levels. The right-hand diagram shows the case where one of the protons is replaced by a neutron. By the exclusion principle, this extra neutron cannot go into one of the lowest two energy levels as these are already filled. It must go into the next energy level so that the energy of such a nucleus in its ground state is larger

---

[3] A brief derivation of this result is given in Appendix 3.

Therefore, (neglecting the very small mass difference between the proton and the neutron), the ground-state energy for Z protons and N neutrons in a volume, V, is proportional to

$$\frac{Z^{5/3} + N^{5/3}}{V^{2/3}}.$$

The difference between the number of protons and the number of neutrons is small compared with the atomic mass number and so we get a good approximation by writing $Z = \frac{1}{2}(A + (Z - N))$, $N = \frac{1}{2}(A - (Z - N))$ and expanding this expression up to quadratic order $(Z - N)$, the difference between the number of protons and the number of neutrons (for fixed A). Using the fact that the volume is proportional to A, we obtain an expression for the contribution to the binding energy from the asymmetry (discrepancy) between the number of protons and the number of neutrons of the form

$$-a_A \frac{(Z - N)^2}{A}.$$

Of course, the assumption that the protons and neutrons are non-interacting is incorrect. Nevertheless, the above model of a Fermi gas of free spin-$\frac{1}{2}$ particles gives the correct dependence of this term on the numbers of protons and neutrons.

5. **Pairing term:** Identical spin-$\frac{1}{2}$ particles bind tightly in pairs with opposite spin. This means that the binding energy of an even number of protons (neutrons) is larger than the binding energy for an odd number for which at least one particle must remain unpaired. This leads to a further term in the expression for nuclear binding energy that is positive if the number of protons and the number of neutrons are both even, so that *both* all the protons and all the neutrons can form pairs, negative if they are both odd, so that *neither* all the protons *nor* all the neutrons are paired, and zero if one number is even and the other is odd, for which *either* all the protons *or* all the neutrons are paired.
We may express such a term in the form

$$a_P \frac{\left((-1)^Z + (-1)^N\right)}{2} A^{k_P}.$$

The dependence on the atomic mass number is some power, k_P, of A. There is no good theoretical determination of this power but comparison with experimental data favours the value $k_P = -\frac{1}{2}$ (although in earlier fits the value was taken to be $k_P = -\frac{3}{4}$).
We therefore have a pairing term

$$a_P \frac{\left((-1)^Z + (-1)^N\right)}{2A^{1/2}}.$$

The complete formula is then

$$B(A, Z) = a_V A - a_S A^{2/3} - a_C \frac{Z^2}{A^{1/3}} - a_A \frac{(Z - N)^2}{A} + a_P \frac{((-1)^Z + (-1)^N)}{2A^{1/2}},$$
$$(3.7)$$

where $N \equiv (A - Z)$.

Much work has been done on finding the best sets of parameters for this model, and the best values have changed as experimental data for the binding energies of various different nuclides improves. To date, the most reliable set of parameters have been derived by Guy Royer and Christian Gautier in 2006 [30]. These parameters a_V, a_S, a_C, a_A, a_P are

$$a_V = 15.78 \text{ MeV}$$
$$a_S = 17.90 \text{ MeV}$$
$$a_C = 0.724 \text{ MeV}$$
$$a_A = 23.72 \text{ MeV}$$
$$a_P = 11.0 \text{ MeV.} \qquad (3.8)$$

For all but the lightest nuclei, this simple formula does a very good job of determining the binding energies – for nuclides with $A > 20$, this is usually better than 1%, although there are a few exceptions where the binding energies are considerably larger than expected from the semi-empirical mass formula. This will be discussed in Chap. 4.

For example, we estimate the binding energy per nucleon of $^{80}_{35}\text{Br}$ (bromine), for which $Z = 35$, $A = 80$ ($N = 80 - 35 = 45$) and insert into the above formulae to get

Volume term: $(15.78 \times 80) = 1262.4 \text{ MeV}$

Surface term: $(-17.90 \times (80)^{2/3}) = -332.3 \text{ MeV}$

Coulomb term: $-\left(\frac{0.724 \times 35^2}{(80)^{1/3}} \right) = -205.8 \text{ MeV}$

Asymmetry term: $-\left(\frac{23.72 \times (45 - 35)^2}{80} \right) = -29.7 \text{ MeV}$

Pairing term: $\left(\frac{-11.0}{(80)^{1/2}} \right) = -1.2 \text{ MeV.}$

Note that we *subtract* the pairing term since both N and Z are odd. This gives a total binding energy of 693.4 MeV. The measured (see e.g. [31]) value is 694.2 MeV, so this model gives very accurate estimate of binding energy of $^{80}_{35}\text{Br}$. Although this high level of accuracy is only expected for nuclides with large atomic mass number ($A \gtrsim 20$), even for $^{12}_{6}\text{C}$ (carbon) the error is still only about 5%, whereas for $^{6}_{3}\text{Li}$ (lithium) the error is increased to 15%.

In order to calculate the mass of the nucleus we *subtract* this binding energy (divided by c^2) from the total mass of the protons and neutrons ($m_p = 938.3\,\text{MeV}/c^2$, $m_n = 939.6\,\text{MeV}/c^2$)

$$m_{Br} = 35m_p + 45m_n - 693.2\,\text{MeV}/c^2 = 74426\,\text{MeV}/c^2.$$

The binding energies of certain nuclides have now been measured with extreme precision, in some cases to better than one part in a million.[4] There has therefore been a whole industry in attempting to refine the model in order to increase the accuracy of the fit (one example is given in [30]).

Nuclear masses are nowadays usually quoted in MeV/c^2 but are still sometimes quoted in atomic mass units (a.u.), defined to be $1/12$ of the *atomic* mass of $^{12}_6\text{C}$. The conversion factor is

$$1\,\text{a.u.} = 931.5\,\text{MeV}/c^2.$$

Binding energies are usually expressed in terms of the binding energy per nucleon, which is the energy required to liberate one nucleon from the nucleus.

3.5 Most Stable Isobars

Different isobars may convert into each other by β-decay, which will be discussed in detail in Chap. 7. This is the process in which either a neutron is converted into a proton, emitting an electron and an antineutrino, or a proton is converted into a neutron, emitting a positron (the antiparticle of an electron) and a neutrino. In both cases, sufficient energy must be liberated by the change in the mass of the isobar to create an electron, which requires an energy of $m_e c^2$.

The process

$$^A_Z\{\text{Ch}\} \rightarrow_{(Z+1)} ^A \{\text{Ch}'\} + e^- + \bar{\nu}$$

is kinematically possible provided

$$B(A, Z+1) - B(A, Z) \geq (m_p + m_e - m_n)c^2.$$

Note that it was necessary to take into account the mass difference between a proton and a neutron.[5] Likewise, the process

$$^A_Z\{\text{Ch}\} \rightarrow_{(Z-1)} ^A \{\text{Ch}'\} + e^+ + \nu$$

---

[4]The experimental data on the binding energies of all stable nuclides are tabulated in [31].
[5]The mass of the neutrino is extremely small and may be neglected.

is kinematically possible provided

$$B(A, Z-1) - B(A, Z) \geq (m_n + m_e - m_p)c^2.$$

For a given atomic mass number, A, the most stable isobar has atomic number Z_m, for which the binding energy is maximum. Ignoring the pairing term for the moment, replacing N by $(A - Z)$ in the expression (3.7) and differentiating with respect to Z give

$$\left.\frac{\partial B(A, Z)}{\partial Z}\right|_A = -2a_C \frac{Z}{A^{1/3}} + 4a_A \frac{(A - 2Z)}{A}. \tag{3.9}$$

Setting this to zero, we find a maximum binding energy at

$$Z_m = \frac{A}{2} \frac{1}{\left(1 + \frac{1}{4} a_c A^{2/3} / a_A\right)}. \tag{3.10}$$

For odd values of A, the pairing term is zero since either the number of protons is odd but the number of neutrons is even or vice versa. Thus for odd values of A the most stable isotope occurs for the nearest integer value of Z to the value given by (3.10). For even values of A the isotopes with even values of both Z and N have a larger binding energy due to the pairing term than isotopes with odd values of both Z and N. Thus for even A the most stable isotope has atomic number Z equal to the nearest *even* integer to the value given by (3.10). Unfortunately, the small errors of the semi-empirical mass formula are often larger than the differences in the binding energies of adjacent isobars, so that this procedure sometimes returns the second most stable isobar. For example, for $A = 129$, the semi-empirical mass formula returns $^{129}_{54}\text{Xe}$ (xenon) as the most stable isobar with binding energy per nucleon 8.37 MeV, whereas the experimental value is 8.43 MeV. The most stable isobar is actually $^{129}_{53}\text{I}$ (iodine) with a binding energy per nucleon of 8.44 MeV, whereas the semi-empirical mass formula gives a binding energy per nucleon of 8.36 MeV. The errors are within 1%, but this is sufficient to interchange the most stable isobar with the second most stable isobar.

3.6   Stability of Isotopes

Since different isotopes have different atomic mass numbers, they will have different binding energies and some isotopes will be more stable than others. It turns out (and can be seen by looking for the most stable isotopes using the semi-empirical mass formula) that for the lighter nuclei, the stable isotopes have approximately the same number of neutrons as protons. This arises as a result of the asymmetry term.

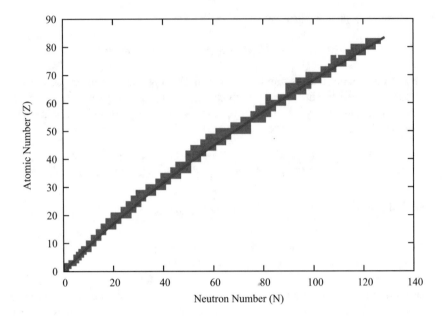

Fig. 3.4 Proton number against neutron number for stable isotopes. The solid blue line is the prediction for the most stable isobar given by (3.10)

However, above $A \sim 20$ the number of neutrons required for stability increases up to about one and a half times the number of protons for the heaviest nuclei (Fig. 3.4).

Qualitatively, the reason for this arises from the Coulomb term. Protons bind less tightly than neutrons because they have to overcome the Coulomb repulsion between them. It is therefore energetically favourable to have more neutrons than protons. For nuclides with a large atomic mass number, this Coulomb effect beats the asymmetry effect that favours equal numbers of protons and neutrons. Quantitatively, this phenomenon can be seen from (3.10), where for small A the maximum binding energy occurs for $Z \approx \frac{1}{2}A$, i.e. the same number of protons and neutrons, but as A increases the maximum binding energy occurs for $Z < \frac{1}{2}A$, i.e. nuclides with more neutrons than protons.

Summary

- Nuclear masses can be measured up to a very high degree of accuracy, by the use of one of three devices:

  1. A Penning trap, in which ions are confined and oscillate in three different modes with frequencies related to the mass-to-charge ratio of the ions

2. A storage ring mass spectrometer, in which ions are stored and rotate around the ring with a frequency that depends on the mass of the ions
3. A multiple reflection time-of-flight spectrometer, in which ions make a large number of trajectories between two electrostatic reflectors and their time of flight is measured

- The mass of a nuclide is smaller than the sum of the masses of the protons and neutrons that comprise the nucleus, by an amount known as the mass defect. This is the binding energy divided by the square of the speed of light.
- For all but the lightest nuclides, and a few other exceptions, the binding energy is given to a high degree of accuracy by the semi-empirical mass formula, which is based on the Liquid Drop Model of the nucleus.
- The semi-empirical mass formula has five terms:

  - A volume term proportional to the atomic mass number, A.
  - A (negative) surface term proportional to $A^{2/3}$.
  - A (negative) Coulomb term, proportional to the square of the atomic number, Z, and inversely proportional to the cube root of the atomic mass number.
  - A (negative) asymmetry term proportional to the square of difference between the number of protons and the number of neutrons. This arises as a result of Pauli's exclusion principle, which states that each energy level can only have up to two identical spin-$\frac{1}{2}$ particles.
  - A pairing term that encodes the fact that spin-$\frac{1}{2}$ particles bind tightly if they pair up – so that an even number of protons or neutrons will have a larger binding energy than an odd number.

- The semi-empirical mass formula can be used to determine whether it is kinematically allowed for isobars to transform into each other via β-decay.
- By maximizing the expression for nuclear binding energies, with respect to the atomic number, Z, one can identify the most stable isobar (the one with the largest binding energy) for a given atomic mass number, A.
- The lighter stable isotopes have approximately the same numbers of protons and neutrons, whereas for the more massive nuclides the stable isotopes have more neutrons than protons. This is because for nuclides with large atomic mass number, the Coulomb repulsion between the protons dominates over the pairing term, which favours equal numbers of protons and neutrons.

Problems

Problem 3.1 Use the semi-empirical mass formula to calculate the binding energies of the nuclides

$$^{238}_{92}\text{U}, \quad ^{145}_{57}\text{La and } ^{90}_{35}\text{Br}.$$

Hence, calculate the energy released in the (spontaneous fission) reaction:

$$^{238}_{92}U \rightarrow {}^{145}_{57}La + {}^{90}_{35}Br + 3n.$$

Problem 3.2 Use the semi-empirical mass formula to calculate the mass of a carbon-12 atom (including the mass of the 6 electrons). Hence, determine the conversion factor between MeV and atomic units. Compare this with the actual value.

Problem 3.3 Find the atomic number of the most stable isobar with atomic mass number 138.

Problem 3.4 By considering the nucleus to be a uniformly charged sphere with electric charge Ze, and radius $1.22A^{1/3}$ fm, calculate the value of the coefficient, a_C, of the Coulomb term and compare this with the value obtained from the best fit to data.

Problem 3.5 By considering a model in which the nucleons are treated as a Fermi gas of non-interacting particles, show that the coefficient, a_A, of the asymmetry term is given by

$$a_A = \frac{\left(3\pi^2\right)^{1/3} \hbar^2}{8m_p r_0^2},$$

where the radius of a nucleus of atomic mass number A is taken to be equal to $r_0 A^{1/3}$.
[This gives an underestimate by about a factor of 2, indicating that the non-interactive Fermi gas model is only qualitatively useful.]

Appendix 3 – The Energy of a Fermi Gas of Free Particles

In one dimension, a particle confined to a "box" of length L has a momentum, which is restricted such that L is an integer number of de Broglie wavelengths, i.e.

$$p = n\frac{2\pi\hbar}{L}, \quad (n \text{ integer}).$$

The interval, Δp, between adjacent allowed momenta is

$$\Delta p = \frac{2\pi\hbar}{L}.$$

The number of allowed momentum states, dn_s, in an interval dp of momentum is the number of these intervals in a length dp, i.e.

Fig. 3.5 Elementary cell in momentum space for a particle in a cube of side L

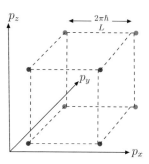

$$dn_s = \frac{dp}{\Delta p} = \frac{L}{2\pi\hbar}dp.$$

In three dimensions, for a particle confined to a cubic box of side L, the momentum vector, p, must be such that L is an integer number of de Broglie wavelengths of each component of p, i.e.

$$p = \frac{2\pi\hbar}{L}\left(n_x, n_y, n_z\right), \quad \left(n_x, n_y, n_z, \text{ integer}\right).$$

The allowed momenta occupy the corners of a cubic cell of length $2\pi\hbar/L$ in momentum space, as shown in Fig. 3.5. The volume, ΔV_p, of such a cell is

$$\Delta V_p = \frac{(2\pi\hbar)^3}{V},$$

where $V = L^3$ is the volume of the box.

Each cell has eight corners, but when the cells are packed together in three dimensions, each corner is shared by eight cells so the volume in momentum space occupied by one momentum point is the volume of one cell, ΔV_p. The number of allowed states in a volume element, d^3p, of momentum space, is then the ratio of this volume element to the volume of one cell, i.e.

$$d^3n_s = \frac{d^3p}{\Delta V_p} = V\frac{d^3p}{(2\pi\hbar)^3}. \tag{3.11}$$

This is known as the *"phase space"* element for one particle.

If we express the momentum vector, p, in terms of spherical polar coordinated with magnitude p, polar angle ϑ and azimuthal angle ϕ, the momentum-space volume element is

$$d^3p = p^2dp \sin\vartheta\, d\vartheta\, d\phi.$$

The number of allowed states, dn_s, with *magnitude* of momentum between p and $p + dp$ is obtained by integrating d^3n_s of (3.11) over all angles to give

$$dn_s = V \frac{1}{2\pi^2\hbar^3} p^2 dp. \tag{3.12}$$

For a free non-relativistic particle of mass m the momentum p is related to the (kinetic) energy, E, of the particle by

$$p = \sqrt{2mE},$$

so that an interval dp in magnitude of momentum corresponds to an interval, dE in energy, where

$$dp = \sqrt{\frac{m}{2E}} dE.$$

This allows us to convert (3.12) so that it tells us the number of allowed states in an energy interval between E and $E + dE$:

$$dn_s = V \frac{(2m)^{3/2}\sqrt{E}}{4\pi^2\hbar^3} dE. \tag{3.13}$$

The number of allowed states with energy less than the Fermi energy, E_F, is

$$n_s = \int_0^{E_F} dn_s = \int_0^{E_F} V \frac{(2m)^{3/2}\sqrt{E}}{4\pi^2\hbar^3} dE = V \frac{(2m)^{3/2}}{6\pi^2\hbar^3} E_F^{3/2}. \tag{3.14}$$

By the Pauli exclusion principle, each such state can be occupied by two spin-$\frac{1}{2}$ fermions, so for a gas of identical non-interacting fermions with spin-$\frac{1}{2}$ and mass m, confined to a box of volume V, the number of fermions (in their ground state) with energies up to E_F is $2n_s$. Accounting for this and inverting (3.14), the Fermi energy for n spin-$\frac{1}{2}$ particles of mass m in a volume V is

$$E_f = \pi^{4/3} \frac{\hbar^2}{2m} \left(\frac{3n}{V}\right)^{2/3}. \tag{3.15}$$

The total energy, E_0, of such a gas in its ground state is given by

$$E_0 = 2\int_0^{E_F} E\, dn_s = \int_0^{E_F} V \frac{(2m)^{3/2}}{2\pi^2\hbar^3} E^{3/2} dE = V \frac{(2m)^{3/2}}{5\pi^2\hbar^3} E_F^{5/2}. \tag{3.16}$$

Substituting for E_F using (3.15), we end up with

$$E_0 = \pi^{4/3} \frac{\hbar^2}{10m} \frac{(3n)^{5/3}}{V^{2/3}}. \tag{3.17}$$

Chapter 4
The Nuclear Shell Model

The semi-empirical mass formula derived from the Liquid Drop Model, discussed in the previous chapter, provides quite a good estimate of the binding energy for many nuclei, but not for all. Let us recall that in order to describe correctly the nuclear mass as a function of A and Z, the asymmetry and pairing terms are required by the semi-empirical formula, and this constitutes a deviation from the classical liquid drop picture of the nucleus. One should also note that the semi-empirical mass formula does not quantitatively describe nuclear excitations or the phenomenon of fission of heavy nuclei into asymmetric lighter nuclei (described in Chap. 9).

4.1 Magic Numbers

It turns out that nuclei have a periodic nature analogous to the periodic nature of atoms described by the Mendeleev periodic table of chemical elements. This feature of nuclei is not described by the semi-empirical mass formula. In particular, it was found that the binding energies predicted by that formula underestimate the actual binding energies of "magic nuclides" for which either the number of neutrons, $N = (A - Z)$, or the number of protons, Z, is equal to one of the following *"magic numbers"*:

$$2, \ 8, \ 20, \ 28, \ 50, \ 82, \ 126, 184.$$

This effect is especially pronounced for the case of "doubly magic" nuclides in which *both* the number of neutrons and the number of protons are equal to magic numbers, for example, for $^4_2$He, $^{16}_8$O, $^{40}_{20}$Ca, $^{208}_{82}$Pb. For $^4_2$He the semi-empirical mass formula predicts a binding energy of 21.69 MeV, whilst the measured value is 28.30 MeV (underestimated by about 30%), for $^{16}_8$O (oxygen) these values are 123.18 and 127.62 MeV, respectively (underestimated by about 3%), for $^{40}_{20}$Ca

© The Author(s), under exclusive license to Springer Nature Switzerland AG 2021
A. Belyaev, D. Ross, *The Basics of Nuclear and Particle Physics*, Undergraduate
Texts in Physics, https://doi.org/10.1007/978-3-030-80116-8_4

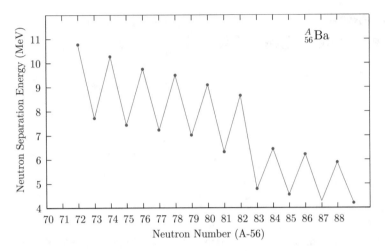

Fig. 4.1   Neutron separation energy for $_{56}^{A}$Ba isotopes

(calcium) they are 338.90 and 342.05 MeV (underestimated by about 1%) and for $_{82}^{208}$Pb (lead) they are 1612.2 and 1636.4 MeV (underestimated by about 1.5%).

Magic nuclides have special features related to their binding energy properties, such as:

- The neutron (or proton) separation energies (the energy required to remove the last neutron (or proton)) peak if N (Z) is equal to a magic number. For example, Fig. 4.1 shows the neutron separation energy for isotopes of $_{56}^{A}$Ba (barium) as a function of neutron number, N. We can see a clear step in the binding energy as the number of neutrons crosses the magic number 82.
- There are more stable isotopes if Z is a magic number and more stable isotones if N is a magic number.
- If N is a magic number, then the cross section for neutron capture is much lower (by a factor of 10–100) than for other nuclides as demonstrated in Fig. 4.2.
- The energies of the excited states are much higher than the ground state if either N or Z or both are magic numbers. As an example, in Fig. 4.3 we present the values of the excitation energies for various isotopes of $_{82}^{A}$Pb, where we can see that the magic nuclide $_{82}^{126}$Pb has an excitation energy of about a factor of 3 higher than the other isotopes.
- Elements with Z or/and N equal to a magic number have a larger natural abundance than those of nearby elements or isotopes with even values of Z or N. Let us take a look, for example, at $_{20}^{40}$Ca. Actually, it makes sense to compare nuclides which differ from magic ones in N or Z by even number, since such nuclides are more stable and therefore have larger natural abundance. One should also note that $_{20}^{40}$Ca is the heaviest stable isotope with $Z = N$. Its abundance amongst other isotopes of Ca is about 97%. The previous nuclide with $Z = N$ (*not* equal to a magic number) is $_{18}^{36}$Ar (argon), which has an abundance of only

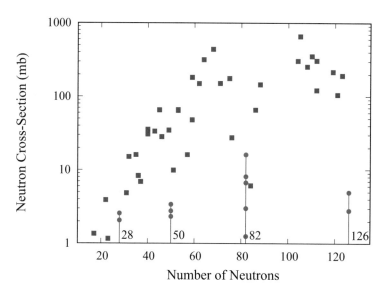

Fig. 4.2 The neutron capture cross section for various nuclei, including nuclei with magic numbers of neutrons (red points)

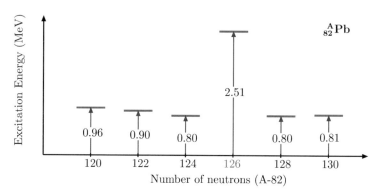

Fig. 4.3 The excitation energies for various isotopes of lead, $^A_{82}\text{Pb}$

0.34% amongst other isotopes of Ar, whilst the $Z = N$ isotope with $Z = 22$, $^{44}_{22}\text{Ti}$ (titanium) is totally absent.

- Elements with magic numbers of protons have zero electric quadrupole moment (their charge distribution is spherically symmetric), reflecting their particular stability.

4.2 Shell Model

The periodicity of the properties of nuclei observed in terms of magic numbers, which is similar to the periodicity of the properties of atoms, gives us a hint about the shell structure of nuclei in analogy to the shell structure of atoms. Indeed magic numbers can be explained in terms of the Shell Model of the nucleus, which considers each nucleon moving in some potential and classifies the energy levels in terms of quantum numbers, n, ℓ and j, in the same way as the wavefunctions of individual electrons are classified in Atomic Physics. However, contrary to the electromagnetic potential in atoms, the potential in nuclei arises from the strong interactions between nucleons, which have different properties from the electromagnetic interactions that bind electrons in atoms, as we discuss below. The Nuclear Shell Model was proposed by Eugene Gapon and Dmitri Iwanenko in 1932 [32] and was later developed by Eugene Wigner [33], Maria Göppert-Mayer [34, 35] and J. Hans D. Jensen [36].

For a spherically symmetric potential, the wavefunction (neglecting spin for the moment) for a nucleon, whose position, $\mathbf{r}$, from the centre of the nucleus is given by polar coordinates (r, θ, ϕ), has the form

$$\Psi_{n\ell m} = R_{n\ell}(r) Y_\ell^m(\theta, \phi), \tag{4.1}$$

where Y_ℓ^m are spherical harmonics, which give the angular part of a wavefunction for a particle moving in a spherically symmetric potential.

The energy eigenvalues depend on the principle quantum number, n, and the orbital angular momentum quantum number, ℓ, but are degenerate in the magnetic quantum number, m. Unlike the case of a Coulomb potential in Atomic Physics, the quantum number, ℓ, is *not* restricted to take values smaller than n.

These energy levels come in "bunches" called "shells" with a relatively large energy gap between each shell. In their ground state, the nucleons fill up the available energy levels from the bottom upwards with two protons (and/or two neutrons), with opposite z-component of spin, in each available proton (neutron) energy level, as required by the Pauli exclusion principle. Thus a state with a given n and ℓ can accommodate up to $2 \times (2\ell + 1)$ protons or neutrons.

Unlike in Atomic Physics, we do not understand, even in principle, the properties of this strong force potential, so we need to make a guess. If we assume a simple harmonic potential (i.e. $V(r) \propto r^2$), then we will get equally spaced energy levels and we would not see the shell structure giving rise to magic numbers.

It turns out once again that a Saxon–Woods potential,

$$V(r) = -\frac{V_0}{1 + \exp\left(((r - R)/\delta)\right)}, \tag{4.2}$$

is a reasonable guess and provides an explanation for the first three magic numbers as shown in Fig. 4.4. For such a potential, it turns out that the lowest level is $1s$ (i.e.

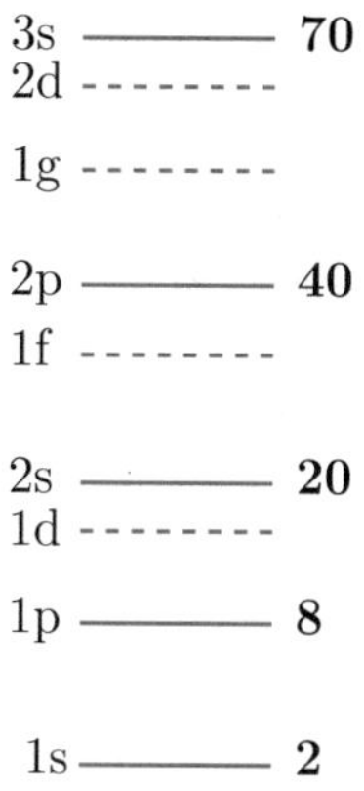

Fig. 4.4 Energy levels of the nucleons for Saxon–Woods model of the strong potential, without spin–orbit coupling

3s ———— 70
2d --------

1g --------

2p ———— 40
1f --------

2s ———— 20
1d --------

1p ———— 8

1s ———— 2

$n = 1$ and $\ell = 0$), which can contain up to 2 protons or neutrons. Next comes $1p$ level ($\ell = 1$), which can contain up to a further 6 protons (neutrons). This explains the first two magic numbers (2 and 8). Then there is the level $1d$ ($\ell = 2$), but this is quite close in energy to $2s$ so that they form the same shell. This allows a further $2 + 10$ protons (neutrons) giving us the next magic number of 20. The next two levels are $1f$ ($\ell = 3$) and $2p$, which are also quite close together and allow a further $6 + 14$ protons (neutrons). This would suggest that the next magic number should be 40 – but experimentally it turns out to be 50. Moreover, there is one additional magic number below 50 and above 20 – namely 28.

The solution to this puzzle lies in the strong spin–orbit interaction. From Atomic Physics, we know that spin–orbit coupling, i.e. the interaction between orbital angular momentum and spin angular momentum, is of magnetic origin and the effect of this interaction generates a small *positive* correction to atomic energy levels. In the case of nuclear binding, the nature of the spin–orbit coupling is "*chromomagnetic,*"[1] and the effect is about 20 times larger. This effect comes from a term in the nuclear potential itself, which is proportional to the scalar product of spin–orbit vectors, $\boldsymbol{L} \cdot \boldsymbol{S}$, and requires modification of the Saxon–Woods potential incorporating a spin–orbit term and the function $W(r)$, which also has the shape of a Saxon–Woods distribution:

$$V(r) \rightarrow V(r) + W(r)\boldsymbol{L} \cdot \boldsymbol{S}. \tag{4.3}$$

This model, suggested in 1949 by two independent groups [35, 36], had a great success. As in the case of Atomic Physics (in the j–j coupling scheme), the orbital and spin angular momenta of the nucleons combine to give a total angular momentum j, which can take the values $j = \ell + \frac{1}{2}$ or $j = \ell - \frac{1}{2}$. The spin–orbit coupling term leads to an energy shift proportional to

---

[1]The interaction between particles that carry the strong force charge is described by Quantum Chromodynamics (QCD). Chromomagnetic interactions are the QCD analogue of magnetic interactions in Electromagnetism.

$$j(j+1) - \ell(\ell+1) - s(s+1), \quad (s = 1/2). \tag{4.4}$$

A further feature of this spin–orbit coupling in nuclei is that the energy split is in the opposite sense from its effect in Atomic Physics, namely that states with higher j have *lower energy*. This happens because $W(r)$ is always *negative*. From Fig. 4.5 we see that this spin–orbit effect can sometimes be so large that levels of different n, ℓ can cross over, and in some cases the splitting between energies of states with the same n, ℓ but different j is sufficiently large for them to end up in different shells. For example, the gap above the third magic number, 28, originates from the splitting of the level $1f$ ($\ell = 3$), where the larger j level, $1f_{\frac{7}{2}}$, has the lower energy and the 8 allowed protons (neutrons) complete the lower shell giving a total of 28 in this shell, whereas the higher energy level, $j = \frac{5}{2}$, belongs to the next shell. For the next magic number, note that the state above the $2p$ state is $1g$ ($\ell = 4$), which splits into $1g_{\frac{9}{2}}$ ($j = \frac{9}{2}$) and $1g_{\frac{7}{2}}$ ($j = \frac{7}{2}$). The energy of the $1g_{\frac{9}{2}}$ state is sufficiently low that it joins the shell below, so that this fifth shell consists of $2p_{\frac{3}{2}}$, $1f_{\frac{5}{2}}$, $2p_{\frac{1}{2}}$ and $1g_{\frac{9}{2}}$. The maximum occupancy of this state is, respectively, $\Sigma(2j+1) = 4 + 6 + 2 + 10 = 22$, which added to the previous magic number, 28, gives the next observed magic number of 50. Further up, it is the $1h$ state that undergoes a large splitting into $1h_{\frac{11}{2}}$ and $1h_{\frac{9}{2}}$, with the $1h_{\frac{11}{2}}$ state joining the shell below.

4.3   Spin and Parity of Nuclear Ground States

Nuclear states have an intrinsic spin and a well-defined parity, $\eta = \pm 1$, determined by the behaviour of the wavefunction of all the nucleons under mirror reversal $(r \rightarrow -r)$, with the centre of the nucleus at the origin.

$$\Psi(-r_1, -r_2 \cdots - r_A) = \eta \Psi(r_1, r_2 \cdots r_A). \tag{4.5}$$

The spin and parity of nuclear ground states can usually be determined from the Shell Model. Protons and neutrons tend to pair up so that the total angular momentum of each pair is zero and each pair has even parity ($\eta = 1$). Therefore, the unpaired neutron and/or proton define nuclear spin and parity. Thus, we have the following:

- Even–even nuclides (both Z and A even) have zero intrinsic spin and even parity.
- Odd-A nuclei have one unpaired nucleon. The spin of the nucleus is equal to the j-value of that unpaired nucleon and the parity is $(-1)^\ell$, where ℓ is the orbital angular momentum of the unpaired nucleon. For example, $^{47}_{22}$Ti has an even number of protons and 25 neutrons. Twenty of the neutrons fill the shells up to magic number 20 and there are 5 in the $1f_{\frac{7}{2}}$ state ($\ell = 3$, $j = \frac{7}{2}$). Four of

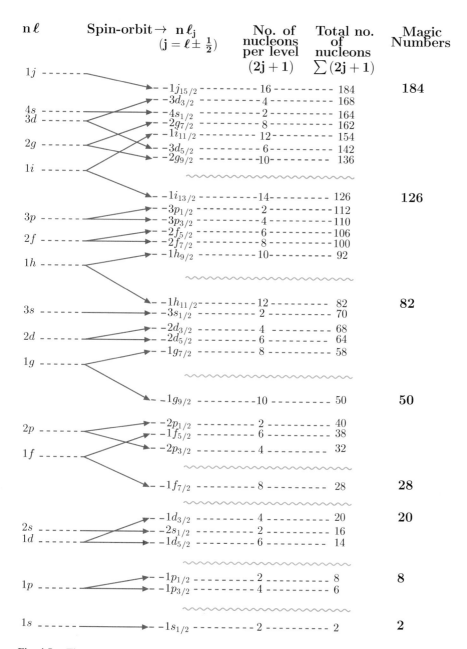

Fig. 4.5 The nuclear shell structure predicted by a Saxon–Woods potential with spin–orbit coupling which agrees with experimental data. In spectroscopic notation the letters $(s, p, d, f, g, h, f, i, j)$ correspond to orbital angular momentum $\ell = (0, 1, 2, 3, 4, 5, 6, 7, 8)$, respectively

these form pairs, and the remaining one leads to a nuclear spin of $\frac{7}{2}$ and parity $(-1)^3 = -1$.

- Odd–odd nuclei: In this case, there is an unpaired proton whose total angular momentum is j_1 and an unpaired neutron whose total angular momentum is j_2. The total spin of the nucleus is the (vector) sum of these angular momenta and can take values between $|j_1 - j_2|$ and $|j_1 + j_2|$ (in unit steps). The parity is given by $(-1)^{(\ell_1 + \ell_2)}$, where ℓ_1 and ℓ_2 are the orbital angular momenta of the unpaired proton and neutron, respectively.

  For example, ^6_3Li (lithium) has 3 neutrons and 3 protons. The first two of each fill the $1s$ level and the third is in the $1p_{\frac{3}{2}}$ level. The orbital angular momentum of each is $\ell = 1$, so the parity is $(-1) \times (-1) = +1$ (even), but the spin can take any value between 0 and 3.

4.4   Magnetic Dipole Moments

Since nuclei with an odd number of protons and/or neutrons have intrinsic spin, they also generally possess a magnetic dipole moment.

The unit of the magnetic dipole moment for a nucleus, the *"nuclear magneton"*, is defined as

$$\mu_N = \frac{e\hbar}{2m_p}, \tag{4.6}$$

which is analogous to the Bohr magneton but with the electron mass replaced by the proton mass. It is defined such that the magnetic moment of a proton due to its orbital angular momentum ℓ is $\mu_N \ell$. Experimentally it is found that the magnetic moment of the proton due to its spin is

$$\mu_p = 2.79\mu_N = 5.58\mu_N s, \quad \left(s = \frac{1}{2}\right) \tag{4.7}$$

and that of the neutron is

$$\mu_n = -1.91\mu_N = -3.82\mu_N s, \quad \left(s = \frac{1}{2}\right). \tag{4.8}$$

If we apply a magnetic field in the z-direction to a nucleus, then the unpaired proton with orbital angular momentum ℓ, spin s and total angular momentum j will give a contribution to the z-component of the nuclear magnetic moment

$$\mu_z = (5.58s_z + \ell_z)\mu_N. \tag{4.9}$$

As in the case of the *"Zeeman effect"*, we can use the vector model approach to project s and ℓ onto the total momentum, j, and then project j onto the z-axis yielding

$$\mu_z = \frac{(5.58\langle s \cdot j\rangle + \langle \ell \cdot j\rangle)}{\langle j^2\rangle} j_z\, \mu_N. \tag{4.10}$$

The expectation values (denoted by $\langle\ \rangle$) of the scalar products, $s \cdot j$ and $\ell \cdot j$, may be expressed in terms of the expectation values of s^2, ℓ^2 and j^2:

$$\langle s \cdot j\rangle = \frac{1}{2}\left(\langle j^2\rangle + \langle s^2\rangle - \langle \ell^2\rangle\right)$$

$$\langle \ell \cdot j\rangle = \frac{1}{2}\left(\langle j^2\rangle + \langle \ell^2\rangle - \langle s^2\rangle\right). \tag{4.11}$$

Recall that the nuclear states are eigenstates of s^2, ℓ^2 and j^2 with eigenvalues $s(s+1)\hbar^2$, $\ell(\ell+1)\hbar^2$ and $j(j+1)\hbar^2$, respectively, so that

$$\langle s \cdot j\rangle = \frac{\hbar^2}{2}\left(j(j+1) + s(s+1) - \ell(\ell+1)\right)$$

$$\langle \ell \cdot j\rangle = \frac{\hbar^2}{2}\left(j(j+1) + \ell(\ell+1) - s(s+1)\right)$$

$$\langle j^2\rangle = j(j+1)\hbar^2. \tag{4.12}$$

Substituting (4.12) into (4.10), we end up with expression for the contribution to the nuclear magnetic moment of a proton, $\mu_p(\ell, j)$, in a state $|n\ell j\rangle$,

$$\mu_p(\ell, j) = \left[\frac{5.58\,(j(j+1) + s(s+1) - \ell(\ell+1))}{2j(j+1)}\right.$$
$$\left. + \frac{(j(j+1) + \ell(\ell+1) - s(s+1))}{2j(j+1)}\right] j\, \mu_N. \tag{4.13}$$

Likewise, for the magnetic moment of the neutron, $\mu_n(\ell, j)$, in a state $|n\ell j\rangle$ (orbital angular momentum ℓ and the total angular momentum j), we have

$$\mu_n(\ell, j) = -\frac{3.82\,(j(j+1) + s(s+1) - \ell(\ell+1))}{2j(j+1)} j\, \mu_N, \tag{4.14}$$

where we used the fact that there is no contribution from the orbital angular momentum because the neutron is uncharged, so only $\langle s \cdot j\rangle$ contributes to μ_n.

Thus, for example, for the nuclide $^{7}_{3}\text{Li}$, for which there is an unpaired proton in the $2p_{\frac{3}{2}}$ state ($\ell = 1$, $j = \frac{3}{2}$), the estimate of the magnetic moment is

$$\mu_p = \frac{5.58\left(\frac{3}{2} \times \frac{5}{2} + \frac{1}{2} \times \frac{3}{2} - 1 \times 2\right) + \left(\frac{3}{2} \times \frac{5}{2} + 1 \times 2 - \frac{1}{2} \times \frac{3}{2}\right)}{2 \times \frac{3}{2} \times \frac{5}{2}} \times \frac{3}{2}$$

$$= 3.79\mu_N. \tag{4.15}$$

The measured value is $3.26\mu_N$, so the estimate is not too good. For heavier nuclei, the estimate from the Shell Model gets even worse.

The origin of the non-trivial value of the magnetic dipole moment of the proton and neutron lies in their composite nature, since they are not elementary particles, as we will discuss in detail in Chap. 15. This is especially evident for the neutron, since if it were an elementary neutral particle, its dipole moment would be zero, which is not the case. The origin of the magnetic dipole moment of the proton and neutron is related to the strong dynamics of their internal components and the nuclear potential which is not well-understood. Therefore, the precise determination of nuclear magnetic moments is not possible, and, in particular, it cannot always even be approximately determined from the Shell Model. For example, for the nuclide $^{17}_{9}\text{F}$, the measured value of the magnetic moment is $4.72\mu_N$, whereas the value predicted from the Shell Model is $-0.26\mu_N$.

4.5 Excited States

As in the case of Atomic Physics, nuclei can be in excited states, which decay via the emission of a photon (γ-ray) back to their ground state (either directly or indirectly).

In the case of some of these excited states, neutrons or protons from the outer shell are promoted to a higher energy level. However, unlike in Atomic Physics, it is also sometimes possible that it is energetically cheaper to promote a nucleon from an *inner* closed shell (rather than a nucleon from an outer shell) into a higher energy state. Moreover, excited states in which more than one nucleon is promoted above its ground state are much more common in Nuclear Physics than in Atomic Physics. Thus the nuclear spectrum of states is very rich indeed, but very complicated and cannot be easily understood in terms of the Shell Model.

Most of the excited states decay so rapidly that their lifetimes cannot be measured. There are some excited states, however, which are metastable because they cannot decay without violating the selection rules (discussed in Chap. 8). These excited states are known as *"isomers"*, and their lifetimes can be measured.

4.6 The Collective Model

The Shell Model has its shortcomings. In spite of its great success, the usefulness of the Shell Model should not be overstated. It has a limited range of validity – it can explain phenomena mainly relevant to the light spherical nuclei, but even in this case one observes discrepancies between the predictions of the model and experiment. These discrepancies are even larger for heavier nuclei. We have already seen that the Shell Model does not predict magnetic dipole moments or the spectra of excited states very well.

One further failing of the Shell Model is the prediction of electric quadrupole moments. The Shell Model predicts very small values for these. However, for heavier nuclei with A in the range of 150–190 and for $A > 220$, these electric quadrupole moments are found to be rather large. The failure of the Shell Model to correctly predict electric quadrupole moments arises from the assumption that the nucleons move in a spherically symmetric potential.

A model that generalizes the Shell Model is the Collective Model, which considers the effect of a non-spherically symmetric potential (leading to substantial deformations for large nuclei and consequently large electric quadrupole moments) and takes into account interactions between nucleons. One of the most striking consequences of the Collective Model is the explanation of low-lying excited states of heavy nuclei. These excitations are of two types:

- **Rotational States.** A nucleus whose nucleon density distributions are spherically symmetric (zero quadrupole moment) cannot have rotational excitations (this is analogous to the application of the principle of equipartition of energy to monatomic molecules for which there are no degrees of freedom associated with rotation). On the other hand, a nucleus with a non-zero quadrupole moment can have excited levels due to rotation perpendicular to the (rotational) axis of symmetry.

  For an **even–even** nucleus whose ground state has zero spin, these states have energies

$$E_{\text{rot}} = \frac{I(I+1)\hbar^2}{2\mathcal{I}},$$

(4.16)

where $\mathcal{I}$ is the moment of inertia of the nucleus about an axis through the centre perpendicular to the axis of rotational symmetry as shown in Fig. 4.6, and the integer I is the quantum number of rotational angular momentum.

It turns out that the rotational energy levels of an even–even nucleus can only take even values of I. For example, the nuclide $^{170}_{72}\text{Hf}$ (hafnium) has a series of rotational states with excitation energies

$$E \text{ (keV)}: \quad 100, \quad 321, \quad 641.$$

(4.17)

Fig. 4.6 A nucleus rotating about an axis perpendicular to its rotational axis of symmetry

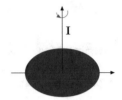

These are almost exactly in the ratio $2 \times 3 : 4 \times 5 : 6 \times 7$, meaning that these are states with rotational spin equal to $2, 4, 6$, respectively. The relation is not exact because the moment of inertia changes slightly as the spin increases.

We can extract the moment of inertia for each of these rotational states from (4.16). We could express this in SI units, but more conveniently nuclear moments of inertia are quoted in $\text{MeV}/c^2\text{fm}^2$.

Therefore, the moment of inertia of the $I = 2$ state, whose excitation energy is 0.1 MeV, is given (inserting $I = 2$ into (4.16)) by

$$\mathcal{I} = 2 \times 3 \times \frac{\hbar^2 c^2}{2c^2 E_{\text{rot}}} = 3 \times \frac{197.3^2}{0.1} = 1.17 \times 10^6 \, \text{MeV}/c^2 \, \text{fm}^2$$

($\hbar c = 197.3 \, \text{MeV fm}$).

For **odd-A** nuclides for which the spin of the ground state I_0 is non-zero, the excitation energy levels related to their rotation are given by

$$E_{\text{rot}} = \frac{1}{2\mathcal{I}} \left(I(I + 1) - I_0(I_0 + 1) \right) \hbar^2, \tag{4.18}$$

where I can take the values I_0+1, I_0+2 *etc.* For example, the first two rotational excitation energies of $^{143}_{60}\text{Nd}$ (neodymium), whose ground state has spin $\frac{7}{2}$, have energies 128 and 290 keV. They correspond to rotational levels with nuclear spin $\frac{9}{2}$ and $\frac{11}{2}$, respectively. The ratio of these two excitation energies, which is equal to 2.27, is almost exactly equal to that described by (4.18)

$$\frac{\frac{11}{2} \times \frac{13}{2} - \frac{7}{2} \times \frac{9}{2}}{\frac{9}{2} \times \frac{11}{2} - \frac{7}{2} \times \frac{9}{2}} = 2.22.$$

- **Shape oscillations.** These excitations are related to the modes of vibration in which the shape of the nucleus oscillates – the electric quadrupole moment oscillates about its mean value. It could be that this mean value is very small, in which case the nucleus is oscillating between an oblate and a prolate spheroidal shape as shown in Fig. 4.7(left). There can also be oscillations in the octupole or even higher multipole moments, which generate more complicated shape oscillation, such as the one shown in Fig. 4.7(right) for octupole oscillations. The small oscillations about the equilibrium shape perform simple harmonic motion. The energy levels of such modes are equally spaced. Thus an observed sequence of equally spaced energy levels within the spectrum of a nuclide is interpreted as a manifestation of such shape oscillations.

Fig. 4.7 The shape
oscillation of nuclei for
quadrupole (left) and
octupole (right) excitations

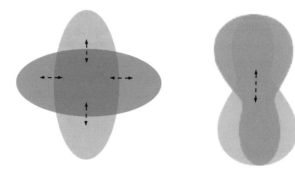

Summary

- The Nuclear Shell Model makes an important step beyond the Liquid Drop Model in describing the properties of nuclei. In analogy with Atomic Physics, it assumes that nuclear energy levels are arranged in closed shells. But contrary to the electromagnetic potential in atoms, the potential in the nuclei is the strong force potential which has different properties.
- The Shell Model is based on a spherically symmetric potential, with energy eigenvalues defined by the principle quantum number, n, and the orbital angular momentum, ℓ. These energy levels come in "bunches" known as "shells" with a large energy gap above each shell.
- A vital feature of the Shell Model is the strong spin–orbit interaction, which necessitates the addition of an extra term proportional to $\boldsymbol{L} \cdot \boldsymbol{S}$, to the Saxon–Woods potential. The orbital and spin angular momenta of the nucleons combine to give a total angular momentum j, which can take the values $j = \ell + \frac{1}{2}$ or $j = \ell - \frac{1}{2}$. An important feature of the spin–orbit coupling in nuclei is that states with higher j have *lower energy* (contrary to the case of Atomic Physics), because the spin–orbit coupling term is *negative*. This model successfully explains magic numbers. The first six levels of the Shell Models are

$$1s_{\frac{1}{2}}, \ 1p_{\frac{3}{2}}, \ 1p_{\frac{1}{2}}, \ 1d_{\frac{5}{2}}, \ 2s_{\frac{1}{2}}, \ 1d_{\frac{3}{2}}$$

  and explain first three magic numbers. The total angular momentum is half-integer, and the maximum occupation of protons or neutrons for a given j is an even number: $(2j + 1)$ (j is half-odd-integer). Detailed information about the nuclear shell structure can be found in Fig. 4.5.
- The spin and parity of nuclear ground states can usually be determined from the Shell Model. Protons and neutrons tend to pair up so that the total angular momentum of each pair is zero and each pair has even parity ($\eta = 1$). Therefore, the unpaired neutron and/or proton, in nuclides with an odd number of protons or neutrons, define nuclear spin and parity as $\eta = (-1)^{\ell}$ for one unpaired nucleon or $\eta = (-1)^{\ell_1 + \ell_2}$ for two unpaired nucleons.

- The Shell Model has shortcomings mainly arising from the fact that it is based on a spherically symmetric potential and does not take into account interactions of nucleons with each other. These effects are taken into account in the Collective Model that generalizes the Shell Model. One of the most striking consequences of the Collective Model is the explanation of low-lying excited states of heavy nuclei coming from rotational excitations and shape deformation excitations.

Problems

Problem 4.1 Use the Shell Model to determine, wherever possible, the spins and parities of the ground states of the following nuclides:

$$^{14}_{6}C, \quad ^{17}_{8}O, \quad ^{16}_{8}O, \quad ^{33}_{16}S, \quad ^{31}_{15}P, \quad ^{30}_{15}P, \quad ^{32}_{15}P.$$

Where it is not possible to determine the spins and/or parities exactly, give the possible values that these quantities can take.

Problem 4.2 Given that the first two shells in the Shell Model are $1s$ and $1p$ states, determine the spin and parities of the following nuclides in their ground state: $^{13}_{7}N$, $^{8}_{3}Li$ and $^{7}_{4}Be$. In the case of several possible spins, list all possibilities.

Problem 4.3 Find the approximate value of the neutron contribution to the magnetic moments of the following nuclides: $^{2}_{1}H$, $^{3}_{2}He$, $^{4}_{2}He$, $^{7}_{3}Li$ and $^{10}_{5}B$.

Problem 4.4 An even–even nucleus with A=170 has a sequence of excited states with energies above the ground state:

$$E\,(keV)\quad 97, \quad 321, \quad 678, \quad 1164.$$

(a) Explain why we can interpret these as rotational states.
(b) Calculate the moment of inertia of the nucleus and estimate the uncertainty.
(c) Taking the nuclear radius to be $R = 1.22 A^{1/3}$ fm, compare this with the moment of inertia about an axis through the centre of a sphere of mass equal to the mass of the nucleus and radius equal to the radius of the nucleus.

(The moment of inertia of a sphere of mass M and radius R about an axis through its centre is $\frac{2}{5}MR^2$.)

Chapter 5
Radioactivity

In 1896, Henri Becquerel [37], whilst conducting an experimental investigation into phosphorescence, placed a sample of a uranium salt on top of a photographic plate wrapped in black (light-proof) paper and kept in darkness. Nevertheless, the plate became exposed. Since the sample had not been exposed to light, as required for the observation of phosphorescence, Becquerel concluded that this was caused by direct emission of radiation from the uranium. Later, Marie Skłodowska Curie named this phenomenon *"radioactivity"*.

5.1 Types of Radioactivity

Some nuclides have a higher binding energy than some of their neighbours. When this is the case, it is often energetically favourable for a nuclide with a lower binding energy (*"parent nuclide"*) to decay into one with a higher binding energy (*"daughter nuclide"*) plus another particle or particles associated with different types of radioactive decay. In many cases, a given element will have two or more naturally occurring radioactive isotopes in addition to two or more stable isotopes.

There are three main types of radioactivity corresponding to three different types of emitted particles. The sum of the kinetic energies of the final-state particles is the difference in the binding energy between the parent and daughter nuclides and is usually a few MeV.

- *"α-decay"* – The emission of an α-particle, which is a $^4_2$He (helium) nucleus. In such decays the daughter nuclide has an atomic number which is two less than that of the parent and an atomic mass number four less than that of the parent. The rest energy of the α-particle, $m_\alpha c^2$, is much larger than its kinetic energy, so that it travels at a speed much lower than the speed of light. It is also doubly charged and therefore easily ionizes media through which it travels. It can be shown that if the electrons in an absorbing material are treated as free particles

© The Author(s), under exclusive license to Springer Nature Switzerland AG 2021
A. Belyaev, D. Ross, *The Basics of Nuclear and Particle Physics*, Undergraduate
Texts in Physics, https://doi.org/10.1007/978-3-030-80116-8_5

(a reasonable approximation as their binding energy is certainly small compared with the kinetic energy of the emitted α-particles), then the energy loss per unit distance travelled through the material is inversely proportional to the square of the velocity. This means that as the α-particles have a small velocity when they enter the absorbing material, they lose most of their initial energy very rapidly and can only penetrate a thin sheet of paper and are absorbed by a single layer of human skin.

- "β-decay" – The emission of an electron or a positron.

  Positron emission is accompanied by the emission of a very low-mass particle called a "neutrino". Electron emission is accompanied by the antiparticle of the neutrino (the antineutrino). The daughter nuclide has atomic number one more (in the case of electron emission) or one less (in the case of positron emission) than that of the parent but has the same atomic mass number. The difference in the binding energies is equal to the total energy of the decay products. The electron rest energy is usually small in comparison with the energy with which they are emitted in β-decay so that they travel ultra-relativistically, i.e. with almost the speed of light. As such they have much higher penetrating power than α-particles and can penetrate as much as a few millimetres of aluminium or a few centimetres of human tissue.

- "γ-decay" – The emission of a very short wavelength photon.

  This type of radioactivity takes place when a nucleus in a metastable excited state (known as a "nuclear isomer") decays directly or indirectly to its ground-state, emitting one or more high energy photons, in the γ-ray range of the electromagnetic spectrum. This is the nuclear counterpart of atomic radiation, but whereas atomic energy levels are of the order of a few electron volts (eV), giving rise to the emission of photons with wavelengths of a few hundred nanometres (i.e. in the visible spectrum), nuclear energy levels are of an order of MeV, so that these γ-rays have wavelengths of the order of $100\,\text{fm}$ (10^{-13} m).[1] γ-rays have far higher penetrating power than α particles or electrons from β decay and require several centimetres of lead to absorb them. They easily penetrate human tissue.

Apart from these main types of radioactivity, there are also radioactive decays involving the capture of an atomic electron, and decays involving proton or neutron emission.

5.2 Detecting Radioactivity

Methods for detecting radiation are based on the physical principles of the interaction mechanism of radioactive particles with matter. Radiation detectors can be classified according to the type of detection medium: gas-filled detectors, scintillation detectors, and semiconductor (or solid-state) detectors.

---

[1] By convention a γ-ray is defined as a photon with energy above $100\,\text{keV}$, corresponding to a wavelength of less than $\sim 2 \times 10^{-12}$ m.

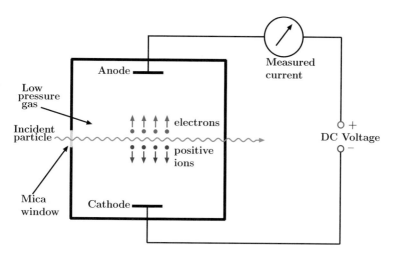

Fig. 5.1 An ionization chamber

5.2.1 Ionization Chamber

In 1898, J. J. Thompson demonstrated that X-rays could make air conductive [38], i.e. they produced electrical charges in the air. Together with Rutherford, he hypothesized that X-rays ionized the air by stripping off small negatively charged particles, later called electrons. Further studies of the ionization characteristics in 1899 led Thomson to the idea that these charges could be collected by producing an electric field across that volume of air. This marked the birth of the ionization chamber.

When a radiation particle (i.e. a particle from a radioactive decay) passes close to an atom it scatters off the electrons in that atom. For α-particles and β-particles, which carry electric charge, this is *"Coulomb scattering"* and for γ-rays it is *"Compton scattering"*. In both cases, momentum from the incident particle is imparted to the electron. Since the binding energy of an electron in an atom is only of the order of a few electron volts, particles from the radioactive decay, which have energies of order MeV, are sufficiently energetic to strip one or more electrons from the atom and ionize it.

One of the simplest devices for the detection and measurement of radioactivity is the ionization chamber. A diagram of such a chamber is shown in Fig. 5.1. It consists of a chamber filled with a gas under low pressure with two electrodes across which there is a DC voltage. The particle enters the chamber and interacts with atoms in the gas, producing free electrons and positively charged ions. The electrons drift towards the positively charged *"anode"*, and the ions drift towards the negatively charged *"cathode"*, generating an electric current which is proportional to the energy flux from the radiation particles entering the chamber. Unfortunately, a chamber of this design cannot be used to determine the energies of

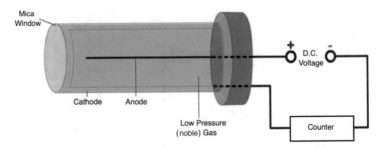

Fig. 5.2 Geiger–Müller Counter. The tube is called a "Geiger–Müller tube", and it is connected to an electronic counter, which counts the electric discharges. [*Source*: https://i.ytimg.com/vi/A8EEH5Fyc8k/maxresdefault.jpg (edited)]

the individual incident particles. However, since the particles emitted in the different types of radioactivity have different specific ionization, the ionization chamber *can* distinguish between these different types. The type of gas, the gas pressure, the voltage and the geometric design of the chamber are selected to optimize the detection of the different types of radioactivity.

5.2.2 Geiger Counter

A modified type of ionization chamber was invented in 1928 by Hans Geiger and Walter Müller [39]. It is known as a Geiger counter (or Geiger–Müller counter). A diagram of such a device is shown in Fig. 5.2. The anode is a cable along the axis of a cylindrical tube and the cathode is the metallic cylindrical surface. The particle from a radioactive decay enters a mica window at the end of the tube and ionizes the low pressure gas inside the tube. Mica is a commonly used material since, owing to its low density, it allows the particle to enter, but nevertheless seals the tube sufficiently to prevent the escape of the gas.

The high voltage across the tube is sufficient to induce an "avalanche" in which the electrons from the initial ionization scatter off the electrons in other atoms of the gas, ionizing those atoms and the emitted electrons ionize further atoms, *etc.* This is known as the *"Townsend avalanche"* phenomenon [40]. In this way, a single radiation particle can induce an electrical discharge between the anode and cathode, which is registered by a counter, thereby enabling the counting of individual radioactive events.

The voltage in a Geiger counter is between 500 and 1000 V. A lower voltage would be insufficient to induce the required Townsend avalanche, whereas a higher voltage could give rise to spontaneous ionization of the gas inside the tube, thereby generating a "false positive" (i.e. erroneously counting a radioactive decay). Usually a mixture of noble (inert) gases is used, chosen so that even at the high voltages required for the avalanche effect, spontaneous ionization of the gas atoms is very unlikely.

The time taken for the electrons and ions to drift to the electrodes is only a few microseconds. However, following each discharge there is a "dead time" which is the time required for the electrodes to regain their operating voltage. In modern Geiger counters, this dead time is 10–$100\,\mu s$, meaning that a Geiger counter can count a radioactive decay rate 10^4–10^5 decays per second.

A Geiger counter is almost 100% efficient in the detection of α-particles or β-particles, but only 1–2% efficient in the detection of γ-rays. The fact that a Geiger counter relies on an avalanche has its disadvantages – the output pulses from different radioactive particles are of about the same size in the region of avalanche, and therefore provide no direct information about the type and energy of radiation. Therefore Geiger–Muller counters can only be used to measure count rates, but not to measure the energies of the incident particles.

5.2.3 Scintillation Counters

The second major type of detectors utilized in radiation detection are scintillation counters, invented by Sam Curran in 1944 [41] whilst working on the "Manhattan project". These detectors record light produced when radiation interacts with materials that are luminescent. These materials, called "scintillators", may be liquid or solid. Gaseous scintillators are also in use, usually for detection of heavy charged particles. Unlike a Geiger counter, a scintillation counter is the most effective device for the detection of γ-rays, although such a device can also be usefully employed for the detection of α- and β-particles.

A diagram of a scintillation counter is shown in Fig. 5.3. The radiation particle is incident on a crystal of scintillating material. The interaction causes some of the electrons in the crystal to be promoted to an excited state. They then make a spontaneous transition back to their ground state, emitting photons, usually within the visible spectrum. These photons are incident on a photomultiplier tube. This consists of an electrode of photoelectric material kept at a high negative voltage, called a "photocathode". The emitted photoelectrons are accelerated down the tube, making collisions with "dynodes" (intermediate electrodes), which emit several electrons for each incident electron. These emitted electrons themselves travel down the tube, making collisions with other dynodes, producing yet more electrons, such that at the end of the tube a macroscopic electric pulse is detected and measured. The magnitude of this pulse carries information about the intensity of the original scintillator light flash, which in turn is proportional to the energy of the incident particle. Therefore, unlike a Geiger counter, this device can also be used to measure the energy of the incident particle. The response time is also very short. The time taken for the photons to travel down the photomultiplier tube is of the order of 1 ns

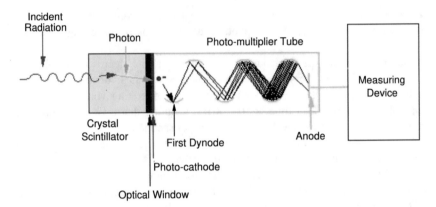

Fig. 5.3 A schematic diagram of a scintillation counter [*Source*: https://qph.fs.quoracdn.net/main-qimg-17dc0e28c40e192fdf798d6266179aae (edited)]

and the recovery time of the photomultiplier tube is a few tens of nanoseconds, so that a scintillation counter can count up to ten million events per second.

The following types of scintillators are in common use: inorganic and organic crystals, organic plastics, and liquids. Solid inorganic crystals are characterized by a high density and high atomic number. They are therefore particularly effective for detecting high energy γ-rays (above 1 MeV) due to their greater stopping power compared with lower density materials with lower atomic number. There are many different materials used for the scintillator crystal, each with somewhat different properties, designed to detect specific particles more efficiently, making them more useful for different types of detection. Caesium iodide is often used for counters designed to detect α-particles, whereas sodium iodide, doped with small amounts of thallium, is found to be more suitable for γ-ray detection.

There are also organic crystals which are hydrocarbons containing benzene rings such as naphthalene or anthracene. Compared with inorganic scintillators, all types of organic scintillators have lower density and significantly shorter times of luminescence. This is one of the important characteristics of the luminescence process in a given material. In order to permit the detection of fast signals, luminescence times should be as short as possible. Detectors based on organic crystals have a good time resolution, usually 10^{-8}–10^{-10} s, but lower detection efficiency for γ-rays. They are therefore mostly used for the detection of α- and β-particles and also for neutron radiation.

A variation of this technique, useful for the detection of low-energy α- and β-particles, is the liquid scintillation counter, in which the radioactive material together with a scintillating material, such as zinc sulphide, is dissolved in an organic solvent. The radioactive particle first transfers energy to the solvent molecule which is then promoted to an excited state. The excitation energy is then transferred to the scintillator, which in turn emits photons. These photons are detected using a photomultiplier tube as in the case of the crystal scintillation

Fig. 5.4 A p-n junction semiconductor diode in reverse diode mode

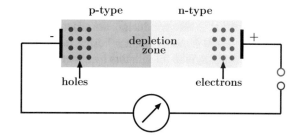

counters. One of the advantages of liquid scintillators is their use in large-volume detectors, which significantly increases the efficiency of registering particles such as neutrinos, which have a small interaction cross section with matter.

5.2.4   Semiconductor Detectors

The most efficient modern-day detector is the semiconductor detector. The basic operating principle of semiconductor detector is similar to an ionization chamber, but the medium is now a solid semiconductor material instead of gas. This is a semi-conductor diode consisting of adjacent strips of p-type and n-type semiconductors. Near the junction between the two, there is a depletion zone with no free charge carriers – neither electrons nor holes.

This depletion zone (the depth of which determines the sensitive region and hence the performance of the detector) is enhanced by placing the diode between electrodes in *"reverse bias mode"*, i.e. the n-type semiconductor is connected to the positive electrode and the p-type to the negative, as shown in Fig. 5.4. When a radiation particle enters the depletion zone and its energy is absorbed, a large number of electrons are promoted from the *"valence band"* to the *"conduction band"*, leaving positively charged holes in the valence band. These pairs of charge carriers diffuse around the circuit giving rise to an electric impulse whose height is proportional to the energy of the incident particle. The pulse rate is equal to the rate of incidence of radiation particles so that these detectors can measure both the rate of incidence and the energies of incident particles. The energy required to create an electron–hole pair is a few eV, which is an order of magnitude less than a typical ionization energy in a gas. A radiation particle with energy 1 MeV can create hundreds of thousands of pairs of charge carriers. Semiconductor detectors have extremely good energy resolution and can also be used for the detection of relatively low-energy particles.

The first radiation measurement using a semiconductor detector was carried out by Pieter Van Heerden in 1945 [42, 43] using a cooled silver chloride crystal. Nowadays, the semiconductor material is usually silicon or germanium, although cadmium-telluride is also sometimes used. Strips of silicon semiconductor have to be only a few millimetres thick, whereas germanium semiconductor strips can be thicker, with a depletion zone of up to a few centimetres, enabling them to absorb

γ-rays with energies up to a few MeV. Another advantage of the semiconductor detector is that as semiconductors are around 1000 times more dense than gas, they can be made much more compact. Furthermore, semiconductor detectors have better spatial resolution than scintillators by about one order of magnitude and thus have a wide application in high energy physics detectors, in particular at the LHC.

5.3 Decay Rates

The probability of a parent nucleus decaying in 1 s is called the *"decay constant"*, (or *"decay rate"*) λ. If we have $N(t)$ parent nuclei at time, t, then the number of decays expected per second is $\lambda N(t)$. The number of parent nuclei decreases by this amount and so we have

$$\frac{dN(t)}{dt} = -\lambda N(t). \tag{5.1}$$

This differential equation has a simple solution – the number of parent nuclei decays exponentially:

$$N(t) = N(0)e^{-\lambda t} . \tag{5.2}$$

The time taken for the number of parent nuclei to fall to $1/e$ of its initial value is called the *"mean lifetime"* (or simply *"lifetime"*), τ, of the radioactive nucleus, and we can see from (5.2) that

$$\tau = \frac{1}{\lambda}. \tag{5.3}$$

More often one talks about the *"half-life"*, $\tau_{\frac{1}{2}}$, of a radioactive nuclide, which is the time taken for the number of parent nuclei to fall to one-half of its initial value. From (5.2) we can also see that

$$\tau_{\frac{1}{2}} = \frac{\ln 2}{\lambda} = \tau \ln 2. \tag{5.4}$$

For example, the uranium isotope $^{238}_{92}$U has a half-life of 4.47 billion years, equivalent to 1.41×10^{17} s. From (5.3) and (5.4), this gives a decay constant

$$\lambda = \frac{\ln 2}{1.41 \times 10^{17} \, [\text{s}]} = 4.92 \times 10^{-18} \, \text{s}^{-1}.$$

Using the semi-empirical mass formula, the mass of one atom of $^{238}_{92}$U is 2.22×10^5 MeV/c^2, equivalent to 3.97×10^{-25} kg. Thus we would expect a (pure) sample of 1 g of this isotope to give an average number of radioactive counts of

$$\bar{n} = \frac{10^{-3} \, [\text{kg}]}{(3.97 \times 10^{-25} \, [\text{kg}])} \times (4.92 \times 10^{-18} \, \text{s}^{-1}) = 12,380 \, \text{Bq}.$$

One decay event per second is the unit of radioactivity called a *"Becquerel"* (Bq) – named after the discoverer of radioactivity. Radioactivity is more often measured in Curies. One Curie is the number of decays per second of 1 g of $^{226}_{88}$Ra (radium), and corresponds to 3.7×10^{10} Bq. This unit is named after Marie Curie, who is famous for having developed the theory of radioactivity and discovering techniques for separating radioactive isotopes. Together with Pierre Curie, she discovered two new radioactive elements – radium and polonium [44].

5.4 Random Decay

A nucleus is a sub-microscopic object to which Quantum Physics must be applied. It is therefore not possible to determine exactly when a given radioactive nucleus will decay. The best we can do is determine the *probability* that it will decay in unit time (the decay constant, λ).

This means that whereas the "expected" number of decays in a sample of N nuclei is λN per second, this does not mean that there will always be precisely this number of decays per second.

The average number of decays over several measurements of duration 1 s is given by

$$\bar{n} = \lambda N,$$

but there will be random fluctuations around this value. A measure of the size of these fluctuations over a set of $\mathcal{N}$ measurements, n_i, with average value $\bar{n}$, is given by the *"standard deviation"*, which is determined as follows:

- For each measurement, determine the deviation of the number of counts, n_i from the average value $\bar{n}$.
- Since this number can be positive or negative with an average value of zero, this quantity is squared – the square of the deviation is always positive.
- Take the average of the square of the deviation.
- The standard deviation, σ, is the square root of this quantity. For this reason, the standard deviation is also known as the *"root-mean-square (r.m.s.) deviation"*.

$$\sigma = \sqrt{\frac{1}{\mathcal{N}} \sum_{i=1}^{\mathcal{N}} (n_i - \bar{n})^2}. \tag{5.5}$$

More precisely, if the expected number of decays in a particular time period is $\bar{n}$, then the probability, $P(n)$, that there will be n decays in that period is given by the *"Poisson distribution"* [2]

---

[2]This result is derived in Appendix 5.

Fig. 5.5 A Poisson
distribution around an
average of 100

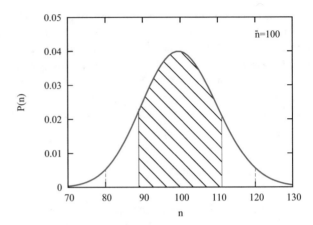

$$P(n) \ = \ \frac{\overline{n}^{n}}{n!} e^{-\overline{n}}. \tag{5.6}$$

It is shown in Appendix 5 that for such a distribution, the standard deviation is

$$\sigma \ = \ \sqrt{\overline{n}} \ . \tag{5.7}$$

The exact meaning of the standard deviation is that we expect about 68% of
the measurements to yield a result between $\overline{n} - \sigma$ and $\overline{n} + \sigma$. Only 5% of the
measurements yield a result with $n > \overline{n} + 2\sigma$ or $n < \overline{n} - 2\sigma$.

The Poisson distribution is plotted in Fig. 5.5 for the case of $\overline{n} = 100$. From
(5.7) this gives a standard deviation of 10. 68% of the measurements will yield
results between 90 and 110, (the shaded area) whereas only 5% will yield results
greater than 120 or less than 80 (outside of the 2σ range).

An immediate consequence of this is that if we want to measure the decay
constant (or half-life) to within an accuracy of ϵ, we need to collect at least $1/\epsilon^2$
decays.

Table 5.1 shows a typical set of 100 readings (in Bq) of a sample of 8.1 mg
of $^{238}_{92}$U, for which the average number of counts per second is 100 Bq (1 g yields
12,380 Bq as shown above). We see that the results are distributed around 100, but
only four of the measurements yield precisely 100. 68% of the measurements yield
a result between 90 and 110, and only 5 have either a count greater than 120 or less
than 80. On the other hand, the total number of counts is close to 10,000 (actually
9976). The expected error on this total is $\sqrt{10000} = 100$, so the sum of these
measurements is sufficient to determine the half-life up to an accuracy of 1%. Once
again note that we can never be sure that this measurement is accurate to 1%. What
is actually meant is that there is a probability of about 68% that the true count rate

Table 5.1 Typical set of 100 readings of the radioactivity of a sample of 8.1 mg of $^{238}_{92}$U

102	97	90	114	103	99	105	103	112	103
91	95	92	110	105	99	92	97	99	108
99	103	109	100	99	101	104	98	89	85
100	121	115	89	103	108	98	114	90	111
84	96	109	103	122	121	93	102	77	89
96	100	99	106	119	94	87	96	97	97
113	85	99	99	103	83	104	94	97	88
112	102	103	83	97	100	105	88	100	113
95	94	102	102	107	107	94	94	86	105
109	94	87	79	88	113	111	92	101	108

of a sample of $^{238}_{92}$U of this size is between 99 and 101 Bq. The enthusiastic reader can check that the r.m.s. deviation of this set of readings is close to 10 (actually 9.4), as expected.

5.5 Radiometric Dating

One of the useful applications of radioactivity is the *"radiometric dating"* method for determining the age of rock samples. The concentrations (number of nuclei per unit volume) of different nuclides in a rock sample can be measured using a mass spectrometer. Suppose the concentration of a radioactive nuclide, whose decay constant is λ, is $X(t)$, where t is the age of the rock sample, and the concentration of its daughter nuclide is $Y(t)$. When the rock was formed these concentrations were $X(0)$ and zero, respectively, where for the moment, we have assumed that the rock contained none of the daughter nuclide when the rock was formed.

From (5.2) we have

$$X(t) = X(0)e^{-\lambda t} \tag{5.8}$$

and

$$Y(t) = X(0)\left(1 - e^{-\lambda t}\right), \tag{5.9}$$

so that the age of the rock is given by

$$t = \frac{1}{\lambda}\ln\left(1 + \frac{Y(t)}{X(t)}\right). \tag{5.10}$$

However, in most cases there was a primordial concentration of the daughter nuclide at the time of formation of the rock, i.e. $Y(0) \neq 0$, so that (5.9) is modified to

$$Y(t) \; = \; Y(0) + X(0)\left(1 - e^{-\lambda t}\right).\tag{5.11}$$

This is known as the *"age equation"*. One of the most effective ways of determining
the primordial concentration of the daughter nuclide is to use the fact that there is
usually another stable isotope of the daughter nuclide present in the rock sample,
which is not involved in the radioactive process and whose concentration is therefore
time-independent. Let us call the concentration of this other isotope W. We have the
relation

$$Y(0) \; = \; \rho W,$$

where ρ is the relative abundance of the two isotopes at the time of rock formation.
Usually the ratio of primordial concentrations of the daughter nuclide and its
isotopes are different in different parts of the rock sample, but the primordial relative
abundance, ρ, should be the same everywhere in the sample. If this is the case and
we measure the concentrations, $X_i(t)$, $Y_i(t)$ and W_i at several different places, i, of
the rock sample, then the ratios obey the linear relation

$$\frac{Y_i(t)}{W_i} \; = \; \rho + \frac{X_i(t)}{W_i}(e^{\lambda t} - 1).\tag{5.12}$$

The ratios $Y_i(t)/W_i$ plotted against $X_i(t)/W_i$ lie on a straight line, whose slope is
a simple function of the age of the rock sample.

   The optimal choice of radioactive nuclide to examine depends on the age of the
rock sample under investigation. If the sample is much older than the half-life of the
radioactive nuclide, then almost all of it would have decayed and it is very difficult
to measure its remnant concentration. On the other hand if the half-life of the nuclide
is much larger than the age of the sample, the concentration of the daughter nuclide
will only be slightly larger than the primordial concentration, and therefore also
difficult to measure with sufficient accuracy.

   One of the most accurate examples of radiometric dating is that of uranium. This
is because it has two radioactive isotopes – $^{238}_{92}U$, which decays into $^{206}_{82}Pb$ with a
half-life of 4.47 billion years and $^{235}_{92}U$ which decays into $^{207}_{82}Pb$, with a half-life of
703.8 million years. Measurement of the ratios of lead to uranium concentrations
for *both* these isotopes can serve as a check, and an accuracy of better than 0.1%
has been achieved with this dating method.

5.6   Radiocarbon Dating

A variant of radiometric dating, invented by Willard Libby in 1946 [45], which can
be used for determining the age of organic fossils, is *"radiocarbon dating"*. The

carbon isotope, $^{14}_{6}$C decays into $^{14}_{7}$N (nitrogen), via β-decay with a half-life, $\tau_{\frac{1}{2}}$, of 5730 years.

Living organisms absorb the radioactive isotope of carbon $^{14}_{6}$C, which is created in the atmosphere by cosmic ray activity. The production of $^{14}_{6}$C from cosmic ray bombardment exactly cancels the rate at which that isotope of carbon decays so that the global concentration of $^{14}_{6}$C remains constant.

A sample of carbon taken from a living organism has a relative abundance, ρ, equal to about one part in 10^{12}. This isotope is being continually circulated by exchanging carbon with the environment (either by photosynthesis or by eating plants which have undergone photosynthesis or by eating other animals that have eaten such plants), so all living organisms – plants or animals – have the same small abundance, ρ, of $^{14}_{6}$C.

On the other hand, a sample of carbon from a dead object does not exchange its carbon with the environment, and therefore its concentration of $^{14}_{6}$C is not replenished as it decays radioactively. Such fossils therefore have a smaller concentration, $\rho'(t)$, of $^{14}_{6}$C, where $\rho'(t)$ depends on the age, t, of the fossil,

$$ t = \frac{\tau_{\frac{1}{2}}}{\ln 2} \ln \left(\frac{\rho'(t)}{\rho} \right) . \tag{5.13} $$

This means that from a measurement of the abundance of $^{14}_{6}$C in a fossil sample, one can determine its age. It is not necessary to measure directly the concentration of $^{14}_{6}$C but simply to ascertain the total mass of the carbon in a given sample. Then, provided that any other radioactive nuclides are present in such small quantities that the radioactivity from them is negligible, it is sufficient merely to measure the radioactivity rate from the fossil sample.

Radiocarbon dating is unreliable for fossils whose age is much larger than the half-life of $^{14}_{6}$C, although it has been successfully used for fossils up to 50,000 years old, and modern techniques for carbon dating are accurate to within 1%.

The reliability of this method of dating assumes that the relative abundance of $^{14}_{6}$C in the atmosphere has remained constant over time. Comparison with the carbon dating method with other dating methods, such as the analysis of tree rings, suggests that this has indeed been the case until the advent of the industrial era at the beginning of the nineteenth century. From that time on, the burning of coal and other fossil fuels has reduced the relative abundance of $^{14}_{6}$C, because in fossil fuels the $^{14}_{6}$C has almost all decayed away. The carbon emitted into the atmosphere from such fossil fuels is almost entirely $^{12}_{6}$C. The relative concentration of $^{14}_{6}$C in the atmosphere decreased by around 30% between 1800 and 1950. An even more disruptive change (in the opposite direction) occurred between 1945 and 1963 as a result of the testing of atomic bombs in the atmosphere. These tests caused the emission of a large quantity of radioactive material into the atmosphere, which increased the quantity of $^{12}_{6}$C converted to $^{14}_{6}$C by several tonnes. The relative abundance of $^{14}_{6}$C had increased by 75% in the northern hemisphere by 1964. It has since dropped off considerably, but is still responsible for an increase in abundance of a few percent. All this serves to make future carbon dating for the twentieth century almost impossible.

5.7 Multi-Modal Decays

A radioactive nuclide can sometimes decay into more than one *"channel"* , each of which has its own decay constant. Thus a parent nuclide P can decay radioactively into one of two or more daughter nuclides, D_i, each with its separate decay constant, λ_i.

The number, $N(t)$, of parent nuclei at time t has an exponential decay factor for each of the possible decay channels, i.e.

$$N(t) = N(0) \prod_i e^{-\lambda_i t} = N(0) e^{-\sum_i \lambda_i t}. \tag{5.14}$$

The half-life of the parent is therefore given by

$$t_{1/2} = \frac{\ln 2}{\sum_i \lambda_i}. \tag{5.15}$$

The daughter nuclides are produced in quantities which are proportional to their individual decay constants. This means that the fraction, B_j, of decays into the daughter nuclide D_j, known as the *"branching ratio"* into D_j is

$$B_j = \frac{\lambda_j}{\sum_i \lambda_i}. \tag{5.16}$$

If the half-life of the parent is known together with the fractions of daughter nuclides produced, then the individual decay constants can easily be determined from (5.16).

An example of this is $^{212}_{83}\text{Bi}$ (bismuth) which can either decay as

$$^{212}_{83}\text{Bi} \rightarrow \ ^{208}_{81}\text{Ti} + \alpha$$

$$\text{or } ^{212}_{83}\text{Bi} \rightarrow \ ^{212}_{84}\text{Po} + e^- + \bar{\nu}$$

$$\text{or } ^{212}_{83}\text{Bi} \rightarrow \ ^{208}_{82}\text{Pb} + e^- + \bar{\nu} + \alpha.$$

The half-life is 3630 s, and the fractions of the Bi which decay into Ti, Po and Pb are 64.05%, 35.94% and 0.014%, respectively.

From these data we can deduce that the total decay constant, λ, is given by

$$\lambda = \frac{\ln 2}{3630[\text{s}]} = 1.91 \times 10^{-4}\,\text{s}^{-1}.$$

The individual decay rates, λ_1, λ_2 and λ_3 from (5.16) for the decay channels into Ti, Po and Pb respectively are

$$\lambda_1 = 0.6405 \times 1.91 \times 10^{-5}\,\mathrm{s}^{-1} = 1.22 \times 10^{-4}\,\mathrm{s}^{-1}$$
$$\lambda_2 = 0.3594 \times 1.91 \times 10^{-4}\,\mathrm{s}^{-1} = 6.866 \times 10^{-5}\,\mathrm{s}^{-1}$$
$$\lambda_3 = 0.00014 \times 1.91 \times 10^{-4}\,\mathrm{s}^{-1} = 2.67 \times 10^{-8}\,\mathrm{s}^{-1}.$$

5.8 Decay Chains

It is possible that a parent nuclide decays, with decay constant λ_1, into a daughter nuclide, which is itself radioactive and decays (either into a stable nuclide or into another radioactive nuclide) with decay constant λ_2. An example of this is

$$^{212}_{83}\mathrm{Bi} \xrightarrow{\beta} {}^{210}_{84}\mathrm{Po} \xrightarrow{\alpha} {}^{206}_{82}\mathrm{Pb}$$

The half-life for the first stage of decay is 5 days, and the half-life for the second stage is 138 days.

If at time t we have $N_1(t)$ nuclei of the parent nuclide and $N_2(t)$ nuclei of the daughter nuclide, then for $N_1(t)$ we simply have

$$\frac{dN_1(t)}{dt} = -\lambda_1 N_1(t) \tag{5.17}$$

and therefore,

$$N_1(t) = N_1(0)e^{-\lambda_1 t}, \tag{5.18}$$

whereas for N_2 there is a production mechanism which contributes a rate of increase of N_2 equal to the rate of decrease of N_1. In addition there is a contribution to the rate of decrease of N_2 from its own decay process, so we have

$$\frac{dN_2(t)}{dt} = \lambda_1 N_1(t) - \lambda_2 N_2(t). \tag{5.19}$$

Inserting (5.18) into (5.19) gives

$$\frac{dN_2(t)}{dt} = \lambda_1 N_1(0)e^{-\lambda_1 t} - \lambda_2 N_2(t). \tag{5.20}$$

This is an inhomogeneous differential equation whose solution with zero initial concentration, i.e. $N_2(0) = 0$, is given by

$$N_2(t) \;=\; N_1(0)\frac{\lambda_1}{(\lambda_1 - \lambda_2)}\left(e^{-\lambda_2 t} - e^{-\lambda_1 t}\right). \tag{5.21}$$

The concentrations $N_1(t)$ and $N_2(t)$ are plotted in Fig. 5.6, as a function of time. Initially the daughter nuclide is completely absent. As the parent decays, the quantity of the daughter nuclide grows faster than it decays, but after some time the available quantity of the parent nuclide is depleted so the production rate decreases and the decay rate of the daughter nuclide begins to dominate. If the half-life of the daughter is more than half the half-life of the parent ($\lambda_2 < 2\lambda_1$) then at some point in time, t_1, the concentration of the daughter will exceed that of the parent, i.e.

$$N_2(t) \;>\; N_1(t), \quad (t > t_1),$$

as shown in the left-hand graph of Fig. 5.6. On the other hand if the half-life of the daughter is less than half the half-life of the parent, the daughter nuclei decay too fast for the concentration of the daughter ever to exceed that of the parent, as shown in the right-hand graph of Fig. 5.6. This can be seen from the fact that the time, t_1, at which the concentrations are equal is obtained by equating $N_1(t) = N_2(t)$ in (5.18) and (5.21). Solving for t gives

$$t_1 \;=\; \frac{1}{(\lambda_1 - \lambda_2)}\ln\left(2 - \frac{\lambda_2}{\lambda_1}\right).$$

We see that this only has a (real) solution for $\lambda_2 < 2\lambda_1$. Note that for $\lambda_2 = 0$, meaning the daughter is stable, t_1 becomes equal to the half-life of the parent nuclide, i.e. the time at which there are as many daughter nuclei as parent nuclei.

Some heavy nuclides have a very long decay chain, decaying at each stage to another unstable nuclide before eventually reaching a stable nuclide. An example of this is ${}^{238}_{92}\mathrm{U}$, shown in Fig. 5.7 which decays in no fewer than 14 stages – eight by

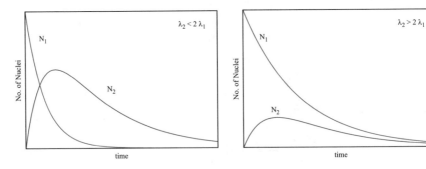

Fig. 5.6 Concentrations of first two nuclides in a two-stage radioactive decay chain. The left-hand graph is for the case where the half-life of the daughter nuclide is more than half the half-life of the parent. The right-hand graph is for the case where the half-life of the daughter nuclide is less than half the half-life of the parent

Fig. 5.7 $^{238}_{92}$U decay chain

Decay Mode	Nuclide	Half-Life
α	Uranium-238	4.5×10^9 years
β	Thorium-234	24.1 days
β	Protactinium-234	1.17 minutes
α	Uranium-234	2.45×10^5 years
α	Thorium-230	7.54×10^5 years
α	Radium-226	1600 years
α	Radon-222	3.82 days
α	Polonium-218	3.04 minutes
β	Lead-214	26.8 minutes
β	Bismuth-214	19.7 minutes
α	Polonium-214	1.64×10^{-4} seconds
β	Lead-210	22.3 years
β	Bismuth-210	5 days
α	Polonium-210	138 days
	Lead-206	STABLE

α-decay and six by β-decay before reaching a stable isotope of Pb. The lifetimes for the individual stages vary from around 10^{-4} s to 10^9 years.

In such cases, if the first parent is very long-lived, so that the number of parent nuclei does not decrease much, it is possible to reach what is known as *"secular equilibrium"*, in which the quantities of various daughter nuclei remain unchanged. This happens when the numbers of nuclei in the chain N_A, N_B, N_C, ... are in the ratio

$$\lambda_A N_A = \lambda_B N_B, \quad etc.,$$

where λ_A, λ_B, ... are the decay rates for these nuclides. In other words, the concentrations of the nuclides in the chain are all proportional to their half-lives. What is happening here is that the rate of production of daughter B, is the rate of decay of A, which is $\lambda_A N_A$ and this is equal to $\lambda_B N_B$, the rate of decay of B, so the quantity of B nuclei remains unchanged.

5.9 Induced Radioactivity

It is possible to convert a nuclide which is not radioactive into a radioactive one by bombarding it with neutrons or other particles. The stable nuclide (sometimes) absorbs the projectile to become an unstable, radioactive nucleus.

For example, bombarding $^{23}_{11}\text{Na}$ (sodium) with neutrons can convert the nuclide to $^{24}_{11}\text{Na}$, which is radioactive and decays via β-decay to $^{24}_{12}\text{Mg}$ (magnesium).

In this case, if we assume that the rate at which the radioactive nuclide (with decay constant λ) is being generated is R, then the number of such nuclei is given by the differential equation

$$\frac{dN(t)}{dt} = R - \lambda N.$$

If at time $t = 0$ the number of these nuclei is zero (i.e. we start the bombardment at $t = 0$), then the solution to this differential equation is

$$N(t) = \frac{R}{\lambda}\left(1 - e^{-\lambda t}\right).$$

This starts at zero and then grows so that asymptotically (as $t \to \infty$)

$$R = \lambda N,$$

which is the equilibrium state in which the production rate R is equal to the decay rate λN. An example of such a process which takes place in Nature is the production of $^{14}_{6}\text{C}$ (relevant to the carbon dating discussed above) from cosmic rays interacting with $^{12}_{6}\text{C}$.

5.10 Neutron Emission

Some naturally occurring isotopes are neutron rich, i.e. they contain more neutrons than their stable isotopes. Such isotopes can spontaneously emit neutrons.

Summary

- There are three main types of radioactivity: α-decay – the emission of nuclei of $^{4}_{2}\text{He}$, β-decay – the emission of a positron and a neutrino, or an electron and an antineutrino, and γ-decay – the emission of short wavelength photons.
- Radioactive decays can be detected and counted using a Geiger counter, in which atoms of a low pressure gas are ionized by the radiation particles and induce an

electric discharge, a scintillation counter, in which the incidence of a radioactive particle on a scintillating crystal produces a light burst which is amplified by a photomultiplier tube, or a semiconductor detector in which incident radiation particles excite electron–hole pairs in a semiconductor diode.

- The decay of a radioactive particle is exponential, with exponent $-\lambda t$, where λ is the decay constant. The half-life, $\tau_{\frac{1}{2}}$ is the time taken for the concentration of the parent nuclide to fall to one-half of its original value and is equal to $\ln 2/\lambda$.
- The average number of decays in a short time interval Δt, of a radioactive sample of N nuclei is $\bar{n} = \lambda N \Delta t$. But the decays are random and so the number of decays observed in a given time interval Δt will fluctuate around this average value. The size of these fluctuations is determined by the r.m.s. deviation, which is equal to $\sqrt{\bar{n}}$. The probability of observing n counts in one such time period is given by a Poisson distribution.
- The age of rock samples can be determined by comparing the concentration of a radioactive nuclide with a long half-life and the concentration of the daughter nuclide. The concentration of the daughter nuclide at the time of the rock formation is obtained by measuring the concentrations of the daughter nuclide and another stable isotope of the daughter element at different places in the rock sample, where the primordial ratios of parent element to daughter element differ.
- Fossilized organisms can be dated by measuring the radioactivity from the $^{14}_{6}C$ in order to compare the ratio of concentrations of $^{14}_{6}C$ and $^{12}_{6}C$ with that of atmospheric carbon. The relative abundance of $^{14}_{6}C$ in the atmosphere was kept in constant equilibrium (at least until the start of the industrial era) by the bombardment with cosmic rays in the outer atmosphere, which induces the regeneration of $^{14}_{6}C$ at the same rate as its radioactive decay.
- Some radioactive nuclides can decay radioactively into two or more different channels. The concentrations of the different radioactive daughters are in the ratio of the decay constants into the different channels.
- Many radioactive nuclides decay into daughters which are themselves radioactive, thereby forming a radioactive chain, which can be quite long. If the parent is long-lived, it is possible to reach secular equilibrium in which the concentrations of all the nuclides in the chain remain constant and are proportional to their half-lives.
- Stable nuclides can be converted into radioactive nuclides by bombarding them with neutrons or other particles.

Problems

[A year is 3.156×10^7 s.]

Problem 5.1 Identify the mode of radioactive decay (i.e. α-decay, β-decay) for the following radioactive transitions:

(a) $^{214}_{84}\text{Po} \rightarrow ^{210}_{82}\text{Pb}$,

(b) $^{238}_{92}\text{U} \rightarrow ^{234}_{90}\text{Th}$,

(c) $^{234}_{90}\text{Th} \rightarrow ^{234}_{91}\text{Pa}$,

(d) $^{40}_{19}\text{K} \rightarrow ^{40}_{18}\text{A}$,

(e) $^{40}_{19}\text{K} \rightarrow ^{40}_{20}\text{Ca}$.

In the case of β-decay indicate whether an electron or a positron is emitted.

Problem 5.2 One gram of carbon from a relic found in an Egyptian tomb has a measured activity of 0.16 Bq. The ratio of $^{14}_{6}\text{C}$ to $^{12}_{6}\text{C}$ in living plants is 1.3×10^{-12}. The half-life of $^{14}_{6}\text{C}$ is 5731 years. How old is the relic? [1 a.m.u. is equivalent to 1.66×10^{-27} kg.]

Problem 5.3 $^{235}_{92}\text{U}$ has a radioactive half-life of 703.8 ± 0.1 million years. For how long does one have to observe the radioactivity from 1 mg of $^{235}_{92}\text{U}$ in order to be able to determine its lifetime to this accuracy? [The mass of an atom of $^{235}_{92}\text{U}$ may be taken to be 3.90×10^{-25} kg.]

Problem 5.4 One of the oldest meteorites to have fallen to Earth is the meteorite NWA 7325 found in Erfoud, Morocco in 2012. It was examined using the long-lived radioactive isotope $^{177}_{71}\text{Lu}$, which decays via β-decay into $^{177}_{72}\text{Hf}$ with a half-life of 37.8 billion (10^9) years. Hafnium also has a naturally occurring stable isotope – $^{176}_{72}\text{Hf}$.

The table below gives the ratios of the concentrations of ^{177}Hf to ^{176}Hf and of ^{177}Lu to ^{176}Hf at four different parts of the meteorite. Plot these and from the measured slope of the fitted straight line, determine the age of the meteor.

$^{177}\text{Hf}/^{176}\text{Hf}$	$^{177}\text{Lu}/^{176}\text{Hf}$
0.2822	0.0250
0.2827	0.0327
0.2832	0.0377
0.2836	0.0434

What was the ratio of the concentrations of the isotopes $^{177}_{72}\text{Hf}$ to $^{176}_{77}\text{Hf}$ when the meteorite was formed?

Problem 5.5 A radioactive nuclide, A, decays with a decay constant λ_1 into a daughter nuclide, B, which itself decays into a stable nuclide, C with decay constant λ_2.

(a) Show that the time, t_2, at which the number of nuclei of B is greatest, is given by

$$t_2 = \frac{1}{(\lambda_1 - \lambda_2)} \ln\left(\frac{\lambda_1}{\lambda_2}\right).$$

(b) Write down an expression for the total radioactive decay count from A and B combined. Show that, provided $\lambda_1 < \lambda_2$, this total count has a maximum at time, t_3, where

$$t_3 = \frac{1}{(\lambda_1 - \lambda_2)} \ln \left(1 - \frac{(\lambda_1 - \lambda_2)^2}{\lambda_2^2} \right).$$

Explain what happens if $\lambda_1 > \lambda_2$.

Appendix 5: Poisson Distribution

The probability that a particular nucleus will undergo radioactive decay in a time interval Δt is $\lambda \Delta t$, where λ is the decay constant. Likewise, the probability that it will *not* decay is $(1 - \lambda \Delta t)$.

If we have a sample of N nuclei and $N \gg \lambda \Delta t$, then the probability, $P(n)$, that exactly n nuclei will decay in time interval Δt is the probability that n nuclei will decay, multiplied by the probability that $(N - n)$ nuclei will *not* decay, multiplied by $^N C_n$, the number of ways of selecting n decaying nuclei out of a total of N nuclei.

$$P(n) = {}^N C_n (\lambda \Delta t)^n (1 - \lambda \Delta t)^{N-n}, \tag{5.22}$$

$$\text{where } {}^N C_n = \frac{N!}{n!(N - n)!}.$$

The average number of nuclei that will decay in this time interval is given by

$$\bar{n} = N \lambda \Delta t, \tag{5.23}$$

so we may rewrite (5.22) as

$$P(n) = \frac{N!}{n!(N - n)!} \left(\frac{\bar{n}}{N} \right)^n \left(1 - \frac{\bar{n}}{N} \right)^{N-n}. \tag{5.24}$$

$P(n)$ is negligible except in the region $n \sim \bar{n}$, so we can always take $n \ll N$. In this limit, we can make the standard approximations

$$\left(1 - \frac{\bar{n}}{N} \right)^{N-n} \approx \left(1 - \frac{\bar{n}}{N} \right)^N \approx e^{-\bar{n}},$$

and

$$(N - n)! \approx \frac{N!}{N^n}.$$

Using this approximation $P(n)$ reduces to the Poisson distribution:

$$P(n) = \frac{\bar{n}^n}{n!} e^{-\bar{n}}. \tag{5.25}$$

We note that the total probability, $\sum_{n=0}^{N} P(n)$, is equal to 1, and that the average value,

$$\bar{n} = \sum_{n=0}^{N} n P(n), \tag{5.26}$$

as expected.

We now calculate the average value of $n(n-1)$, using the relation

$$n(n-1)\bar{n}^n = \bar{n}^2 \frac{d^2}{d\bar{n}^2} \left(\bar{n}^n \right),$$

so that

$$\overline{n(n-1)} = \sum_{n=0}^{N} n(n-1) P(n) = \bar{n}^2 e^{-\bar{n}} \sum_{n=0}^{N} \frac{d^2}{d\bar{n}^2} \left(\frac{\bar{n}^n}{n!} \right) \tag{5.27}$$

For $N \gg \bar{n}$, we can take the limit $N \to \infty$ to obtain

$$\overline{n(n-1)} = \bar{n}^2 e^{-\bar{n}} \frac{d^2}{d\bar{n}^2} \left(e^{\bar{n}} \right) = \bar{n}^2. \tag{5.28}$$

This gives us an expression for the mean of the square, $\overline{n^2}$:

$$\overline{n^2} = \bar{n}^2 + \bar{n}. \tag{5.29}$$

The mean square deviation is

$$\overline{(n - \bar{n})^2} = \sum_{n=0}^{N} \left(n^2 - 2n\bar{n} + \bar{n}^2 \right) P(n) = \overline{n^2} - \bar{n}^2, \tag{5.30}$$

(where we have used (5.26)).

From (5.29) and (5.30), we see that the standard deviation is

$$\sigma = \sqrt{\overline{(n - \bar{n})^2}} = \sqrt{\bar{n}}. \tag{5.31}$$

Chapter 6
Alpha Decay

Alpha (α)-decay is the radioactive emission of an α-particle – a nucleus of $_2^4$He, consisting of two protons and two neutrons. This is a very tightly bound nucleus as it is doubly magic. The daughter nuclide has two protons and four nucleons fewer than the parent nuclide, and α-decay is described by the following process:

$$_Z^A\{Ch\}_P \rightarrow {}_{(Z-2)}^{(A-4)}\{Ch\}_D + \alpha, \tag{6.1}$$

which can occur only if the binding energy of the daughter nuclide, $_{(Z-2)}^{(A-4)}\{Ch\}_D$, plus the binding energy of the α-particle exceeds the binding energy of the parent nuclide, $_Z^A\{Ch\}_P$. Since the binding energy per nucleon of an α-particle is so large, this can happen in very many nuclides. On the other hand, the decay of a parent nuclide into a daughter plus a nuclide with larger atomic number than an α-particle is far less likely, although it does happen for some of the heavier nuclides due to fission (discussed in Chap. 9). With the exception of $_4^8$Be, which decays into two α-particles, only nuclides with atomic number $Z \geq 52$ (tellurium) exhibit α-decay.

6.1 Kinematics

The "*Q-value*" of the decay, Q_α, is defined to be the difference between the mass of the parent and the sum of the mass of the daughter and the α-particle, multiplied by c^2.

$$Q_\alpha = (m_P - m_D - m_\alpha)c^2.$$

This is equal to the difference of the total binding energies mentioned above. The binding energies of the parent and daughter nuclides can usually be estimated quite accurately from the semi-empirical mass formula, but this formula underestimates

© The Author(s), under exclusive license to Springer Nature Switzerland AG 2021
A. Belyaev, D. Ross, *The Basics of Nuclear and Particle Physics*, Undergraduate
Texts in Physics, https://doi.org/10.1007/978-3-030-80116-8_6

the binding energy of the α-particle, as explained in Chap. 4. Here we need to use the experimentally measured value of 28.3 MeV. The values of Q_α are usually of the order of a few MeVs.

The α-particle emerges with a kinetic energy T_α, which is slightly below the value of Q_α. This is because if the parent nucleus is at rest before the decay, there must be some recoil of the daughter nucleus in order to conserve momentum. The daughter nucleus therefore has kinetic energy T_D such that

$$Q_\alpha = T_\alpha + T_D. \tag{6.2}$$

The momentum, p, of the α-particle is equal and opposite to the recoil momentum of the daughter nucleus (assuming that the parent nucleus is at rest). The kinetic energies of the α-particle and daughter nucleus are given by

$$T_\alpha = \frac{p^2}{2m_\alpha} \tag{6.3}$$

and

$$T_D = \frac{p^2}{2m_D}, \tag{6.4}$$

respectively. Inserting (6.3) and (6.4) into (6.2) gives

$$Q_\alpha = \frac{p^2}{2\tilde{m}_\alpha}, \tag{6.5}$$

where $\tilde{m}_\alpha$ is the *"reduced mass"* of the α-particle-daughter system:

$$\tilde{m}_\alpha = \frac{m_\alpha m_D}{m_\alpha + m_D}, \tag{6.6}$$

and the kinetic energy of the α-particle is

$$T_\alpha = \frac{\tilde{m}_\alpha}{m_\alpha} Q_\alpha. \tag{6.7}$$

Neglecting the binding energies in comparison with the masses of the particles this may be written as

$$T_\alpha = \frac{A_D}{A} Q_\alpha,$$

where $A_D \equiv (A - 4)$ is the atomic mass number of the daughter nuclide.

As an example, let us consider the decay of $^{214}_{84}\text{Po}$ to $^{210}_{82}\text{Pb}$

$$^{214}_{84}\text{Po} \rightarrow {}^{210}_{82}\text{Pb} + \alpha.$$

The binding energy of $^{214}_{84}$Po is 1666.03 MeV, and the binding energy of $^{210}_{82}$Pb is 1645.56 MeV. The Q-value for the decay is

$$Q_\alpha = 1645.56 + 28.3 - 1666.03 = 7.83 \text{ MeV}.$$

The kinetic energy of the α-particle is then given by

$$T_\alpha = \frac{210}{214} \times 7.83 \simeq 7.68 \text{ MeV}.$$

Sometimes the α-particles emerge with kinetic energies that are somewhat lower than this prediction. Such α-decays are accompanied by the emission of γ-rays. What is happening is that the daughter nucleus is being produced in one of its excited states, so that there is less energy available for the α-particle (or the recoil of the daughter nucleus).

For example, consider the decay of $^{228}_{90}$Th (thorium) to $^{224}_{88}$Ra (radon)

$$^{228}_{90}\text{Th} \rightarrow \,^{224}_{88}\text{Ra} + \alpha.$$

The binding energy of $^{220}_{90}$Th is 1743.09 MeV, and the binding energy of $^{224}_{88}$Ra is 1720.31 MeV. The Q-value is therefore

$$Q_\alpha = 1720.31 + 28.3 - 1743.09 = 5.52 \text{ MeV}.$$

The kinetic energy of the α-particle is then given by

$$T_\alpha = \frac{224}{228} \times 5.52 \simeq 5.42 \text{ MeV}.$$

α-particles are observed with this kinetic energy, but also with kinetic energies 5.34, 5.21, 5.17 and 5.14 MeV. From this we can conclude that there are excited states of the daughter nuclide, $^{224}_{88}$Ra, with excitation energies of 0.08, 0.21, 0.25 and 0.28 MeV. The α-decay is therefore accompanied by γ-rays (photons) with energies equal to the differences of these energies.

It is sometimes possible to find an α-particle whose energy is *larger* than that predicted from the Q-value. This occurs when the parent nuclide is itself a product of a decay from a further ("grand")parent. In this case the parent α-decaying nucleus can be produced in one of its excited states. In most cases, this state will decay to the ground state by emitting γ-rays before the α-decay takes places. But in some cases where the excited state is relatively long-lived and the decay constant for the α-decay is sufficiently large, the parent in the excited state can α-decay before returning to its ground state. The Q-value for such a decay is larger than for decay from the ground state by an amount equal to the excitation energy.

In the above example of α-decay from $^{214}_{84}$Po the parent nuclide is actually unstable and is produced by β-decay of $^{214}_{83}$Bi, states with energies 0.61, 1.41, 1.54,

1.66 MeV above the ground state. Therefore as well as an α-decay with Q-value 7.83 MeV, calculated above, there are α-decays with Q-values of 8.44, 9.24, 9.37 and 9.49 MeV.

6.2 Decay Mechanism

The mean lifetime of α-decaying nuclides varies from the order of 10^{-7} s to 10^{10} years. We can understand this by investigating the mechanism for α-decay.

What happens is that two protons from the highest proton energy levels and two neutrons from the highest neutron energy levels combine to form an α-particle inside the nucleus – this is known as a *"quasi-bound state"*. It acquires an energy that is approximately equal to Q_α (neglecting the small correction due to the recoil of the nucleus).

The α-particle is bound to the nucleus in a potential well created by the strong, short-range nuclear forces. There is also a Coulomb repulsion between this "quasi-" α-particle and the rest of the nucleus. Since the strong-interaction force is rapidly attenuated at distances from the centre exceeding the nuclear radius, R, we can make the approximation that the nuclear potential well can be taken to be a spherical well, of radius R.

The Coulomb repulsion between the α-particle, (charge $2e$), and the daughter nucleus is described by the potential (written in terms of the fine structure constant, α)

$$V_c = \frac{2Z_D\alpha\hbar c}{r}, \tag{6.8}$$

where $Z_D \equiv (Z - 2)$ is the atomic number of the daughter nuclide. It is convenient to rewrite this as

$$V_c = \frac{Q_\alpha r_b}{r}, \tag{6.9}$$

where r_b is the radial distance at which the kinetic energy of the α-particle inside the nucleus is equal to the Coulomb energy i.e.

$$r_b = \frac{2Z_D\alpha\hbar c}{Q_\alpha}. \tag{6.10}$$

The nuclear potential well combined with the Coulomb potential forms a potential barrier as shown in Fig. 6.1.

Classically, whenever r_b exceeds the nuclear radius, R, the α-particle does not have enough kinetic energy to pass over the potential barrier from inside the nucleus,

Fig. 6.1 The potential barrier formed by the potential well of the nuclear force and the Coulomb repulsion between the α-particle and the daughter nucleus

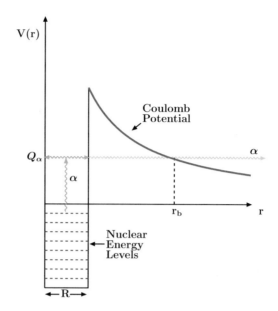

$r < R$, to free space, $r \gg r_b$. In Quantum Physics, on the other hand, quantum tunnelling can occur with probability, T, which is given approximately by[1]

$$T = e^{-2S(R,r_b)/\hbar},\tag{6.11}$$

with

$$S(R, r_b) = \int_R^{r_b} \sqrt{2m_\alpha \left(U(r) - Q_\alpha\right)}dr \;,\tag{6.12}$$

where $U(r)$ is the effective potential that includes the centrifugal repulsion term for an α-particle with orbital angular momentum denoted by the quantum number ℓ given by

$$U(r) \equiv V_c(r) + \frac{\ell(\ell + 1)\hbar^2}{2m_\alpha r^2}.\tag{6.13}$$

Finally, we need to multiply the transition probability by the number of times per second, N_{coll}, that the α-particle "tries" to escape, i.e. how many times per second it makes a collision with the potential barrier. This is the inverse of the time taken for the α-particle to travel from the centre to the edge of the nucleus and back, given approximately by

---

[1] An outline derivation of this result is given in Appendix 6.

$$\frac{v}{2R},$$

where $v = \sqrt{2Q_\alpha/m_\alpha}$ is the velocity of the α-particle inside the nucleus. This varies for different nuclides and it is also a very rough approximation. Further refinements of this estimate suggest that a reasonable value for N_{coll} is 10^{21} s^{-1}.

We therefore end up with an approximate expression for the α-decay rate, λ:

$$\lambda \approx N_{\text{coll}} \exp\left\{-\frac{2}{\hbar}S(R, r_b)\right\}. \tag{6.14}$$

This can easily be converted into an expression for the $\log_{10}$ of the half-life,[2] $\tau_{\frac{1}{2}}$:

$$\log_{10}\left(\tau_{\frac{1}{2}}\right) \approx \frac{2}{\hbar \ln(10)}S(R, r_b) - \log_{10}(N_{\text{coll}}). \tag{6.15}$$

This expression for the half-life in α-decay was originally derived by George Gamow in 1928 [46], and the exponential term in (6.14) is known as the *"Gamow factor"*, which is the probability that the α-particle will tunnel through the Coulomb barrier.

For even–even nuclei the ground states of the parent and daughter nuclides both have zero spin, so by conservation of angular momentum the α-particle has zero orbital angular momentum ($\ell = 0$). In this case, $S(R, r_b)$ may be written as

$$S(R, r_b) = \sqrt{2m_\alpha Q_\alpha} \int_R^{r_b} \sqrt{\left(\frac{r_b}{r} - 1\right)}dr. \tag{6.16}$$

The integral in (6.16) can be performed exactly and then simplified by expanding in the small quantity $\sqrt{R/r_b}$:

$$\int_R^{r_b} \sqrt{\left(\frac{r_b}{r} - 1\right)}dr = r_b\left(\arccos\sqrt{x} - \sqrt{x(1-x)}\right) \simeq r_b\left(\frac{\pi}{2} - 2\sqrt{x}\right), \tag{6.17}$$

where $x = R/r_b$ and for the last step we have used the expansion for $x \ll 1$ around zero. Using (6.10) and (6.17), $S(R, r_b)$ simplifies to

$$S(R, rb) \simeq \sqrt{\frac{2m_\alpha}{Q_\alpha}}\left(\pi\alpha\hbar c Z_D - 2\sqrt{2Q_\alpha\alpha\hbar c r_0 Z_D}A_D^{1/6}\right), \tag{6.18}$$

where we have used the radius of the daughter nucleus, $R = r_0 A_D^{1/3}$.

Thus we obtain the following approximation for the logarithm of the half-life:

---

[2]We neglect here the very small correction in the logarithm due to the ratio of the half-life to the mean lifetime.

$$\log_{10}\left(\tau_{\frac{1}{2}}\right) \approx \frac{\sqrt{8m_\alpha c^2}\pi\alpha}{\ln(10)}\frac{Z_D}{\sqrt{Q_\alpha}} - \frac{8}{\ln(10)}\sqrt{\frac{\alpha m_\alpha c r_0}{\hbar}}\sqrt{Z_D}A_D^{1/6}$$
$$+ \cdots - \log_{10}(N_{\text{coll}}) \ . \tag{6.19}$$

Inserting numerical values ($m_\alpha = 3727\,\text{MeV}/c^2$, $r_0 = 1.22\,\text{fm}$) this becomes

$$\log_{10}\left(\tau_{\frac{1}{2}}\right) \approx 1.72\frac{Z_D}{\sqrt{Q_\alpha}} - 1.42\sqrt{Z_D}A_D^{1/6} - 21. \tag{6.20}$$

In this (Gamow) formula, the half-life, $\tau_{\frac{1}{2}}$, is in seconds. The first term on the RHS of (6.20) explains the huge variation of half-lives over the range of atomic number and Q-value. The ratio of the atomic number to the square root of the Q-value (in MeV) varies from 27 to 43, and this gives us a range of half-lives from less than a microsecond to 10^{18} s.

An improvement to this estimate comes from accounting for the recoil of the daughter nucleus. This is achieved by replacing the mass of the α-particle by its reduced mass. However, the expression (6.20) often gives values that are different from the experimentally measured half-life by a few orders of magnitude, so that this improvement makes a very small difference in comparison with such errors.

6.3 Empirical Formulae

In 1911 Hans Geiger and John Nuttall [47] observed that the logarithm of the half-lives of different α-decays was a linear function of the inverse square root of the Q-value:

$$\log_{10}\left(t_{\frac{1}{2}}\right) \approx \frac{f}{\sqrt{Q_\alpha}} + g. \tag{6.21}$$

Comparing this with the theoretical expression, (6.20), the coefficient f has a Z dependence given by

$$f = 1.72Z_D.$$

From (6.20), the parameter g is given by

$$g = -1.42\sqrt{Z_D}A_D^{1/6} - 21.$$

This is *not* a constant but varies over the range of α-decaying nuclides from around -46 to -58. It turns out, however, that using the Geiger–Nuttall formula with g fixed at the value of -54 gives a better fit to data. This fit is shown in Fig. 6.2. One can see that this fit is very approximate and sometimes has large errors of up to 3 in the logarithm of the half-life (i.e. 3 orders of magnitude in the half-life). The purely theoretical prediction (6.20) can produce even larger errors for the absolute values of the half-lives. On the other hand it reproduces the dependence on the Q-values

Fig. 6.2 The half-lives of several α-decaying nuclides over the full range of $Z_D/\sqrt{Q_\alpha}$. The blue line is the Geiger–Nuttall formula (6.21) with the slope set to the theoretical value of 1.72 and the green line is the best fit through the data

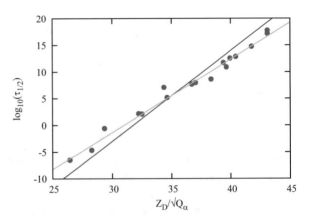

reasonably well, although it is clear from Fig. 6.2, comparing the best fit to data (green line) with the theoretical slope (blue line), that the theoretical slope is around 10% too large.

An empirical formula that is based on the theoretical expression was proposed by Guy Royer in 2000 [48]. The form is identical to that of (6.20) but with coefficients that are free parameters

$$\log_{10}\left(\tau_{\frac{1}{2}}\right) = a + b\sqrt{Z}A^{1/6} + \frac{cZ}{\sqrt{Q_\alpha}}. \tag{6.22}$$

In these empirical fits, Z and A are taken to refer to the parent nuclide. Good fits were found, but it was necessary to fit the set of parameters differently for even–even, odd–even, even–odd and odd–odd nuclides. In the case of even–even nuclides the best-fit parameters were

$$a = -25.3, \quad b = -1.15, \quad c = 1.59.$$

We note that these differ a little from their theoretical values, but this results in a very good fit to data. For a range of 356 even–even nuclides, the standard deviation (in the logarithm of the half-lives) is 0.28.

There have been many proposals for improved fits. One notable empirical formula was proposed by Victor Viola and Glenn Seaborg [49] in 1966, which involved four parameters

$$\log_{10}\left(t_{\frac{1}{2}}\right) = \frac{aZ + b}{\sqrt{Q_\alpha}} + Zc + d. \tag{6.23}$$

A very good fit [50] using this model has been found with

$$a = 1.64, \quad b = -8.55 \quad c = -0.194, \quad d = -33.9.$$

This formula reproduces the half-lives of a range of 64 even–even nuclides with a standard deviation as low as 0.143.

6.4 Angular Momentum

So far we have only considered the case of an α-particle with zero angular momentum tunnelling through a purely Coulomb potential barrier with no centrifugal term. There are cases in which the α-particle is formed inside the nucleus with non-zero orbital angular momentum.

For even–even nuclides the ground states of both the parent and daughter nuclides are 0^+. However, the parent could also decay to one of the excited states of the daughter nuclide, which has non-zero spin. In order to conserve angular momentum the α-particle emitted in such a decay to an excited state will carry the orbital angular momentum equal to the spin of the excited state of the daughter nuclide. An example is the decay of $^{244}_{96}$Cm (curium) to $^{240}_{94}$Pu (plutonium), which has a series of rotational excited states 2^+, 4^+, $\cdots$ with excitation energies 0.043 MeV, 0.142 MeV etc. The Q-value of the decay to the ground state is 5.901 MeV. The decays to the excited states will have Q-values that are lower by an amount equal to the excitation energy.

The decay rates in which the α-particle has non-zero orbital angular momentum are suppressed because the effective potential, (6.13), is enhanced by the centrifugal term. In such a case the integral $S(R, r_b)$ in (6.12) cannot be expressed in terms of elementary functions. However, since the correction due to the orbital angular momentum is small, it is sufficient to expand the integrand to first order in $\ell(\ell + 1)$. We may also make the same approximation as was done to obtain (6.18), and expand in the small quantity $\sqrt{R/r_b}$. This yields

$$S(R, r_b) = S(R, r_b)_{\ell=0} + \frac{\ell(\ell + 1)\hbar^2}{\sqrt{2m_\alpha Q_\alpha}} \left[\frac{1}{\sqrt{Rr_b}} - \frac{1}{2}\frac{R}{r_b^{3/2}} + \cdots \right] + \cdots \quad (6.24)$$

Keeping only the leading term, using (6.10) for r_b, and again using $R = r_0 A_D^{1/3}$, we find that due to the orbital angular momentum, the half-life is increased by a factor

$$\exp\left\{ \left(\frac{\hbar}{m_\alpha r_0 c Z_D A_D^{1/3}} \right)^{1/2} \ell(\ell + 1) \right\}. \quad (6.25)$$

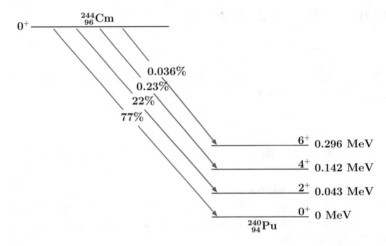

Fig. 6.3 The branching ratios for the decay of $^{244}_{96}$Cm to the ground state and the first 3 rotational excited states of $^{240}_{94}$Pu

Another way of describing this is to say that the decay rate is suppressed by the same factor. This allows us to determine the branching ratios for the decays into various excited states of the daughter nuclide. Note that as well as the suppression of the decay rate due to the orbital angular momentum, there is an enhancement owing to the fact that decays to the excited states have a lower Q-value. If we take this into account as well, then, for example, the branching ratios for the decay of $^{244}_{96}$Cm to the different rotational excited states of $^{240}_{94}$Pu are as shown in Fig. 6.3. 77% of the decays are to the 0^+ ground state, 22% to the first excited, 2^+, state and 0.23% to the second excited 4^+ state.

α-particles with orbital angular momentum can also be emitted from parent nuclides with an odd number of protons or neutrons or both. In such cases the ground states of both the parent and daughter nuclides will have non-zero spin, I_P and I_D, respectively, and conservation of angular momentum allows for the emission of an α-particle with orbital angular momentum, ℓ ranging from $|I_P - I_D|$ to $I_P + I_D$. Not all of these values of orbital angular momentum are permitted because it is necessary in α-decay to conserve parity. The wavefunction of an α-particle with orbital angular momentum ℓ has parity $(-1)^\ell$ and so the parities, π_P and π_D, of the parent and daughter states must satisfy the relation

$$\pi_P = (-1)^\ell \pi_D. \tag{6.26}$$

As an example, we consider the decay of the nuclide $^{237}_{93}$Np (neptunium) to $^{233}_{91}$Pa (protactinium). The ground state of $^{237}_{93}$Np has spin–parity $\frac{5}{2}^+$, whereas the ground state of $^{233}_{91}$Pa has spin–parity $\frac{3}{2}^-$. From angular momentum conservation the allowed values of ℓ for the α-particle are $\frac{5}{2} - \frac{3}{2}$ to $\frac{5}{2} + \frac{3}{2}$, i.e. $\ell = 1, 2, 3$ or 4.

However, the ground states of the parent and daughter have opposite parity and therefore ℓ has to be odd – we can have $\ell = 1$ or $\ell = 3$. The $\ell = 1$ decay will dominate since the suppression of the decay constant is smaller for the lower value of ℓ. Recall that decays to excited states of the daughter nuclide are accompanied by the emission of a γ-ray as the daughter nuclide makes a transition (either directly or indirectly) back to the ground state. The emitted γ-rays will have frequencies corresponding to the differences between the excited state energy levels.

6.5 Hindrance Factor

We have been concentrating mainly on decays from even–even nuclides, For odd-A or odd–odd nuclides, the half-lives are considerably longer than those from neighbouring even–even nuclides.

The reason for this lies in the probability of creating an α-particle inside the nucleus. For an even–even nuclide a spin-zero pair of protons and a spin-zero pair of neutrons can be extracted from the highest shell and easily form an α-particle. In Quantum Physics language we say that the overlap integral between the wavefunction of the parent nuclide and the wavefunction of the daughter nuclide plus an α-particle is large. Strictly speaking the factor N_{coll}, which counts the rate of collisions with the potential barrier, includes a factor that accounts for the overlap integral, i.e. it includes a factor P_{form}, which gives the probability of creating an α-particle.

For nuclides with an odd number of protons and/or neutrons, the unpaired nucleon has a spin and usually an orbital angular momentum, which are not aligned with any of the other nucleons to give a state of zero total angular momentum. In such cases the overlap integral of the parent wavefunction with that of the daughter plus an α-particle will be much smaller than the even–even case. Sometimes the unpaired nucleon is not used at all to form the α-particle; rather it is formed from the zero angular momentum nucleon pairs in levels below that of the unpaired nucleon. In that case the daughter nuclide finds itself in an excited state since there are empty energy levels below the level of the unpaired nucleon.

For the Viola–Seaborg model this hindrance is encoded by adding a term h (known as the "*hindrance factor*") to the RHS of (6.23). It is zero for even–even nuclides and has different non-zero values for even–odd, odd–even or odd–odd nuclides. The values of h are $\mathcal{O}(1)$, implying an enhancement of the half-life by around a factor of ten. In the case of the Royer model, the high quality fit is achieved by allowing all 3 of the parameters to take different values for all of these different types of nuclides.

Summary

- The energy released in an α-decay process (the "Q-value") is the difference between the mass of the parent nucleus and the combined mass of the daughter nucleus and the α-particle.
- Owing to the recoil of the daughter nucleus, the kinetic energy of the emitted α-particle is smaller than the Q-value by a factor equal to the ratio of the reduced mass of the α-particle to its actual mass.
- The mechanism of α-decay is quantum tunnelling through the potential barrier formed by the combination of the strong attractive nuclear force and the Coulomb repulsion between the α-particle and the daughter nucleus. This can be estimated by using the WKB approximation to calculate the tunnelling probability and multiplying this by the number of collisions per second between the α-particle with the potential barrier. This product determines the decay rate and consequently the half-life.
- The estimated half-life has a dependence on the Q-value that agrees fairly well with the dependence that was observed by Geiger and Nuttall.
- Although the theoretical prediction reproduces approximately the dependence on the Q-value, there are usually very large discrepancies (up to three orders of magnitude) between theory and experiment. A model proposed by Royer maintains the form of the theoretical prediction but with coefficients that are taken to be free parameters. The Viola–Seaborg model is a four-parameter fit with a more generalized dependence on the atomic number. Both of these models have been tuned to reproduce the experimental data on the half-lives over a large range of α-decaying nuclides, with a high degree of accuracy.
- For even–even nuclides decaying to the ground state of the daughter nuclide, the α-particle has zero orbital angular momentum. However, for decays to excited states of the daughter nuclide, or from excited states of the parent nuclide, and for decays of nuclides with an odd number of protons or neutrons (or both), the α-particle can, in general, have orbital angular momentum. This adds a centrifugal barrier term to the nuclear potential barrier and substantially decreases the tunnelling probability. Consequently, the decay rates in which the α-particle has orbital angular momentum are much smaller.
- Nuclides with an odd number of protons and/or neutrons have a considerably longer half-life. This arises because for such nuclides the probability of formation of an α-particle inside the decaying nucleus is very much smaller.

Problems

Problem 6.1 Write down an expression for the Q-value in terms of the masses of the parent, daughter and α-particles. For α-decay of $P \rightarrow D + \alpha$, derive the relation between the Q-value of the energy released and the binding energy of the parent particle, B_P^E, the binding energy of the daughter particle, B_D^E, and the binding energy of the α-particle, B_α^E.

Problem 6.2 The binding energy of $^{240}_{94}\text{Pu}$ is 1813.45 MeV, and that of $^{236}_{92}\text{U}$ is 1790.40 MeV. The binding energy of ^4_2He is 28.30 MeV.
Calculate (to an accuracy of 10 keV) the kinetic energy of the α-particle emitted in the decay

$$^{240}_{94}\text{Pu} \rightarrow ^{236}_{92}\text{U} + \alpha.$$

For some of these decays the α-particle is accompanied by a 50 keV photon. Explain how this occurs and find the kinetic energy of the emitted α-particle for these events.

Problem 6.3 The nuclide $^{256}_{102}\text{No}$ decays to $^{252}_{100}\text{Fm}$ with a Q-value of 8.58 MeV. Calculate the half-life for this decay using Gamow's theoretical approximation as well as the Viola–Seaborg and Royer empirical expressions. Which gives a result closest to the experimental half-life of 2.9 s?

Problem 6.4 The isotope of polonium, $^{210}_{84}\text{Po}$, decays into $^{206}_{82}\text{Pb}$ with a half-life of 138 days. The Q-value for the decay is 5.4 MeV. The isotope $^{212}_{84}\text{Po}$ decays into $^{208}_{82}\text{Pb}$ with a half-life of 0.3 μs. From the ratio of these two half-lives estimate the Q-value for the decay of $^{212}_{84}\text{Po}$.

Appendix 6: Quantum Tunnelling

Consider a particle of mass m with (kinetic) energy E incident upon a spherically symmetric potential $V(r)$ as shown in Fig. 6.4. r_a and r_b are the radial distances at which the potential energy is equal to the kinetic energy. If the particle has angular momentum quantum number ℓ, then the radial part of the Schroedinger equation for the particle is

$$-\frac{\hbar^2}{2mr^2}\frac{\partial}{\partial r}\left(r^2\frac{\partial \Psi(r)}{\partial r}\right) + U(r)\Psi(r) = E\Psi(r), \tag{6.27}$$

where the effective potential, $U(r) \equiv V(r) + \frac{\ell(\ell+1)\hbar^2}{2mr^2}$, includes the centrifugal potential due to the orbital angular momentum.
    We seek a solution of the form

$$\Psi(r) = \frac{r_a}{r}\Psi(r_a)e^{-S(r_a,r)/\hbar}.$$

Fig. 6.4 A particle of energy
E enters a spherically
symmetric potential barrier
$U(r)$ at $r = r_a$ and leaves at
$r = r_b$. r_a and r_b are the
values of r for which the
kinetic energy of the particle
is equal to the potential
energy

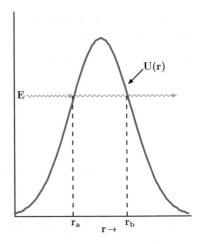

Inserting this into the Schroedinger equation (6.27), we get a differential equation
for $S(r_a, r)$:

$$\left(\frac{dS(r_a, r)}{dr}\right)^2 - \hbar \frac{d^2 S(r_a, r)}{dr^2} = 2m \left(U(r) - E\right), \qquad (6.28)$$

with boundary value $S(r_a, r_a) = 0$.

Using the leading order WKB approximation, we drop the term proportional to
$\hbar$ in (6.28) and obtain

$$S(r_a, r) = \int_{r_a}^{r} \sqrt{2m \left(U(r') - E\right)} \, dr'. \qquad (6.29)$$

For $r < r_a$ or $r > r_b$, $E > U(r)$ (see Fig. 6.4), $S(r_a, r)$ is imaginary so that $\Psi(r)$
is an oscillatory function with modulus that decreases as $1/r$. On the other hand, for
$r_a < r < r_b$, $S(r_a, r)$ is real so that in this region $\Psi(r)$ decreases as

$$\Psi(r) \sim \frac{1}{r} e^{-S(r_a, r)/\hbar}.$$

There is also an exponentially increasing solution, but careful matching of the
wavefunction to that of an outgoing plane wave as $r \to \infty$ shows that this term
may be neglected provided $S(r_a, r_b) \gg \hbar$. Comparing the wavefunctions at $r = r_a$
and $r = r_b$ gives us

$$r_b \Psi(r_b) = r_a \Psi(r_a) e^{-S(r_a, r_b)/\hbar}.$$

For $r \ll r_a$, the (effective) potential is negligible, so that the particle momentum is $\sqrt{2mE}$ and the incident particle flux, F_i, is given by

$$F_i(r) = \frac{2E}{m} \frac{r_a^2}{r^2} |\Psi(r_a)|^2 .$$

Similarly, the outgoing particle flux for $r \gg r_b$ is

$$F_o(r) = \frac{2E}{m} \frac{r_b^2}{r^2} |\Psi(r_b)|^2 = \frac{2E}{m} \frac{r_a^2}{r^2} |\Psi(r_a)|^2 e^{-2S(r_a,r_b)/\hbar} .$$

The incident particle flux at $r = r_a$ is the rate of arrival of particles at the barrier, *per unit area*. Likewise the outgoing flux at $r = r_b$ is the rate *per unit area* at which the particles leave the barrier, so that the tunnelling probability, T, defined by

$$T = \frac{\text{Probability for outgoing particle to cross a surface at } r > r_b}{\text{Probability for incoming particle to cross a surface at } r < r_a},$$

is given by

$$T = \frac{4\pi r^2 F_o(r)}{4\pi r^2 F_i(r)} = e^{-2S(r_a,r_b)/\hbar}, \tag{6.30}$$

where

$$S(r_a, r_b) = \int_{r_a}^{r_b} \sqrt{2m(U(r) - E)} dr. \tag{6.31}$$

Chapter 7
Beta-Decay

Beta (β)-decay is the radioactive decay of a nuclide in which an electron or a positron is emitted. These decays are described by the following processes:

$$^A_Z\{P\} \rightarrow {^A_{Z+1}}\{D\} + e^- + \bar{\nu}, \quad (\beta^--\text{decay}), \tag{7.1}$$

or

$$^A_Z\{P\} \rightarrow {^A_{Z-1}}\{D\} + e^+ + \nu, \quad (\beta^+-\text{decay}). \tag{7.2}$$

The atomic mass number is unchanged so that these reactions occur between isobars (nuclides with the same atomic mass number). The electron (or positron) does not exist inside the nucleus but is created either in the reaction

$$n \rightarrow p + e^- + \bar{\nu} \tag{7.3}$$

or

$$p \rightarrow n + e^+ + \nu. \tag{7.4}$$

In fact, the neutron has a mass that exceeds the sum of the masses of the proton plus the electron, so that a free neutron can undergo this decay with a half-life of 10.2 min. On the other hand, the proton is stable, so reaction (7.4) is not proton decay but proton conversion in the presence of other particles that make this process energetically possible. The system of forces inside the nucleus that effect β-decay are not the same forces that are responsible for the binding of nucleons within a nucleus. Rather, they are a set of very short-range forces (with a range of order 3×10^{-3} fm), known as *"weak interactions"*.

© The Author(s), under exclusive license to Springer Nature Switzerland AG 2021
A. Belyaev, D. Ross, *The Basics of Nuclear and Particle Physics*, Undergraduate Texts in Physics, https://doi.org/10.1007/978-3-030-80116-8_7

7.1 Kinematics

A β-decay event is energetically permitted provided that the mass of the parent exceeds the mass of the daughter plus the mass of an electron.

$$M(Z, A) > M((Z+1), A) + m_e, \tag{7.5}$$

for electron emission, and

$$M(Z, A) > M((Z-1), A) + m_e, \tag{7.6}$$

for positron emission. In the latter case a proton is converted into a neutron (with larger mass), but the binding energy of the daughter is sufficiently large that the total nuclear mass of the daughter is less than that of the parent by more than the electron mass, m_e. In these reactions the mass of the neutrino is neglected since it is very small (as we discuss later) in comparison even with the electron mass.

The mass of the electron can be included directly by comparing atomic masses, $\mathcal{M}(A, Z)$, since a neutral atom always has Z electrons. Thus we require

$$\mathcal{M}(Z, A) > \mathcal{M}((Z+1), A) \tag{7.7}$$

for electron emission. The atomic (as opposed to nuclear) mass includes the masses of the electrons.

One should note, however, that (7.7) will not work for positron emission, for which Z *decreases* by one unit. These decays are often not energetically allowed because of the difference in the binding energies of the parent and daughter nuclides. When a neutron is converted into a proton, the Coulomb repulsion between the nucleons increases – thereby decreasing the binding energy. Moreover, there is a pairing term in the semi-empirical mass formula that favours even numbers of protons and neutrons and a symmetry term that tells us that the number of protons and neutrons should be roughly equal.

For nuclides with even A it turns out that, because of the pairing term in the binding energy, nuclides with odd numbers of protons and neutrons (odd–odd nuclides) are nearly always unstable against β-decay. On the other hand, even–even nuclides can also sometimes be unstable against β-decay if the number of neutrons in a particular isobar is too large or too small for stability.

For example, consider the isobars for $A = 100$, whose atomic masses are shown in Fig. 7.1. We note that all the odd–odd nuclides (blue) have a larger atomic mass than one of the adjacent even–even nuclides (red) and that for the case of $Z = 43$, *both* electron and positron emissions are energetically allowed so that this nuclide, $^{100}_{43}$Tc, (technetium), can decay either by electron emission to $^{100}_{44}$Ru (ruthenium) or by positron emission to $^{100}_{42}$Mo. (molybdenum). Moreover, the even–even nuclide $^{100}_{40}$Zr (zirconium) has a binding energy that is less than its odd–odd neighbour $^{100}_{41}$Nb (niobium), so it can decay into it by electron emission.

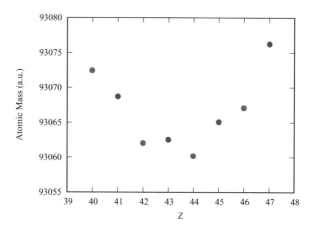

Fig. 7.1 Atomic masses (in a.u.) of isobars with atomic mass number $A = 100$

For nuclides with odd A there is either an even number of neutrons *or* an even number of protons. In this case the pairing term does not change from isobar to isobar and the question of stability relies on the balance between the symmetry term that prefers equal numbers of protons and neutrons and the Coulomb term that prefers fewer protons. For such nuclides, there is only one stable isobar, with a certain atomic number, Z_A. This means that the isobars with atomic number $Z > Z_A$ have too many protons to be stable and therefore can always β-decay emitting a positron, whereas isobars with $Z < Z_A$ have too many neutrons, and can undergo β-decay emitting an electron. The value of Z_A for a given A can be obtained by minimizing the *atomic mass* (including the masses of the electrons) from the semi-empirical mass formula. This gives

$$Z_A = A \frac{2a_A + (m_n - m_p - m_e)c^2/2}{4a_A + a_C A^{2/3}}, \tag{7.8}$$

where a_A and a_C are the coefficients of the asymmetry term and Coulomb term in the semi-empirical mass formula (3.8).

The difference between the mass of the parent nucleus, m_P, and the mass of the daughter, m_D, plus the electron gives the Q-value for the decay,

$$Q = (m_P - m_D - m_e)c^2, \tag{7.9}$$

and in this case the recoil of the daughter can be neglected because the electron is so much lighter than the nucleus.

Unlike the case of α-decay, the Q-value is not given only in terms of binding energies because a neutron is converted into a proton (or vice versa), which has a slightly different mass, and furthermore, energy equal to the rest energy of an electron is required to produce an electron or positron.[1] Thus for electron emission from a parent with atomic number Z and atomic mass number A, the Q-value is given by

$$Q_{\beta^-} = B(A, Z+1) - B(A, Z) + (m_n - m_p - m_e)c^2$$
$$= B(A, Z+1) - B(A, Z) + 0.782 \, \text{MeV} \qquad (7.10)$$

and for positron emission

$$Q_{\beta^+} = B(A, Z-1) - B(A, Z) + (m_p - m_n - m_e)c^2$$
$$= B(A, Z-1) - B(A, Z) - 1.804 \, \text{MeV}. \qquad (7.11)$$

Also, a nuclide with atomic number Z is unstable against β^- decay if its binding energy is less than the sum of the binding energy of its isobar with atomic number $(Z + 1)$ plus 0.782 MeV, and is unstable against β^+ decay if its binding energy is less than that of its isobar with atomic number $(Z - 1)$ by more than 1.804 MeV.

7.2   Neutrinos

For two-body decay (i.e. two particles in the final state) of the parent nucleus into the daughter plus the electron (or positron), we would expect the Q-value to be equal to the kinetic energy of the emitted electron, but what is observed is a spectrum of electron energies up to a maximum value that is equal to this Q-value. For example the intensity of electrons with different energies from the β-decay of $^{210}_{83}\text{Bi}$ [51] is shown in Fig. 7.2.

There is a further puzzle. Since the number of spin-$\frac{1}{2}$ nucleons is the same in the parent and daughter nuclei, the difference in the spins of the parent and daughter nuclei must be an integer. But the electron has spin-$\frac{1}{2}$, so there appears to be a violation of conservation of angular momentum here.

The solution to both of these puzzles was provided in 1930 by Wolfgang Pauli who postulated (in an unpublished open letter) the existence of a massless[2] neutral particle with spin-$\frac{1}{2}$ that always accompanies the electron in β-decay. This was called a *"neutrino"* (ν). Nowadays the particle that accompanies an emitted electron in β-decay is called an "antineutrino"($\bar{\nu}$)) – the antiparticle of the neutrino. Neutrinos and antineutrinos have almost all the same properties and so in what follows we will often use the term "neutrino" to refer to either.

---

[1] Particles always have exactly the same mass as their antiparticles, so electrons and positrons have the same mass.

[2] We now know that neutrinos have a tiny mass of order much less than one millionth of the mass of an electron.

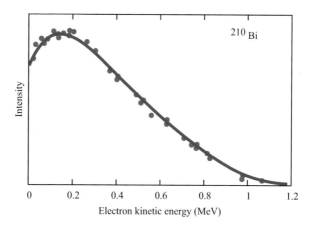

Fig. 7.2 The energy spectrum of electrons from $^{210}_{83}Bi$

Intensity

^{210}Bi

0   0.2   0.4   0.6   0.8   1   1.2

Electron kinetic energy (MeV)

Neutrinos interact very weakly with matter. Most cosmic neutrinos pass though the Earth without interacting. They are therefore very difficult to detect and their existence was not confirmed until 1956, by Clyde Cowan and Frederick Reines [52]. They used a very large flux of 5×10^{13} antineutrinos per square centimetre per second from a nearby nuclear reactor. The apparatus consisted of two tanks containing 200 litres of cadmium chloride solution. These tanks were sandwiched between three liquid scintillator detectors and buried deep underground in order to shield the apparatus from cosmic radiation. The same weak-interaction force that is responsible for β-decay allows antineutrinos to scatter off protons in the solution, converting them into neutrons and emitting a positron:

$$\bar{\nu} + p \rightarrow n + e^+. \tag{7.12}$$

The emitted positron annihilates with an electron in the water molecules, giving off two γ-rays,

$$e^+ + e^- \rightarrow \gamma + \gamma, \tag{7.13}$$

which are then detected by a liquid scintillator. These events were detected at a rate of three per hour, corresponding to a cross section, σ, for antineutrino scattering off a proton

$$\sigma_{\bar{\nu}p \rightarrow ne^+} = 6.3 \times 10^{-44} \, cm^2.$$

This tiny cross section is much smaller even than the lowest cross section of order 10^{-41} (0.01 fb), currently being measured at the Large Hadron Collider (LHC) at CERN. Nevertheless, this cross section was approximately equal to the expected cross section from theoretical calculations, thereby confirming Pauli's hypothesis.

The fact that the neutrino has spin-$\frac{1}{2}$ means that the total angular momentum can be conserved (provided the electron–neutrino system has the required orbital angular momentum) and the Q-value is the sum of the energies of the electron and antineutrino. The kinetic energy of the electron can vary from zero, where all the Q-values are taken by the antineutrino (the momentum being conserved by the small recoil of the daughter nucleus), up to the Q-value, in which case the energy carried off by the antineutrino is negligible.

Electrons and neutrinos are examples of *"leptons"*, which are particles that do not interact under the strong nuclear forces – they are not found inside nuclei. By convention, electrons and neutrinos are assigned a "lepton number" of $+1$, which means that positrons and antineutrinos have a lepton number of -1. Lepton number is conserved since it is actually an antineutrino that is emitted together with electron emission β-decay and a neutrino that is emitted together with a positron.

7.3 Fermi Theory

In 1934 Enrico Fermi developed his theory of β-decay [53], which provided an explanation of the decay spectrum. It was based on Fermi's golden rule (derived in Appendix 7), which tells us that the transition rate, λ, from an initial state $|i\rangle$ to a final state $|f\rangle$ from a perturbing potential H' is given by

$$\lambda = \frac{2\pi}{\hbar} \left| \langle f|H'|i\rangle \right|^2 \rho(E_f), \tag{7.14}$$

where $\rho(E_f)$ is the density of states with final-state energy E_f, and $\langle f|H'|i\rangle$ is the *"matrix element"* of H' between the initial and final states.

Applying this to β-decay, we seek the quantity, $\lambda(T_e)$, where $\lambda(T_e)dT_e$ is the rate of β-decay with an electron whose kinetic energy lies between T_e and $T_e + dT_e$. The density of states, $\rho(E_f)$, is replaced by $\rho(Q, T_e)$, where $\rho(Q, T_e)dQdT_e$ is the number of electron and antineutrino[3] states with the electron kinetic energy between T_e and $T_e + dT_e$ and the total energy between Q and $Q + dQ$.

7.3.1 Density of States

We have shown in Appendix 3 that the number of allowed momentum states for a particle confined to a box of volume V, with magnitude of momentum between p and $p + dp$, is given by

---

[3]In this section we will assume that we are considering β-decays with electron and antineutrino emission. The same considerations apply for positron and neutrino emission unless stated otherwise.

$$n(p) = V \frac{p^2 dp}{2\pi^2 \hbar^3},\tag{7.15}$$

so the number of allowed states with an electron momentum between p_e and $p_e + dp_e$ and an antineutrino with momentum between $q_{\bar{\nu}}$ and $q_{\bar{\nu}} + dq_{\bar{\nu}}$ is given by

$$n_{e\bar{\nu}}(p_e, q_{\bar{\nu}}) = n(p_e)n(q_{\bar{\nu}}) = \left(\frac{V}{2\pi^2 \hbar^3}\right)^2 p_e^2 dp_e q_{\bar{\nu}}^2 dq_{\bar{\nu}}.\tag{7.16}$$

Let us now convert this into the number of states, $n_{e\bar{\nu}}(T_e, E_{\bar{\nu}})$, with electron kinetic energy between T_e and $T_e + dT_e$ and antineutrino energy between $E_{\bar{\nu}}$ and $E_{\bar{\nu}} + dE_{\bar{\nu}}$. Neglecting the neutrino mass, the antineutrino energy and momentum are simply related by

$$q_{\bar{\nu}} = \frac{E_{\bar{\nu}}}{c}, \quad dq_{\bar{\nu}} = \frac{dE_{\bar{\nu}}}{c}.\tag{7.17}$$

For electrons and neutrinos we need to use the relativistic relation between momentum and kinetic energy, since very often the electron will have kinetic energy of the order of or greater than its rest energy, $m_e c^2$:

$$p_e = \frac{1}{c}\sqrt{T_e(T_e + 2m_e c^2)}, \quad dp_e = \frac{(T_e + m_e c^2)}{c^2 p_e} dT_e.\tag{7.18}$$

Inserting (7.17) and (7.18) into (7.16) we get the expression for $n_{e\bar{\nu}}(T_e, E_{\bar{\nu}})$,

$$n_{e\bar{\nu}}(T_e, E_{\bar{\nu}}) = \frac{V^2}{4\pi^4 \hbar^6 c^5} E_{\bar{\nu}}^2 \left(T_e + m_e c^2\right) p_e dT_e dE_{\bar{\nu}}.\tag{7.19}$$

Note, however, that the electron and antineutrino energies are *not* independent but are related by

$$Q = E_{\bar{\nu}} + T_e.\tag{7.20}$$

For fixed electron kinetic energy, T_e, we may replace $dE_{\bar{\nu}}$ by dQ and $E_{\bar{\nu}}$ by $(Q - T_e)$ and obtain an expression for the density of states, $\rho(Q, T_e)$ with electron kinetic energy between T_e and $T_e + dT_e$, and therefore we have

$$\rho(Q, T_e)dT_e dQ = \frac{V^2}{4\pi^4 \hbar^6 c^5} (Q - T_e)^2 \left(T_e + m_e c^2\right) p_e dT_e dQ.\tag{7.21}$$

7.3.2 Decay Matrix Element

Let us consider now the matrix element of the weak-interaction Hamiltonian

$$\langle f \,|H_{\mathrm{WK}}|\, i \rangle \;=\; \int \Psi_{\bar{\nu}}^{*}(r)\Psi_{e}^{*}(r)\Psi_{D}^{*}(r)H_{\mathrm{WK}}\Psi_{P}(r)d^{3}r, \qquad (7.22)$$

where we have used the fact that the interactions are very short range so that we can take the wavefunctions for all the particles at the same point, r.

The wavefunctions (normalized within a volume V) of the leptons are outgoing plane-wave functions

$$\Psi_{e}(r) \;=\; \frac{e^{i\,p_{e}\cdot r/\hbar}}{\sqrt{V}} \quad\text{and}\quad \Psi_{\bar{\nu}}(r) \;=\; \frac{e^{i\,q_{\bar{\nu}}\cdot r/\hbar}}{\sqrt{V}}.$$

Since the nuclear wavefunctions become negligible for $r \gg R$, where R is the nuclear radius and the momenta of the leptons are such that

$$\frac{p_{e}\cdot r}{\hbar}, \;\; \frac{q_{\bar{\nu}}\cdot r}{\hbar} \;\ll\; 1 \text{ for } r \,\leq\, R, \qquad (7.23)$$

we may approximate these wavefunctions by

$$\Psi_{e}(r), \;\Psi_{\bar{\nu}}(r) \;\approx\; \frac{1}{\sqrt{V}}. \qquad (7.24)$$

The weak Hamiltonian can be written as

$$H_{\mathrm{WK}} \;=\; G_{F}\hbar^{3}c^{3}\mathcal{O}_{\mathrm{WK}}(r), \qquad (7.25)$$

where the operator $\mathcal{O}_{\mathrm{WK}}$ acts only on the nuclear wavefunctions, Ψ_{P} of the parent nucleus and Ψ_{D} of the daughter nucleus. Its matrix element between these states,

$$\mathcal{M}_{DP} \;\equiv\; \int \Psi_{D}^{*}(r)\mathcal{O}_{\mathrm{WK}}(r)\Psi_{P}(r)d^{3}r, \qquad (7.26)$$

is a dimensionless quantity of order unity.

By dimensional counting we find that G_{F}, known as the "*Fermi coupling constant*", has a dimension of inverse energy squared. It takes the value

$$G_{F} \;=\; 1.166 \times 10^{-5}\,\mathrm{GeV}^{-2}.$$

7.3.3 Decay Rate

Finally, we must introduce a factor, called the *"Fermi factor"*, $F(Z_D, T_e)$, which accounts for the Coulomb attraction between the daughter nucleus and the emitted electron (or the Coulomb repulsion between the daughter nucleus and the emitted positron). For electrons this function is given by

$$F(Z_D, T_e) = \frac{2\pi\eta}{1 - \exp(-2\pi\eta)}, \tag{7.27}$$

where

$$\eta = Z_D \alpha \sqrt{\frac{m_e c^2}{T_e}}.$$

We note that for small electron energies this is a large factor, indicating that the electron intensity is larger for small kinetic energies, because the electrons are slowed down by the Coulomb attraction.

For the case of positron and neutrino emission, the Fermi function, which now accounts for the Coulomb repulsion between the daughter nucleus and the positron, is given by

$$F(Z_D, T_e) = \frac{2\pi\eta}{\exp(2\pi\eta) - 1}. \tag{7.28}$$

It is small for small positron energy as the Coulomb repulsion tends to speed up the positrons.

Piecing everything together by inserting (7.26), (7.25), and (7.21) into (7.14) and including the Fermi factor we end up with the final expression for $\lambda(T_e)$:

$$\lambda(T_e) = \frac{G_F^2 c}{2\pi^3 \hbar} |\mathcal{M}_{DP}|^2 (Q - T_e)^2 \left(T_e + m_e c^2\right) p_e F(Z_D, T_e). \tag{7.29}$$

7.3.4 Kurie Plots

A useful quantity to plot is the Kurie function

$$K(T_e) \equiv \sqrt{\frac{\lambda(T_e)}{\left(T_e + m_e c^2\right) p_e F(Z_D, T_e)}}, \tag{7.30}$$

against T_e. This is called a *"Kurie plot"* [54]. As can be seen from (7.29) this is expected to be a straight line with intercept at $T_e = Q$.

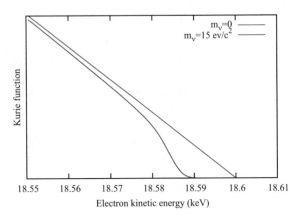

Fig. 7.3 The end of the Kurie plot for the β-decay of tritium ($Q = 18.6\,\text{keV}$) (red line) and the predicted plot assuming a neutrino mass of $15\,\text{eV}/\text{c}^2$ (blue line)

This holds provided that the assumption of zero neutrino mass is valid. Accounting for a non-zero neutrino mass, m_ν, introduces a correction factor to the Kurie function of

$$\left(1 - \frac{m_\nu^2 c^4}{(Q - T_e)^2}\right)^{1/4} \tag{7.31}$$

and this affects the end point of the Kurie plot. Figure 7.3 displays the end point of the Kurie plot for the decay of tritium, ^3_1H, which has a Q-value of $18.6\,\text{keV}$, and shows how this would deviate if the neutrino had a mass of $15\,\text{eV}/\text{c}^2$. If this were the case, the maximum value of electron kinetic energy would be $(Q - m_\nu c^2)$ $= 18.585\,\text{keV}$, whereas for a massless neutrino its total (relativistic) energy can be arbitrarily small and the electron can carry energy up to the Q-value. An ongoing experiment by the KATRIN collaboration [55] has reduced the upper limit on the neutrino mass to $1.1\,\text{eV}/\text{c}^2$ and expects to reduce this upper limit down to $0.2\,\text{eV}/\text{c}^2$.

It is now known that neutrinos *do* have a tiny mass. This was confirmed by experiments at the Kamiokande neutrino detector in Japan in 1999 [56]. The mass of the neutrino is almost certainly smaller than $0.1\,\text{eV}/\text{c}^2$ (much smaller than the electron mass of $0.511\,\text{MeV}/\text{c}^2$) and therefore highly unlikely to be detected in a β-decay experiment. For the remaining part of this chapter neutrinos are considered to be massless.

7.4   Selection Rules

Since the electron and antineutrino emitted in β-decay are both spin-$\frac{1}{2}$ particles, their total spin is either zero or one. Decays in which the total lepton spin is zero are called *"Fermi transitions"*, whereas decays in which the total lepton spin is one are called *"Gamow–Teller transitions"*.

The spins of the parent nuclide, I_P, and daughter nuclide, I_D, differ by S where

$$|I_P - I_D| \leq S \leq I_P + I_D.$$

If both I_P and I_D are zero, then only Fermi transitions can occur, since S then also has to be zero. If $I_P = 1$ and $I_D = 0$ or vice versa, then only Gamow–Teller transitions can occur since S then has to be one. If I_D and I_P are both non-zero and differ by one, then S can be either zero or one and both Fermi transitions and Gamow–Teller transitions are permitted. Such transitions are known as "mixed transitions".

If I_P and I_D differ by more than one, then the decay is called a *"forbidden transition"*. Such decays are not really forbidden, but only suppressed. Angular momentum can still be conserved provided that the leptons have one or more units of orbital angular momentum. A β-transition that requires orbital angular momentum ℓ is called an "ℓth forbidden transition".

Transitions with non-zero orbital angular momentum are only possible if we go beyond the approximation (7.24) for the wavefunctions of the leptons and expand them as

$$\Psi_e(r) = \frac{1}{\sqrt{V}} \left[1 + i \frac{p_e \cdot r}{\hbar} + \cdots \right] \tag{7.32}$$

and similarly for the wavefunction of the (anti)neutrino. The second term in the expansion displayed explicitly in (7.32) has an angular dependence proportional to the wavefunction for an $\ell = 1$ state, so that if we include this term then the weak Hamiltonian has a non-zero matrix element with the leptons in an $\ell = 1$ orbital angular momentum state. For larger values of orbital angular momentum, we need to include further terms in the expansion (7.32). As in the case of α-decay, the orbital angular momentum number, ℓ, of the decay particles must be odd if the parities, π_P and π_D, of the parent and daughter nuclides, respectively, are different and must be even if they are the same. All transitions including "forbidden" transitions must obey this selection rule.

As explained in (7.23), higher order terms in (7.32) become increasingly smaller, since $p_e \cdot r \ll \hbar$. This leads to increasingly diminished decay rates and consequently significantly increased half-lives, as the orbital angular momentum, ℓ, increases. For example, the allowed (Gamow–Teller) transition

$$^{22}_{11}\text{Na} \left(3^+ \right) \rightarrow {}^{22}_{10}\text{Ne} \left(2^+ \right) + e^+ + \nu$$

has a half-life of 2.6 years, whereas the second forbidden decay ($\ell = 2$)

$$^{10}_{4}\text{Be} \left(0^+ \right) \rightarrow {}^{10}_{5}\text{B} \left(3^+ \right) + e^- + \bar{\nu}$$

has a half-life of 1.5 million years. Each degree of forbiddenness increases the half-life by between three and five orders of magnitude.

Fig. 7.4 Kurie plot for a
decay with orbital angular
momentum $\ell = 1$ (red line)
and the straight line for an
allowed decay (blue line).
The functions are normalized
to be equal at $T_e = 0$

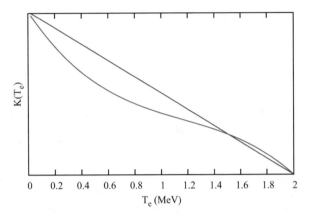

Moreover, the momentum dependence of the lepton wavefunction introduces a
shape factor to the Fermi estimate of the transition rate for a given electron kinetic
energy. These shape functions, $S^\ell(T_e)$, for an ℓth forbidden transition are of the
form

$$S^\ell(T_e) \propto a_0 (q_{\bar{\nu}})^{2\ell} + a_2 (q_{\bar{\nu}})^{2\ell-2} p_e^2 + \cdots + a_{2\ell} p_e^{2\ell}. \tag{7.33}$$

This means that for forbidden transitions the Kurie plot is no longer linear. Figure
7.4 shows a comparison between the Kurie plot for an allowed transition and the
Kurie plot with a shape factor for a first forbidden transition ($\ell = 1$).

As in the case of α-decay, it is often favourable (i.e. less forbidden) for the
parent nuclide to decay into a daughter nuclide in an excited state. Such decays
are followed by a γ-ray emission as the daughter nuclide returns to its ground state.

7.5 Electron Capture

Nuclei that undergo β^+-decay emitting a positron and a neutrino can also decay by
another mechanism.

$$_Z^A\{P\} + e^- \;\; \rightarrow \;\; _{Z-1}^A\{D\} + \nu. \tag{7.34}$$

What happens here is that an atom can absorb an electron from one of the inner
shells (usually the innermost shell, which is called the *"K-shell"*). The captured
electron interacts with a proton, transforming it into a neutron, so that the atom is
converted into an atom with one lower atomic number. The energy is entirely carried
away by the neutrino and is nearly always undetected because neutrinos interact so
weakly with matter.

This mechanism was proposed in 1934 by Gian Carlo Wick [57] and first detected by Luis Alvarez in 1937 [58] in the decay of $^{49}_{23}V$ (vanadium) to $^{49}_{22}Ti$

$$^{49}_{23}V + e^- \rightarrow \, ^{49}_{22}Ti + \nu, \tag{7.35}$$

with a half-life of 330 days. In a normal β-decay one would expect an outgoing positron, but none was observed.

7.6 Parity Violation

β-decay exhibits a further remarkable property. It violates conservation of parity.

Parity violation in the weak interactions was first proposed by Chen-Ning Yan and Tsung-Dao Lee in 1956 [59] and verified in 1957 by Chien-Shiung Wu [60]. She performed a seminal experiment on the decay of radioactive $^{60}_{27}Co$ (cobalt), whose ground state has spin-parity 5^+ into an excited state of $^{60}_{27}Ni$, with spin-parity 4^+

$$^{60}_{27}Co(5^+) \rightarrow \, ^{60}_{28}Ni^*(4^+) + e^- + \bar{\nu}. \tag{7.36}$$

This is an allowed Gamow–Teller transition. The nickel nucleus returns to its ground state, emitting two γ-rays. The cobalt sample was placed in a strong magnetic field in order to align the direction of the nuclear spin and cooled to almost absolute zero so that thermal fluctuations did not destroy the spin polarization. From the properties of the electromagnetic interactions responsible for the emission of the γ-rays, the correlation between the angular distribution of the γ-rays and the direction of the spin of the $^{60}_{27}Co$ nucleus was known. The measurement of this angular distribution was used to determine the degree of polarization of the nuclear spin, which was found to be around 60%.

The rate of emission of electrons in the direction of the magnetic field was measured and the direction of the magnetic field was reversed in order to measure the rate of emission in the direction opposite to that of the magnetic field. What was observed was an excess of electrons (around 20%) emitted in the opposite direction to the magnetic field (i.e. opposite to the direction of polarization of the nuclear spin), as shown schematically in diagram (a) of Fig. 7.5. After a few minutes this excess disappeared, owing to the warming of the sample so that thermal fluctuations erased the spin polarization.

This anisotropic distribution is a signal of parity violation. Under a parity transformation the position vector transforms as $r \rightarrow -r$ and likewise momentum $p \rightarrow -p$. All three components of r change sign under a parity transformation. So if we look at diagram (a) of Fig. 7.5 and turn it upside down and back to front (observe it in a mirror), we will obtain diagram (b) of Fig. 7.5. Note that in both diagrams the direction of spin of the nucleus is the same. This is because angular momentum is an axial vector, which does *not* change sign under parity. This can be seen very easily from the definition of the orbital angular momentum vector, L:

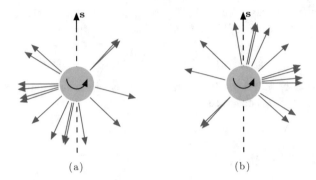

(a)   (b)

Fig. 7.5 (a) The electron distribution from $^{60}_{27}$Co with spin polarized in the upward vertical direction (indicated by the anticlockwise spin as seen from the above). (b) The parity image of (a). In both diagrams the spin, s, of the electron points upwards, meaning that the average value of $\langle s \cdot p_e \rangle$ is negative in diagram (a) and positive in diagram (b)

$$L = r \times p. \tag{7.37}$$

Since r and p both change sign under parity it follows that L remains unchanged. From the fact that the electron distribution shown in diagram (a) is the one that is observed, whereas its parity image, namely the distribution shown in diagram (b), is *not* observed, taken together with the fact that the electron spin must be in the direction of the nuclear spin (by conservation of angular momentum), we conclude that this observed anisotropic distribution implies a violation of conservation of parity in transitions governed by the weak-interaction force.

A full understanding of the action of the weak interactions on electrons and antineutrinos requires a study of the *"Dirac equation"*, which describes the quantum behaviour of relativistic spin-$\frac{1}{2}$ particles. This is beyond the scope of this book, but the upshot is that the action of the weak-interaction Hamiltonian on the wavefunction of an electron with spin s and momentum p_e is proportional to a parity-violating term.

$$\sqrt{T_e + 2m_e c^2} - \frac{2s \cdot p_e / \hbar}{\sqrt{T_e + 2m_e c^2}}. \tag{7.38}$$

Under a parity transformation the first term of (7.38) is unchanged but the second term changes sign (because the momentum, p_e, changes sign, but the spin, s, does not). In Wu's experiment the component of spin of the parent cobalt nucleus is $\hbar$ greater than that of the daughter nickel nucleus in the direction of the spin polarization. By conservation of angular momentum this means that both the electron and the antineutrino must have a component $+\frac{1}{2}\hbar$ in the direction of nuclear spin polarization. The presence of the factor (7.38) in the matrix element of the weak-interaction Hamiltonian then means that there is a larger amplitude (and hence a larger probability) for an event in which $s \cdot p_e$ is negative. This means that for the majority of events the electron is moving in a direction that makes an angle of more than 90° to the direction of polarization of the spin of the cobalt nucleus.

7.6.1   Helicity

The helicity, λ, of a particle of spin-s with momentum $\boldsymbol{p}$ is defined as

$$\lambda \equiv \frac{1}{s\hbar} \boldsymbol{s} \cdot \hat{\boldsymbol{p}}, \tag{7.39}$$

where $\hat{\boldsymbol{p}}$ is a unit vector in the direction of $\boldsymbol{p}$ and $\boldsymbol{s}$ is the spin operator (whose z-component has eigenvalues $\pm\frac{1}{2}\hbar$). Helicity is a measure of the component of its spin in the direction of its motion. For spin-$\frac{1}{2}$ particles it can take the value $\lambda = +1$ (called "right-handed helicity") or $\lambda = -1$ (called "left-handed helicity"). Using this definition together with the relativistic relation (7.18) between the momentum and kinetic energy of an electron, we may rewrite (7.38) as

$$\sqrt{T_e + 2m_e c^2} - \lambda\sqrt{T_e} \,,$$

which tells us that the weak-interaction Hamiltonian acts more strongly on left-handed electrons than on right-handed electrons. More precisely,

$$\frac{\text{weak coupling to left-helicity electrons}}{\text{weak coupling to right-helicity electrons}} = \frac{\sqrt{T_e + 2m_e c^2} + \sqrt{T_e}}{\sqrt{T_e + 2m_e c^2} - \sqrt{T_e}}. \tag{7.40}$$

Actually, the helicity of a massive particle is not a *"relativistically invariant"* quantity since it can always be observed by a boosted observer moving faster than the particle (in the same direction). In the boosted observer's frame, the direction of momentum is reversed, but the spin is not, so the helicity is flipped. A relativistically invariant quantity is the chirality, χ, which is the helicity of a particle observed from a frame in which the particle is moving with almost the speed of light, for which $T_e \gg m_e c^2$. A particle with mass m and kinetic energy T in a state with chirality χ is a superposition of states with helicity, $\lambda = 1$ and helicity, $\lambda = -1$, namely

$$|\chi\rangle = \frac{\sqrt{T + 2mc^2} + \sqrt{T}}{2\sqrt{T + mc^2}} |\lambda = \chi\rangle - \chi \frac{\sqrt{T + 2mc^2} - \sqrt{T}}{2\sqrt{T + mc^2}} |\lambda = -\chi\rangle . \tag{7.41}$$

In the extreme relativistic limit, $T \gg mc^2$ chirality states and helicity states coincide (the coefficient of the state $|\lambda = -\chi\rangle$ in (7.41) vanishes).

We see from (7.41) that the ratio of the coefficients of the helicity states $|\lambda = 1\rangle$ and $|\lambda = -1\rangle$ of the chirality state $|\chi = -1\rangle$ is precisely the same as the ratio of the couplings of the weak interactions to the two electron helicities (7.40). From this we conclude that the weak interactions couple purely negative chirality ($|\chi = -1\rangle$) particle states (also called "left-handed"). For a massive particle this means that the weak interactions couple to a superposition of left-helicity and right-helicity states with coefficients given in (7.41).

The antiparticle of a particle with chirality χ has chirality $-\chi$. Thus for the antineutrino (taken to be massless here) the matrix element of the weak Hamiltonian has a parity violating factor

$$1 + \frac{2c}{\hbar} \frac{\boldsymbol{s} \cdot \boldsymbol{q}_{\bar{\nu}}}{E_{\bar{\nu}}}, \tag{7.42}$$

since for the (massless) antineutrino $E_{\bar{\nu}} = cq_{\bar{\nu}}$. For a massless particle there is no distinction between helicity and chirality. The weak-interaction Hamiltonian acts only on the wavefunction of antineutrinos with positive (i.e. right-handed) chirality. More generally, the weak interactions occur between left-handed leptons or between right-handed antileptons.

Any difference in the action of a Hamiltonian on the wavefunctions of particles with different chiralities is a manifestation of parity violation. The fact that the weak-interaction Hamiltonian *only* couples to leptons with left-handed chirality or antileptons with right-handed chirality means that this parity violation is maximal.

Before the discovery of a small but non-zero mass for neutrinos, it was assumed that only left-handed neutrinos and their right-handed antiparticles existed in nature as these were the only neutrino chiralities that interacted with other matter. The existence of a small mass, however, means that right-handed neutrinos must exist. This will be discussed in more detail in Chap. 22.

7.7 Double Beta-Decay

It can sometimes happen that a particular nuclide, with atomic number Z, is stable against either β^-- or β^+-decay, but its binding energy is less than the binding energy of the isobar with atomic number $(Z+2)$ or exceeds it by less than 1.56 MeV (twice the mass difference between a neutron and a proton plus an electron). In that case it is energetically possible for the nuclide to undergo double β-decay:

$$^A_Z\{P\} \rightarrow {}^A_{Z+2}\{D\} + 2e^- + 2\bar{\nu}, \tag{7.43}$$

emitting two electrons and two antineutrinos.

An example of this is $^{130}_{52}$Te, which has a binding energy of 1095.94 MeV, whereas the binding energies of its neighbouring isobars are smaller than this and by an amount exceeding the rest energy difference $(m_n - m_p - m_e)c^2 = 0.782$ MeV. This nuclide is therefore stable against single β-decay. On the other hand the binding energy of $^{130}_{54}$Xe is 1096.91 MeV so that the double β-decay

$$^{130}_{52}\text{Te} \rightarrow \, ^{130}_{54}\text{Xe} + 2e^- + 2\bar{\nu} \qquad (7.44)$$

is energetically possible. However, the decay rate for such processes is very much suppressed so the half-life is extremely long, making these events extremely difficult (but not impossible) to observe. The half-life of $^{130}_{52}\text{Te}$ is 7×10^{20} years. The relative abundance of $^{130}_{52}\text{Te}$ is 33.8%, so that in a sample of 1 gram of naturally occurring tellurium there will be on average about 2 double β-decay events per year.

Another scenario that can give rise to double β-decay occurs when single β-decay is energetically possible but highly suppressed because the transition is forbidden. An example of this is $^{48}_{20}\text{Ca}$, whose ground state has spin-parity 0^+, and whose binding energy is 415.99 MeV. It is just energetically possible for this nuclide to decay to $^{48}_{21}\text{Sc}$ (scandium) with binding energy 415.48 MeV (lower than that of $^{48}_{20}\text{Ca}$, but by an amount less than 0.782 MeV), but the ground state of $^{48}_{21}\text{Sc}$ has spin-parity 6^+, so such a decay is 5th forbidden. There are two excited states of $^{48}_{21}\text{Sc}$ to which $^{48}_{20}\text{Ca}$ can also decay and these have spin-parities 5^+ and 4^+. This means that such single β-decays are at least 3rd forbidden. On the other hand, the ground state of $^{48}_{22}\text{Ti}$ has spin-parity 0^+, and a binding energy 418.69 MeV. Double β-decay of $^{48}_{20}\text{Ca}$ is an allowed Fermi transition and is energetically permitted. This double β-decay has been observed by the NEMO experiment [61] and the measured half-life is 6×10^{19} years.

7.7.1 Neutrinoless Double Beta-Decay

By assigning lepton number $+1$ to electrons and neutrinos and lepton number -1 to their antiparticles – positrons and antineutrinos – we see that in a β-decay reaction lepton number is always conserved.

However, for double β-decay we no longer need the emission of antineutrinos in order to conserve energy or angular momentum. This presents the possibility of neutrinoless double beta-decay

$$^{A}_{Z}\{P\} \rightarrow \, ^{A}_{(Z+2)}\{D\} + 2e^- . \qquad (7.45)$$

The occurrence of such events would mean a violation of conservation of lepton number. This would have a significant effect on the properties of neutrinos, implying that neutrinos could annihilate each other. This, in turn, would mean that neutrinos were their own antiparticles. Searches for neutrinoless double β-decay are currently being conducted at several laboratories around the world, but so far no such events have been observed.

Summary

- β-decay is the process in which a neutron in a nucleus is converted into a proton, emitting an electron and an antineutrino, or a proton is converted into a neutron, emitting a positron and a neutrino.
- The kinetic energy of the emitted electron or positron can take any value up to the Q-value of the decay. The remaining energy is taken up by an almost massless antineutrino or neutrino, which also has spin-$\frac{1}{2}$, so that angular momentum is conserved.
- The electron spectrum can be derived from Fermi's theory of β-decay, which is based on Fermi's golden rule but has an additional factor (the Fermi factor) that accounts for the Coulomb interaction between the daughter nucleus and the emitted electron or positron.
- The Kurie plot is the plot of the Kurie function – the square root of the transition rate divided by the electron momentum, electron energy, and the Fermi function – against the kinetic energy of the electron. For allowed decays this plot is linear, but for forbidden decays it acquires a shape function, which depends on the degree of forbiddenness. A deviation from linearity near the end point of a Kurie plot for an allowed transition is a signal for a non-zero neutrino mass.
- Allowed β-transitions are those in which the spins of the parent and daughter nuclides differ by at most one unit. Fermi decays are transitions in which the total spin of the electron and antineutrino is zero and Gamow–Teller decays are transitions in which the total spin of the electron and antineutrino is one. Forbidden decays can nevertheless occur if the lepton pair carries off orbital angular momentum, but such decays are strongly suppressed.
- β^{+}-transitions can occur without the emission of a positron in the case of electron capture, in which an inner atomic electron is captured by the nucleus and it interacts with the proton, transforming it into a neutron and emitting a neutrino.
- The interactions responsible for β-decay or β-transitions are the weak interactions. Parity violation was observed in the β-decay of $^{60}_{27}$Co, with polarized spin. The electron distribution was anisotropic, with an excess of electrons in the opposite direction from the spin polarization. This excess of left-handed electrons indicates that weak interactions violate parity and that they couple purely negative chirality states.
- In double β-decay the daughter nuclide has an atomic number that differs from that of the parent nuclide by two units, emitting two electrons or two positrons. This can occur when such a decay is energetically possible but single β-decay is not, or if single β-decay is heavily forbidden.

Problems

Problem 7.1 Indicate whether the following decays are Fermi, Gamow–Teller, mixed, or forbidden transitions. In the case of forbidden transitions write down the minimum orbital angular momentum of the electron or antineutrino:

(a) $^{10}_{4}\text{Be}\left(0^+\right) \rightarrow ^{10}_{5}\text{B}\left(3^+\right) + e^- + \bar{\nu}$

(b) $^{6}_{2}\text{He}\left(0^+\right) \rightarrow ^{6}_{3}\text{Li}\left(1^+\right) + e^- + \bar{\nu}$

(c) $^{37}_{16}\text{S}\left(\frac{7}{2}^-\right) \rightarrow ^{37}_{17}\text{Cl}\left(\frac{3}{2}^+\right) + e^- + \bar{\nu}$

(d) $^{14}_{8}\text{O}\left(0^+\right) \rightarrow ^{14}_{9}\text{N}^*\left(0^+\right) + e^- + \bar{\nu}$

Problem 7.2 The Kurie function, $K(T_e)$, for the β-decay

$$^{114}_{49}\text{In} \rightarrow ^{114}_{50}\text{Sn} + e^- + \bar{\nu}$$

takes the value 5.36 (in arbitrary units) for electron kinetic energy, $T_e = 0.4\,\text{MeV}$ and 4.00 for electron kinetic energy, $T_e = 0.8\,\text{MeV}$. The binding energy of $^{114}_{49}\text{In}$ is 970.365 MeV. What is the binding energy of $^{114}_{50}\text{Sn}$?

Problem 7.3 Take the weak-interaction Hamiltonian to be

$$H_{WK}(r) = G_F \frac{M_W^2 \hbar c^5}{r} e^{-(M_W cr/\hbar)},$$

where M_W is the mass of the W-boson, $M_w = 80.4\,\text{GeV}/c^2$.

(a) Estimate the range of the interaction described by this weak Hamiltonian.

(b) Calculate the magnitude of this Hamiltonian (in GeV) at distances $r = 2 \times 10^{-3}$ fm, $r = 4 \times 10^{-3}$ fm, and $r = 6 \times 10^{-3}$ fm.

(c) Assuming that the wavefunctions $\Psi_i(r)$ and $\Psi_f(r)$ for the initial and final states, respectively, vary very little over the range

$$0 < r < \frac{\hbar}{M_W c},$$

show that the matrix element of the weak-interaction Hamiltonian between an initial state $|i\rangle$ and a final state, $|f\rangle$ is

$$\mathcal{M}_{fi} = 4\pi G_F \hbar^3 c^3 \Psi_f^*(0)\Psi_i(0).$$

Problem 7.4 The shape factor for a first forbidden transition is

$$S^1 = \frac{1}{m_e^2 c^2}\left(p_e^2 + q_{\bar{\nu}}^2\right).$$

Express this in terms of the electron kinetic energy, T_e, and the Q-value, Q.

Appendix 7: Fermi's Golden Rule

The approximate expression for the transition rate for a system due to a perturbing potential is known as Fermi's golden rule, although it was actually first derived by Paul Dirac [62].

If a time-independent perturbing potential, H', is applied to a quantum system in a state $|i\rangle$, energy E_i, at time, $t = 0$, then the amplitude $a_{fi}(t)$ for the system to have made a transition to the state $|f\rangle$, with energy E_f, at time t is given by first order time-dependent perturbation theory to be

$$a_{fi}(t) = 2e^{i\eta} \langle f|H'|i\rangle \frac{\sin\left(\frac{1}{2}(E_i - E_f)t/\hbar\right)}{\left(E_i - E_f\right)}, \tag{7.46}$$

where $\langle f|H'|i\rangle$ is the matrix element of the perturbing Hamiltonian between the initial state $|i\rangle$ and final state $|f\rangle$, and η is a phase.

The probability, $T_{fi}(t)$, for such a transition to have occurred by time t, is then

$$T_{fi}(t) = \left|a_{fi}(t)\right|^2 = 4\left|\langle f|H'|i\rangle\right|^2 \frac{\sin^2\left(\frac{1}{2}(E_i - E_f)t/\hbar\right)}{\left(E_i - E_f\right)^2}. \tag{7.47}$$

The transition rate, λ_{fi}, is given by the derivative of T_{fi} with respect to time

$$\lambda_{fi} = \frac{2}{\hbar}\left|\langle f|H'|i\rangle\right|^2 \frac{\sin\left((E_i - E_f)t/\hbar\right)}{\left(E_i - E_f\right)}. \tag{7.48}$$

To determine the total transition rate, λ, to *any* final state, we sum over all final states $|f\rangle$. However, if these final states are in a continuum, this discrete sum is replaced by an integral over final-state energy, E_f, with a Jacobian factor equal to the density of states, $\rho\left(E_f\right)$ – the number of quantum states per unit energy interval. We then obtain

$$\lambda = \frac{2}{\hbar}\int dE_f \left|\langle f|H'|i\rangle\right|^2 \rho\left(E_f\right) \frac{\sin\left((E_i - E_f)t/\hbar\right)}{\left(E_i - E_f\right)}. \tag{7.49}$$

The function

$$\frac{\sin\left((E_i - E_f)t/\hbar\right)}{\left(E_i - E_f\right)}$$

is very sharply peaked at $E_f = E_i$. This enables us to make the approximation of taking all the other factors outside the integral, under the assumption that the variation of these factors with E_f is much slower. We can then approximate (7.49) by

$$\lambda \approx \frac{2}{\hbar} \left|\langle f|H'|i\rangle\right|^2 \rho\left(E_f\right) \int_{E_f \ll E_i}^{E_f \gg E_i} dE_f \frac{\sin\left((E_i - E_f)t/\hbar\right)}{(E_i - E_f)}. \tag{7.50}$$

The limits on the integral are from far below the peak to far above the peak and so may be replaced by $-\infty$ and ∞. Finally, using the standard integral

$$\int_{-\infty}^{\infty} \frac{\sin(x)}{x} dx = \pi,$$

we end up with Fermi's golden rule:

$$\lambda \approx \frac{2\pi}{\hbar} \left|\langle f|H'|i\rangle\right|^2 \rho\left(E_f\right). \tag{7.51}$$

Chapter 8
Gamma Decay

The emission of γ-rays from nuclei is the nuclear analogue of the atomic emission of photons, which occur when an electron makes a transition from an excited state either to a lower excited state or to the atomic ground state. Similarly, γ-rays are emitted when a nucleus in an excited state makes a transition to a lower state. Atomic excitation energies are typically of the order of a few electron volts (eV), leading to the emission of photons with wavelengths of hundreds of nanometres encompassing the visible spectrum, whereas nuclear excitations are of the order of hundreds of KeV, emitting γ-rays with wavelengths of the order of a picometre (1000 fm), although some nuclear excitation energies are less than 100 keV, so that the emitted photons are strictly classified as X-rays. In contrast to atomic radiation, γ-rays are usually described in terms of their energies, E_γ, rather than their wavelengths.

Most excited states have a very short lifetime – of order $10^{-13} - 10^{-10}$ s. However, there are some excited states which are metastable and therefore have a much longer lifetime. An example of this is the nuclide $^{58}_{27}\text{Co}$, which has a metastable excited state with energy 24.9 keV and half-life of about 9 h. Such excited states are called *"nuclear isomers"* and their decays are called *"isomer transitions"* – often abbreviated to IT.

An excited state with decay rate λ has a mean lifetime $\tau = 1/\lambda$ (see (5.3)). By Heisenberg's uncertainty principle, this implies that the energy of the excited state has an uncertainty $\frac{1}{2}\hbar/\tau$, so that the spectral line of an emitted γ-ray has a half-width, $\frac{1}{2}\Gamma_\gamma$, which is equal to that uncertainty. The line-width is therefore given by

$$\Gamma_\gamma = \frac{\hbar}{\tau} = \hbar\lambda. \qquad (8.1)$$

© The Author(s), under exclusive license to Springer Nature Switzerland AG 2021
A. Belyaev, D. Ross, *The Basics of Nuclear and Particle Physics*, Undergraduate
Texts in Physics, https://doi.org/10.1007/978-3-030-80116-8_8

8.1   Radiation Modes and Selection Rules

As in the case of β-decay, the emitted photon in γ-decay can carry off angular momentum ℓ, which permits a transition between an initial state with spin I_i and a final state with spin I_f, provided that angular momentum is conserved, i.e. that the *vector* sum of $\boldsymbol{I}_f$ and the photon angular momentum, $\boldsymbol{\ell}$, must be equal to $\boldsymbol{I}_i$. The allowed values of ℓ are then given by

$$\left| I_i - I_f \right| \le \ell \le I_i + I_f . \tag{8.2}$$

The interactions responsible for γ-decay are the electromagnetic interactions (as is the case for atomic transitions). There are two types of electromagnetic transitions – electric transitions and magnetic transitions. Electric transitions with angular momentum ℓ=1,2,... are denoted by the symbols E1, E2,.... They are called *"electric 2^l-pole transitions"* – "electric dipole", "electric quadrupole" *etc.* Magnetic transitions with angular momentum $\ell = 1, 2, ...$ are denoted by the symbols M1, M2,.... They are called *"magnetic 2^l-pole transitions"* – "magnetic dipole", "magnetic quadrupole" *etc.* The emitted radiation from such transitions is known as "radiation modes".

Unlike the weak interactions, which mediate β-decay, the electromagnetic interactions are parity conserving. An electric dipole, $\boldsymbol{d}_E = e\boldsymbol{r}$, is odd under parity transformation so that electric dipole transitions are only permitted between initial and final states of opposite parity. On the other hand, a magnetic dipole is proportional to the spin, s, of the nucleon that makes the transition. This is an axial vector and therefore *even* under parity transformations, implying that magnetic dipole transitions are only permitted between initial and final states of the same parity.

More generally, for an electric transition Eℓ, the parities, π, of the initial and final states are related by

$$\pi_i = (-1)^\ell \pi_f, \tag{8.3}$$

whereas for a magnetic transition Mℓ, the parities of the initial and final states are related by

$$\pi_i = (-1)^{(\ell+1)} \pi_f. \tag{8.4}$$

Very often, there are several possible radiation modes which can take place during a transition between two states. For example, let us consider the transition between the excited state of $^{24}_{11}$Na with energy 4.72 MeV and spin–parity 4^+ and its ground state with spin–parity 1^+. The allowed γ-ray modes for this transition, which satisfy (8.2)–(8.4), are M3, E4 or M5. However, the largest decay rate is always the one with the lowest angular momentum, so it is that mode which dominates.

There is a further selection rule which is *always* obeyed. Since the photon has spin one, transitions between an initial state with $I_i = 0$ and a final state with $I_f = 0$ are strictly forbidden. This means that if the ground state has spin zero and the lowest excited state also has spin zero, the excited state cannot decay to the ground state by the emission of a γ-ray. An example of this is the nuclide $^{40}_{20}\text{Ca}$, whose ground state has spin–parity 0^+ and whose first excited state also has spin–parity 0^+ and an excitation energy of 5.35 MeV. It cannot emit a photon, but since the excitation energy is greater than twice the rest energy of an electron, this state can decay to the ground state by creating an electron–positron pair.

8.2 Decay Rates

The decay rates for different radiation modes were estimated by Victor Weisskopf [63] in 1951. A rigorous calculation of transition rates effected by electromagnetic interactions requires *"Quantum Electrodynamics"* (QED), but we can obtain the Weisskopf estimate for electric multipole transitions using Fermi's golden rule (7.50), with the electric interaction Hamiltonian for the emission of a photon with energy E_γ obtained from QED

$$H_{E_\gamma}(r) = \sqrt{\frac{2\pi\alpha\hbar^3 c^3}{E_\gamma}} \Psi^*_{k_\gamma}(r), \tag{8.5}$$

where $\Psi_{k_\gamma}(r)$ is the plane-wave wavefunction for the outgoing photon (in a volume V) with wave number k_γ $(= E_\gamma/\hbar c)$. The decay rate for an electric multipole transition Eℓ of a nuclide with atomic mass number A is then given approximately by [1]

$$\lambda_{E\ell}(A, E_\gamma) \approx \frac{2\alpha c}{r_0} \frac{(\ell+1)}{\ell\,((2\ell+1)!!)^2} \left(\frac{3}{\ell+3}\right)^2 \left(\frac{r_0 E_\gamma}{\hbar c}\right)^{(2\ell+1)} A^{2\ell/3}, \tag{8.6}$$

where the nuclear radius, R, is given by $R = r_0 A^{1/3}$.

The estimate of the decay rates for magnetic transitions involves the nuclear spin. We would expect the magnetic interaction Hamiltonian, H_M, to be proportional to the magnetic moment of the nucleon which makes the transition.[2] Weisskopf estimated that the magnetic interaction Hamiltonian for a nucleus of radius R can be approximated by

---

[1] An outline derivation of this result is given in Appendix 8.

[2] Electric multipole radiation is only produced by transitions of protons which carry electric charge, whereas magnetic multipole radiation can also be effected by neutrons, which although electrically neutral, possess a non-zero magnetic moment.

Table 8.1 Weisskopf estimates for electric and magnetic multipole transitions, in terms of photon energy in MeV and atomic mass number A

ℓ	Multipole	Electric transitions		Magnetic transitions	
		Symbol	Decay rate (s^{-1})	Symbol	Decay Rate (s^{-1})
1	Dipole	E1	$10^{14} E_\gamma^3 A^{2/3}$	M1	$3 \times 10^{13} E_\gamma^3$
2	Quadrupole	E2	$8 \times 10^7 E_\gamma^5 A^{4/3}$	M2	$2 \times 10^7 E_\gamma^5 A^{2/3}$
3	Octupole	E3	$37 E_\gamma^7 A^2$	M3	$11 E_\gamma^7 A^{4/3}$
4	Hexapole	E4	$10^{-5} E_\gamma^9 A^{8/3}$	M4	$4 \times 10^{-6} E_\gamma^9 A^2$

$$H_M \approx \sqrt{10} \frac{\hbar}{m_p c R} H_{E_\gamma} \qquad (8.7)$$

with H_{E_γ} given by (8.5). The decay rate for magnetic transitions is therefore

$$\lambda_{Ml}(A, E_\gamma) \approx 20 \frac{\alpha \hbar^2}{r_0^3 m_p^2 c} \frac{(\ell + 1)}{\ell((2\ell + 1)!!)^2} \left(\frac{3}{\ell + 3}\right)^2 \left(\frac{r_0 E_\gamma}{\hbar c}\right)^{(2\ell+1)} A^{(2l-2)/3}. \qquad (8.8)$$

Inserting numerical values, we obtain decay rates given in Table 8.1. We note the following main features:

- The decay rate increases with atomic mass number and with the energy of the γ-ray.
- The rate of increase with increasing energy and atomic mass number grows with increasing ℓ.
- The decay rates diminish rapidly with increasing ℓ, partly due to the fact that the ℓ-dependence of the coefficients in (8.6) and (8.8) diminishes rapidly with increasing ℓ, but mainly because of the increasing powers of the small quantity $E_\gamma r_0/\hbar c$, which is typically of order 10^{-3}.

These decay rates are very rough estimates and are only good to within a factor of 10. One major source of error arises from the fact that these estimates have assumed a single particle transition, i.e. only one nucleon takes part in the transition, in the same way that in atomic transitions only one electron makes a transition from one energy level to another. However, unlike in the case of Atomic Physics, there are many transitions between nuclear energy levels involving more than one nucleon – this leads to an enhanced transition rate.

Nuclear isomers occur when an excited state has a spin which is significantly different from that of the ground state or other states with lower energy. In such cases, the radiation has a high polarity and consequently a very small decay rate, making the excited state metastable. For example, the nuclide $^{99}_{43}$Tc has an excited state with excitation energy 143 keV and spin–parity $\frac{1}{2}^-$. The ground state has spin–parity $\frac{9}{2}^-$, so that the lowest radiation multipole that can effect the transition from the excited state to the ground state is M4. This excited state is a nuclear isomer with a half-life of about 6 h. Some nuclear isomers have a much longer half-life and are almost stable. An example is $^{180}_{73}$Ta (tantalum), which has an excited state with spin–parity 9^- and an excitation energy of 77 keV. The ground state of this nuclide has spin–parity 1^+, requiring an M8 radiation mode for the transition. This nuclear isomer has a half-life in excess of 7×10^{15} years.

8.3 Small Energy Shifts

There are some very small corrections to the energies of γ-rays due to the interaction of the nucleus with the electrons in the atom in which the nucleus is embedded. There are three main types of these:

- hyperfine shift,
- isomer shift, and
- electric quadrupole shift.

8.3.1 Hyperfine Structure

The magnetic moment (vector) of a nucleus is proportional to its spin and is given by

$$\tilde{\mu}_N = g_I \frac{\mu_N}{\hbar} I, \tag{8.9}$$

where μ_N is the nuclear magneton (4.6), g_I is the nuclear g-factor[3] and I is the nuclear spin vector.

The magnetic moment of the atomic electrons is (analogously)

$$\tilde{\mu}_e = g_J \frac{\mu_e}{\hbar} J, \tag{8.10}$$

where μ_e is the Bohr magneton

$$\mu_e \equiv \frac{e\hbar}{2m_e}, \tag{8.11}$$

where g_J is the atomic g-factor, and J is the total electron angular momentum vector.

These two magnetic moments interact with each other, generating a hyperfine energy shift,

---

[3] g_I is defined through (8.9). It is a number of order unity, and Shell Model estimates of its value were discussed in Chap. 4.

$$\Delta E_{\text{hf}} = \frac{\mu_0}{4\pi} \tilde{\mu}_N \cdot \tilde{\mu}_e \left\langle \frac{1}{r_a^3} \right\rangle = \frac{\mu_0}{4\pi \hbar^2} g_I g_J \mu_N \mu_e \boldsymbol{I} \cdot \boldsymbol{J} \left\langle \frac{1}{r_a^3} \right\rangle, \qquad (8.12)$$

where $\mu_0 (= 1/\epsilon_0 c^2)$ is the permeability of the vacuum, and r_a is the radial distance of the electrons from the nucleus. The nuclear and electron angular momenta combine to produce a total angular momentum with quantum number F, which takes possible values

$$|I - J| \leq F \leq I + J,$$

and using the fact that the entire atomic state is in a simultaneous eigenstate of the operators F^2, I^2 and J^2 with eigenvalues $F(F+1)\hbar^2$, $I(I+1)\hbar^2$ and $J(J+1)\hbar^2$, respectively, we may write

$$\boldsymbol{I} \cdot \boldsymbol{J} = \frac{\hbar^2}{2} \left(F(F+1) - I(I+1) - J(J+1) \right),$$

such that the hyperfine energy shift, ΔE_{hf}, is

$$\Delta E_{\text{hf}} = \frac{\mu_0}{4\pi} \frac{1}{2} g_I g_J \mu_N \mu_e \left\langle \frac{1}{r_a^3} \right\rangle (F(F+1) - I(I+1) - J(J+1))$$

$$= \frac{\alpha}{2} g_I g_J \frac{\hbar^2}{m_p m_e c} \left\langle \frac{1}{r_a^3} \right\rangle (F(F+1) - I(I+1) - J(J+1)), \quad (8.13)$$

where in the last step we have used (4.6), (8.11) and the relation between μ_0 and ϵ_0.

Different nuclear energy states will, in general, have different values of nuclear spin, I, and therefore different values of F, so that the hyperfine energy shift will be different for different states. This introduces a small change in the energy of the emitted γ-ray when the nucleus makes a transition from an excited state to a lower energy state. The atomic radius is of order 10^5 fm so that inserting numbers into (8.13), we find that the hyperfine shift in the emitted γ-rays is of order 10^{-6} eV. This is about 11 orders of magnitude smaller than the typical γ-ray energy, but as we shall see in the next section, this shift can nevertheless be detected.

8.3.2 Isomeric Shift

The wavefunctions for electrons in an s-wave ($\ell = 0$) do not vanish at the origin, $\Psi(0) \neq 0$. This means that s-wave electrons have a small but non-zero probability of being inside the nucleus. When this is the case, the electrostatic potential between the nucleus and these electrons is smaller than that obtained by treating the nucleus as a point particle. It was pointed out by Richard Weiner [64] that since the effective volume of the nucleus is different for different excited states, this would lead to a small correction to the energy of the γ-ray emitted in the transition between two nuclear states.

The shift in energy of a state due to the non-zero volume of a nucleus with charge density $\rho(r)$, interacting with an electron whose wavefunction is $\Psi_e(r)$, is given by

$$\Delta E_{\text{vol}} = \frac{e^2}{4\pi\varepsilon_0} \int d^3r \int d^3r' \, |\Psi_e(r)|^2 \, \rho(r') \left[\frac{1}{|r - r'|} - \frac{1}{|r|} \right]. \tag{8.14}$$

Assuming that the nuclear charge density is spherically symmetric, as well as the s-wave electron wavefunctions, the angular integration in (8.14) can be performed to give

$$\Delta E_{\text{vol}} = \frac{4\pi e^2}{\varepsilon_0} \int r^2 dr \, |\Psi_e(r)|^2 \int_r^\infty dr' \rho(r') \left[r' - \frac{r'^2}{r} \right]. \tag{8.15}$$

If we treat the nuclear charge density as being uniform inside the nuclear radius, R, i.e.

$$\rho(r) = \frac{3Ze}{4\pi R^3}, \quad (r < R)$$
$$= 0 \quad (r > R),$$

the radial integrand is non-zero only for $r < R$. In that region, we can approximate the electron wavefunction by its value at the origin. Radial integration over r and r' then gives

$$\Delta E_{\text{vol}} = \frac{4\pi Z\alpha\hbar c}{10} |\Psi_e(0)|^2 R^2. \tag{8.16}$$

If we approximate the wavefunction at the origin by

$$\Psi_e(0) \approx \sqrt{\frac{3}{4\pi r_a^2}}$$

(r_a being the atomic radius) and take a nuclear radius to be around 7 fm, we find that this correction is of order 10^{-5} eV.

8.3.3 Electric Quadrupole Shift

Nuclei, whose charge distributions are not spherically symmetric, possess an electric quadrupole moment Q (which has the dimension of area) defined in (2.17). This electric quadrupole interacts with the electric field gradient (EFG) given by

Fig. 8.1 Electric quadrupole splitting in a transition from an $I = \frac{1}{2}$ state to an $I = \frac{3}{2}$ state

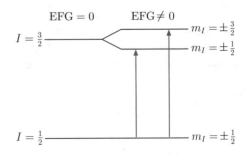

$$V_{ij} \equiv \frac{\partial}{\partial x_i} E_j(\mathbf{x}),$$

which is generated by the electrons in the atom in which the nucleus is embedded and, in the case of a crystal, also by the surrounding ions. The electric field gradient couples to the spin of the nuclear state, denoted by quantum number I, and also to its z-component, $m_I \hbar$.

For nuclear states with spin exceeding $I = \frac{1}{2}$, the contribution to the energy of the nuclear state is[4]

$$\Delta E_{\text{quad}} = \frac{V_{zz} Q}{4I(2I - 1)} \left[3m_I^2 - I(I + 1) \right]. \qquad (8.17)$$

Unlike the Zeeman effect, this depends on the modulus, $|m_I|$, of the quantum number m_I and *not* its sign. Nevertheless, it separates the energies of states with different values of $|m_I|$. So, for example, in a transition from a state with $I = \frac{1}{2}$ to a state with $I = \frac{3}{2}$, the initial state has $m_I = \pm \frac{1}{2}$, but the final state can have either $m_I = \pm \frac{1}{2}$ or $m_I = \pm \frac{3}{2}$. These final states have slightly different energies, and so the γ-ray emitted or absorbed in these transitions will have slightly different energies as shown in Fig. 8.1. These energy differences are only of the order of 10^{-6} eV, but they can nevertheless be detected [65].

These various effects all depend on the atomic electron configuration. Different atomic structures lead to different atomic radii, different values for wavefunctions at the origin and different electric field gradients at the nucleus. This means that there will be a very small difference in the γ-ray energy spectrum of a given nuclide in different chemical environments – for example, in ions with different electric charges or in different crystal structures.

---

[4]For simplicity, we have assumed azimuthal symmetry so that $V_{xx} = V_{yy}$. If this is not the case, then (8.17) acquires an extra term.

8.4 The Mössbauer Effect

In Atomic Physics, it is possible to promote atoms into their excited states by bombarding them with photons with the resonant frequencies, i.e. with energies equal to the energies between the ground state and the excited states.

In nuclei this is not usually possible. The reason for this is to do with the nuclear recoil. The energy of the emitted photon, E_γ, is not exactly equal to the energy difference, $E_{if} = E_i - E_f$, between an initial nuclear state of energy, E_i, and a final state of energy, E_f. The photon carries momentum E_γ/c, and so the recoiling nucleus must have equal and opposite momentum. Consequently, it acquires a recoil kinetic energy of

$$T = \frac{E_\gamma^2}{2m_N c^2}, \tag{8.18}$$

where m_N is the nuclear mass. The excitation energy, E_{if}, is the sum of the photon energy plus this recoil kinetic energy

$$E_{if} = E_\gamma \left(1 + \frac{E_\gamma}{2m_N c^2} \right). \tag{8.19}$$

The photon energy is slightly less than the energy difference between the initial and final nuclear states. On the other hand, in order to absorb a photon and promote the nucleus into a higher energy state, the incident photon needs to have an energy, E_γ', which exceeds the energy difference, E_{if}, in order to provide the energy of the nuclear recoil

$$E_\gamma' = E_{if} + \frac{E_\gamma'^2}{2m_N c^2} \approx E_\gamma \left(1 + \frac{E_\gamma}{m_N c^2} \right). \tag{8.20}$$

For a photon of energy 100 keV and a nucleus with $A = 100$, the energy of the photon required for absorption exceeds that of the emitted photon by about 0.1 eV. This may not seem like a very large energy difference compared with a photon energy of 100 keV. However, the problem is that the γ-ray line-widths are extremely narrow. Even for fast decaying excited states with lifetimes, τ, of about 10^{-12} s, the line-width is given by

$$\Gamma = \frac{\hbar}{\tau} \approx 10^{-3} \, \text{eV},$$

so the difference between the absorbing and emitted photons is much larger than the photon width, thereby making the reabsorption of an emitted photon impossible.

The way out of this was discovered by Rudolf Mössbauer [66] in 1958. If the source and target nuclei are both fixed in a crystal lattice, then the recoil momentum can be taken up by the entire crystal (whose mass is many orders of magnitude larger than that of the nucleus) and the recoil energy is negligible. This is called the *"Mössbauer effect"*.

8.4.1   Measurement of Line-Widths

The Mössbauer effect provides an extremely accurate method for measuring γ-ray line-widths.

Mössbauer [66] used this method to measure the line-width of the decay of the excited state of $^{191}_{77}\text{Ir}$ (iridium) with excitation energy 129 keV. This state is produced in the β-decay of $^{191}_{76}\text{Os}$ (osmium). The source, consisting of osmium embedded in a thin crystal, was moved with a constant velocity $\pm v$ relative to the absorber, which consisted of a crystal of iridium. The motion of the absorber introduces a Doppler shift of the spectral line (in the frame of the absorber)

$$\Delta E_{\text{Doppler}} = E_\gamma \frac{v}{c},$$

so that the absorber is once again slightly off resonance, but as the velocity of the absorber was only a few millimetres per second, this energy shift was only a few parts in 10^{11}.

The results are shown in Fig. 8.2. The intensity, $\mathcal{I}^{\text{Ir}}$, of γ-radiation penetrating the iridium absorber was compared with the intensity, $\mathcal{I}^{\text{Pt}}$, of γ-radiation penetrating an absorber consisting of platinum, which was used as a control. The platinum absorber does not have a resonance around 129 keV, and for sufficiently large source velocities, the penetrating intensities through the two absorbers are the same. The half-width is obtained from the velocity at which the absorption was one-half of the peak absorption. This was found to be

Fig. 8.2 Penetration intensity of a 129 keV γ-ray against velocity of absorber. The vertical axis is the dimensionless quantity $1 - \mathcal{I}^{\text{Ir}}/\mathcal{I}^{\text{Pt}}$, where $\mathcal{I}^{\text{Ir}}$ is the penetration intensity using an iridium absorber and $\mathcal{I}^{\text{Pt}}$ is the penetration intensity for a platinum absorber used as a control. [Reproduced from [66] by permission from Springer Nature]

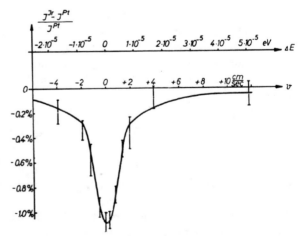

$$v_{1/2} = 0.52 \pm 0.06 \, \text{cm s}^{-1},$$

corresponding to a line-width[5]

$$\Gamma = 2\frac{v_{1/2}}{c}E_\gamma = 4.6 \pm 0.6 \times 10^{-6} \, \text{eV} .$$

8.4.2 Measurement of Isomeric Shift and Quadrupole Splitting

Very small changes in γ-ray energies can be measured using the Mössbauer technique (known as *"Mössbauer spectroscopy"*). This can be used to observe the tiny energy changes discussed in the previous section. The velocity, v, of the absorber relative to the source, which is required to bring a γ-ray of energy E_γ back into resonance, is related to an energy shift, ΔE_γ, between the resonant energy of the absorber and that of the source given by

$$\Delta E_\gamma = E_\gamma \frac{v}{c} . \tag{8.21}$$

If we have two closely spaced lines, which require velocities v_1 and v_2 to bring them back into resonance, then the energy splitting between the lines is given by

$$\left(\Delta E_\gamma\right)_{\text{split}} = (\Delta E_\gamma)_1 - (\Delta E_\gamma)_2 = E_\gamma \frac{(v_1 - v_2)}{c} . \tag{8.22}$$

The absorption of the 14.4 keV line of $^{57}_{26}\text{Fe}$ (iron) is a good example of a line which is split due to the quadrupole interaction. This has been widely studied in many experiments. The excited state has spin $I = \frac{3}{2}$, whereas the ground state has $I = \frac{1}{2}$. This leads to quadrupole splitting. There can also be an isomeric shift if the source and absorber are in different chemical environments.

In one such experiment [65], the source was stainless steel, whereas in the absorber the iron was doubly ionized Fe^{++} embedded in a crystal of $FeSO_4$. The different chemical environments of source and absorber led to an isomeric shift. The results of this experiment are displayed in Fig. 8.3, which shows two absorption lines – one with velocity $+3.4 \, \text{mm s}^{-1}$ and the other with velocity $-0.2 \, \text{mm s}^{-1}$, corresponding to energy shifts $(\Delta E_\gamma)_1 = 1.6 \times 10^{-7} \, \text{eV}$ and $(\Delta E_\gamma)_2 = -9.6 \times 10^{-9} \, \text{eV}$, respectively. These shifts are caused by the sum of the electric quadrupole shift and an isomer shift. Using (8.17) with $I = \frac{3}{2}$, and $m_I = \frac{3}{2}$ and $\frac{1}{2}$, respectively, we have

---

[5]The curve shown in Fig. 8.2 is twice as wide as this because both the source and the absorber have width Γ so that the observed distribution is a convolution of the source and absorber distributions, which leads to a total width of 2Γ.

Fig. 8.3 Mössbauer
spectrogram of the 14.4 keV
line of $^{57}_{26}$Fe with a stainless
steel source and absorber
consisting of crystals of
$FeSO_4$. [Reprinted from [65]
by permission from the
American Physical Society]

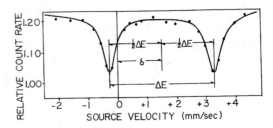

$$(\Delta E_\gamma)_1 = \frac{QV_{zz}}{4} + (\Delta E_\gamma)_{\text{isomer}},$$

$$(\Delta E_\gamma)_2 = -\frac{QV_{zz}}{4} + (\Delta E_\gamma)_{\text{isomer}}.$$

This gives us an isomer shift

$$(\Delta E_\gamma)_{\text{isomer}} = 7.5 \times 10^{-8}\,\text{eV}.$$

For the electric quadrupole shifts, ΔE_{quad}, of the two lines, we find

$$\Delta E_{\text{quad}} = \pm\frac{QV_{zz}}{4} = \pm 8.5 \times 10^{-8}\,\text{eV}.$$

The electric quadrupole moment of $^{57}_{26}$Fe is 0.08 eb (electron-barns). This allows us
to estimate the electric field gradient

$$V_{zz} = 4.25 \times 10^{22}\,\text{V m}^{-2}.$$

8.4.3   The Gravitational Redshift

One of the most spectacular applications of the Mössbauer effect was the measurement of the gravitational shift of a photon as it falls through a gravitational field.
Einstein's General Theory of Relativity predicts that a photon emitted with energy
E, travelling downwards from a height h in a gravitational field with acceleration g,
will increase its energy by

$$\Delta E_{\text{grav}} = \frac{Egh}{c^2}. \tag{8.23}$$

In 1959, Robert Pound and Glen Rebka [67] detected this gravitational shift using
the Mössbauer effect. They placed a source sample of $^{57}_{26}$Fe on the top of the tower

of the Jefferson Laboratory at Harvard University, 22.6 metres above ground level, and an absorber sample at the bottom of the tower, with a scintillation counter below it. From (8.23), the resonant velocity at which the source must move relative to the absorber is

$$v = \frac{gh}{c} = 7.4 \times 10^{-4} \, \text{mm s}^{-1}.$$

This very small velocity, at which absorption was observed, was measured by placing the sample at the top of the tower in the cone of a loudspeaker to which they applied a pure sinusoidal signal with frequency that ranged between 10 and 50 Hz. Absorption occurs once every cycle when the velocity of the loudspeaker is exactly equal to the resonant velocity. The determination of the precise phase of the oscillation at which this absorption is observed allows one to calculate the velocity of the membrane of the loudspeaker at which absorption occurs. In order to improve the accuracy, the experiment was conducted both with the upper sample as the source[6] and the lower sample as the absorber and the other way around.

They obtained a result which was within 10% of the theoretical result. This accuracy was later improved to 1% [68].

8.5   Internal Conversion

As well as decaying by the emission of a γ-ray, a nucleus in an excited state can interact directly with an inner atomic electron and free it from its atomic binding. The electron is then emitted with a fixed energy equal to the energy difference of the excited state and the ground state (or lower excited state) minus the atomic binding energy of the electron.

The emitted electron is not created inside the nucleus as in β-decay, nor is it accompanied by the emission of an antineutrino. This type of electron emission is *not* classified as β-decay and is known as *"internal conversion"*.

An example is shown in Fig. 8.4. An excited state of $^{203}_{81}\text{Tl}$ (thallium), with energy 279 keV above the ground state, is created from the β-decay of $^{203}_{80}\text{Hg}$ (mercury). The β-decay has a Q-value of 214 keV, so that we find the usual spectrum of β-decay electrons with kinetic energy up to 214 keV. Since the daughter nuclide is created in the excited state, these events are followed almost immediately by the emission of a γ-ray with energy 279 keV. In addition, there are two peaks of observed electrons at 195 and 264 keV. These are internal conversion electrons, whose binding energies are 85 keV (K-shell) and 15 keV (L-shell), so that these electrons are emitted with energies $(279-85) = 194$ keV and $(279-15) = 264$ keV, respectively.

---

[6]The source was actually $^{57}_{27}\text{Co}$, which decays to the 14.4 keV excited state of $^{57}_{26}\text{Fe}$ by β-decay.

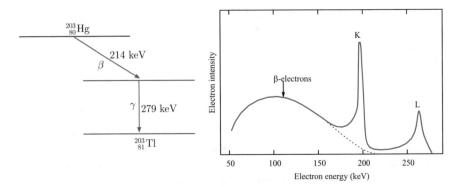

Fig. 8.4 K-shell and L-shell electrons emitted by internal conversion in the β-decay of $^{203}_{80}$Hg to an excited state of $^{203}_{81}$Tl

These electrons are emitted as a result of the Coulomb interaction between the nucleus and the inner atomic electrons. This interaction is largest if the probability of finding the electron near the nucleus is relatively large. The electrons emitted in internal conversion are therefore nearly always in an s-wave state, whose wavefunction does not vanish at the origin.

The ejected electron leaves a vacancy in an inner atomic level, which is then filled by an electron from an outer level. This electron transition is accompanied by the emission of a photon, usually in the X-ray range.

Internal conversion is always possible as an alternative to γ-ray emission. The ratio of the rate of internal electron emissions to the rate of γ-ray emissions is called the *"internal conversion coefficient"* and is denoted by $\alpha_{\{n\}}$ for the emission of an electron from the shell $\{n\}$ ($n = K, L, \ldots$). If λ_γ is the decay rate for γ-decay of an excited state, then the K-shell internal conversion rate for that state is $\alpha_K \lambda_\gamma$ *etc.*, so that the total decay rate for the state is $(1 + \sum_n \alpha_{\{n\}})\lambda_\gamma$. In the example above, the internal conversion coefficient for the K-shell electron is $\alpha_K = 0.163$, meaning that for every 1000 γ-ray emissions, there are 163 internal conversion K-shell electron emissions. As can be seen from Fig. 8.4, the internal conversion factor for the emission of an electron from the L-shell is smaller by about a factor of 3. Internal conversion coefficients increase with atomic number, Z, and decrease with increasing γ-ray energy.

In the case of an excited state with spin zero to another state of spin zero, the transition cannot proceed via the emission of a γ-ray, but the transition can take place via internal conversion.

Summary

- Electromagnetic transitions between nuclear energy levels are classified as electric 2^ℓ-pole transitions or magnetic 2^ℓ-pole transitions, where ℓ is the angular momentum of the emitted γ-ray and can take any value which respects conservation of angular momentum between the spins of the initial and final nuclear states.
- For electric 2^ℓ-pole transitions, the initial and final states have parities which differ by $(-1)^\ell$, whereas for magnetic 2^ℓ-pole transitions they differ by $(-1)^{\ell+1}$.
- γ-ray emission is strictly forbidden in transitions between the initial and final states both of which have nuclear spin $I = 0$.
- The decay rates for different decay modes can be estimated using Fermi's golden rule. The estimates are only accurate to within a factor of 10.
- Decay rates decrease considerably as the angular momentum, ℓ, increases. They increase with increasing atomic mass number and with γ-ray energy. The dependence of the decay rates of atomic mass number and γ-ray energy grows rapidly with increasing angular momentum. Excited states that require a large orbital angular momentum for a transition to a state of lower energy are metastable nuclear isomers.
- There are tiny shifts in γ-ray energies due to the interaction of the nucleus with the electrons in the atom or crystal in which the nucleus is embedded. These small shifts depend on the chemical environment of the nucleus and will differ for different atoms or different ionic structures. There are three main types of such interactions:

  - Hyperfine splitting due to the magnetic interaction between the spin of the nucleus and the total angular momentum of the electrons.
  - Isomeric shift due to the change in the nuclear energy arising from the finite volume of the nucleus. This differs for different excited states of the nucleus which have different effective volumes.
  - Electric quadrupole splitting, proportional to the square of the magnetic quantum number, m_I, of the nucleus and to the electric field gradient provided by the surrounding electrons or ions.

- The Mössbauer effect is the emission of a γ-ray without recoil energy loss due to the fact that the emitting and absorbing nuclei are bound in a crystal. This can be used to measure extremely small differences in γ-ray energies, such as those arising from the interactions of the nucleus with atomic electrons. It has also been used to measure the gravitational redshift predicted by General Relativity.
- Excited states can also decay by internal conversion. In this process the nucleus interacts with an atomic electron in an inner shell, which is then ejected from the atom. There is no γ-ray emission, but this process is accompanied by X-rays as an outer electron fills the vacancy left by the ejected electron.

Problems

Problem 8.1 State the allowed radiation modes for transitions between states with the following spin–parities:

(a)

$$\frac{1^+}{2} \rightarrow \frac{5^+}{2}.$$

(b)

$$3^- \rightarrow 1^+.$$

(c)

$$2^+ \rightarrow 0^+.$$

(d)

$$\frac{7^+}{2} \rightarrow \frac{3^+}{2}.$$

In each case, indicate which is the dominant radiation mode.

Problem 8.2 The nuclide $^{69}_{30}$Zn has a ground state with spin–parity $\frac{1}{2}^-$. Estimate the lifetime of the excited state with excitation energy 439 keV and spin–parity $\frac{9}{2}^+$. Compare this with the observed value of 4.95×10^4 s.

Problem 8.3 Estimate the decay rate of the excited state of $^{132}_{52}$Te with excitation energy 1.77 MeV and spin–parity 6^+ to the ground state which has spin–parity 0^+.

Problem 8.4 The nuclide $^{197}_{97}$Au has a ground state with spin–parity $\frac{3}{2}^+$ and an excited state with spin–parity $\frac{1}{2}^+$, whose excitation energy is 77.4 keV. In a Mössbauer spectroscopy experiment with a gold absorber in a crystal of the salt AuBr and a source of gold foil, two minima of the penetration intensity are observed with source velocity -1.47 mm s^{-1} and $+4.23$ mm s^{-1}.

(a) Calculate the isomeric shift.
(b) Calculate the electric quadrupole energy shifts of the ground state with $|m_I| = \frac{3}{2}$ and $|m_I| = \frac{1}{2}$.
(c) The electric field gradient (in the z-direction), $V_{zz} = 2.75 \times 10^{22}$ V m^{-2}. Determine the electric quadrupole moment of the nucleus.

Problem 8.5 The line-width of the 129 keV line of $^{191}_{77}$Ir is 4.60×10^{-6} eV. The excited state can also decay by internal conversion with internal conversion coefficient 2.04. What is the γ-ray decay constant of the excited state?

Appendix 8: Weisskopf's Decay Rate Estimate

The matrix element of the electric transition Hamiltonian (see (8.5)) between in initial nuclear state with wavefunction $\Psi_i(r)$ and a final state with wavefunction $\Psi_f(r)$ is

$$\langle f|H_{E_\gamma}|i\rangle = \int d^3r \sqrt{\frac{2\pi\alpha\hbar^3 c^3}{E_\gamma}} \Psi_f^*(r)\Psi_{k_\gamma}^*(r)\Psi_i(r), \tag{8.24}$$

where $\Psi_{k_\gamma}^*(r)$ is the photon wavefunction.

In order to extract the 2^ℓ-multipole (Eℓ) part of this matrix element, we need to expand the plane-wave wavefunction of the photon in terms of spherical harmonics, $Y_{\ell m}(\theta, \phi)$. Using the fact that the integrand in (8.24) only has support for r less than or of the order of the nuclear radius, R, (the nuclear wavefunctions are negligible outside this region) and that the wavelengths of the emitted γ-ray is always much larger than the nuclear radius, we can assume that

$$k_\gamma r \ll 1,$$

so that we can keep only the leading power of $k_\gamma r$ in the coefficient of each spherical harmonic in the expansion of the plane-wave photon wavefunction. This expansion then gives (in spherical polar coordinates)

$$\Psi_{k_\gamma}(r, \theta, \phi) = \frac{1}{\sqrt{V}} \sum_{\ell=0}^\infty \sum_{m=-\ell}^\ell \frac{\sqrt{(\ell+1)}}{\sqrt{\ell}(2\ell+1)!!} (k_\gamma r)^\ell Y_{\ell m}(\theta, \phi) \tag{8.25}$$

The matrix element of the interaction Hamiltonian for the transition of an initial nucleon state with wavefunction $\Psi_i(r)$ and a final nucleon state with wavefunction $\Psi_f(r)$, emitting a photon with energy E_γ and orbital angular momentum ℓ is then

$$\langle f|H_{E_\gamma}|i\rangle_l = \frac{1}{\sqrt{V}} \sqrt{\frac{2\pi\alpha\hbar^3 c^3}{E_\gamma}} \sum_{m=-l}^\ell \frac{\sqrt{(\ell+1)}}{\sqrt{\ell}(2\ell+1)!!}$$

$$\times \int d\phi \sin\theta d\theta r^2 dr \, \Psi_f^*(r) (k_\gamma r)^\ell Y_{\ell m}(\theta, \phi)\Psi_i(r), \tag{8.26}$$

Provided that the orbital angular momenta ℓ_i and ℓ_f of the initial and final states are such that

$$|\ell_i - \ell_f| \le \ell \le \ell_i + l_f,$$

the angular part of the integral in (8.26) is approximately

$$\sum_{m=-l}^{\ell} \int_0^{\pi} \sin\theta\, d\theta \int_0^{2\pi} d\phi\, \Psi_f^*(r) Y_{\ell,m}(\theta,\phi) \Psi_i(r) \approx \psi_f^*(r)\psi_i(r),$$

where $\psi_{i/f}(r)$ are the radial parts of the initial and final nuclear wavefunctions.

Finally, we need the radial part of the integral

$$\int r^2 dr\, \psi_f^*(r) r^{\ell} \psi_i(r) .$$

The nuclear wavefunctions extend only over the range $r < R$. If we approximate the radial parts of the nuclear wavefunctions by assuming it takes a constant value inside the nucleus but vanishes outside it:

$$\psi_{i/f}(r) \approx \frac{3}{4\pi R^3}, \ (r < R)$$

$$= 0\ (r > R),$$

then we get

$$\int r^2 dr\, \psi_f^*(r) r^{\ell} \psi_i(r) \approx \frac{3}{(\ell+3)} R^{\ell}. \tag{8.27}$$

Using the expression for the nuclear radius $R = r_0 A^{1/3}$ and piecing this together, the matrix element of the interaction Hamiltonian is approximately given by

$$\langle f|H_{E_\gamma}|i\rangle_l \approx \sqrt{\frac{2\pi\alpha\hbar^3 c^3}{E_\gamma V}} \frac{\sqrt{(\ell+1)}}{\sqrt{\ell}(2\ell+1)!!} \left(\frac{E_\gamma r_0}{\hbar c}\right)^{\ell} A^{\ell/3} \tag{8.28}$$

The factor $E_\gamma r_0/\hbar c$ is typically of order 0.001, so we see that the matrix element of the interaction Hamiltonian decreases rapidly with increasing ℓ.

The density of states for the photon is given by

$$\rho(E_\gamma) = V\frac{E_\gamma^2}{2\pi^2\hbar^3 c^3} . \tag{8.29}$$

Inserting (8.28) and (8.29) Fermi's golden rule, (7.50), the decay rate for electric multipole transitions is

$$\lambda_{El} \approx 2\alpha c\frac{(\ell+1)}{\ell\left((2\ell+1)!!\right)^2} \left(\frac{3}{\ell+3}\right)^2 \left(\frac{E_\gamma}{\hbar c}\right)^{(2\ell+1)} r_0^{2l} A^{2l/3}. \tag{8.30}$$

Chapter 9
Nuclear Fission

9.1 Nuclear Transmutation

In the same way that molecules can interact with each other, exchanging atoms or ions, to produce different molecules, nuclei can interact with each other, exchanging protons and/or neutrons. If the total binding energies of the final-state nuclides is larger than that of the initial nuclides, then energy is liberated in the reaction (the reaction has a positive Q-factor); otherwise, the initial-state nuclei must be accelerated to a minimum kinetic energy before the reaction can take place.

Such a nuclear reaction is called nuclear *"transmutation"*. This term is applied to all nuclear reactions including radioactive decay.

The first demonstration of this transmutation was carried out by John Cockroft and Ernest Walton in 1932 [69]. They built the first particle accelerator, which accelerated protons up to a kinetic energy of 0.7 MeV, using pulsed or AC voltages. The accelerated protons were used to bombard a target of ^7_3Li and set at an angle of $45°$ to the proton beam. This gave rise to the reaction

$$p + {}^7_3\text{Li} \rightarrow {}^4_2\text{He} + {}^4_2\text{He} .$$

The final-state α-particles were observed perpendicular to the direction of the proton beam, using zinc sulphide screens. They were observed to be moving in opposite directions with the same energy and at right angles to the direction of the proton beam. By conservation of momentum, this meant that the final-state particles had equal mass and therefore were different from the initial-state particles. This experiment was popularly described as "splitting the atom".

© The Author(s), under exclusive license to Springer Nature Switzerland AG 2021
A. Belyaev, D. Ross, *The Basics of Nuclear and Particle Physics*, Undergraduate
Texts in Physics, https://doi.org/10.1007/978-3-030-80116-8_9

9.2 Discovery of Fission

In 1934 Enrico Fermi, performed experiments in which he bombarded uranium with neutrons [70], thereby provoking transmutation of the nucleus. Uranium, with atomic number 92, is the heaviest of the naturally occurring elements. Fermi speculated that in some cases the neutron-rich uranium underwent β-decay to create neptunium, the first transuranic element.[1] However, Ida Noddack suggested that rather than the production of neptunium, the neutron bombardment had caused the uranium nucleus to split into two smaller nuclei [71]. The process of such splitting (induced or spontaneous) is called nuclear *"fission"*. We can see from Fig. 3.2, which shows the distribution of binding energies per nucleon, that this process is energetically possible. For the heavier elements, such as uranium, the binding energy per nucleon is considerably lower than for elements in the middle of the periodic table, such as iron or nickel. This means that it is energetically favourable for a heavy nucleus (with atomic mass number greater than about 150) to split into two fragments of smaller nuclei, thereby releasing energy which goes into the kinetic energy of the fragments. A typical fission process releases around 200 MeV.

This hypothesis was verified in 1939 by Otto Hahn and Fritz Strassmann, who identified barium as one of the elements produced during the neutron bombardment of uranium [72].

9.3 Fission Mechanism

The mechanism by which fission occurs was expounded by Lisa Meitner and her nephew Otto Frisch.

The classical picture is displayed in Fig. 9.1. The nucleus starts off (almost) spherical and then becomes distorted into an (azimuthally symmetric) ellipsoid, (b). Further distortion causes the nucleus to develop a "neck" (c). This is known as the *"saddle point"*. After further deformation, the two sides of the neck separate (d) into two different nuclei with smaller atomic number and atomic mass number. The point of separation is called the *"scission point"*.

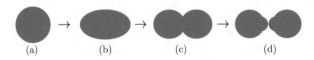

Fig. 9.1 Deformation of a nucleus undergoing fission. The initial spherical shape (**a**) becomes ellipsoidal (**b**) and then develops a "neck" (**c**) before the two fragments separate (**d**)

---

[1]Elements with atomic number greater than 92 are known as *"transuranic elements"*.

We can test for the stability of a nuclide against fission using the Liquid Drop Model, by considering the change in binding energy as the nucleus becomes deformed into an ellipsoid. Since the atomic mass number, A, remains unchanged, the volume term in the semi-empirical mass formula remains unchanged, but the surface term increases in magnitude as an ellipsoid has a larger surface area than a sphere of the same volume. For small deformation, the surface energy term in the binding energy increases (in absolute value) from $-\alpha_S A^{2/3}$ to $-\alpha_S A^{2/3} \left(1 + 2\epsilon^4/45\right)$ [73], where ϵ is the eccentricity of the ellipsoid ($\epsilon \ll 1$).

On the other hand, the Coulomb energy is reduced because the average distance between the protons increases. The Coulomb term changes from $-\alpha_c Z^2/A^{1/3}$ to $-\alpha_c Z^2/A^{1/3} \left(1 - \epsilon^4/45\right)$ [74]. The total change, Δ_{BE}, in the binding energy for an ellipsoid with small eccentricity ϵ is then

$$\Delta_{BE} = \frac{\epsilon^4}{45} \left(\alpha_c \frac{Z^2}{A^{1/3}} - 2\alpha_s A^{2/3}\right) + \mathcal{O}\left(\epsilon^6\right).$$

We therefore see that nuclides for which

$$\Delta_{BE} > 0 \implies \frac{Z^2}{A} > 2\frac{\alpha_s}{\alpha_c} \sim 50$$

are unstable against fission (it is energetically preferable for the nucleus to become deformed and eventually split). Heavy nuclides have a neutron-to-proton ratio of around 1.5, so that the maximum atomic mass number for fission stable nuclei is about 300. Nuclides with a larger atomic mass number than this are *not* found naturally as they fission almost instantaneously into smaller nuclei.

9.4 Spontaneous Fission

For the fission stable nuclides, the energy initially increases as the nucleus becomes less spherical (N.B. a *decrease* in binding energy means an *increase* in the rest energy of the nucleus), as shown in Fig. 9.2.

The energy increases for the ellipsoidal configuration (b) of Fig. 9.1, reaching a maximum at (c) where the neck is formed and then the energy decreases and the two fission fragments separate at (d). Although such a potential makes the nucleus classically stable against fission, spontaneous fission can nevertheless occur via quantum tunnelling in a similar way to the quantum tunnelling, which leads to α-decay. Spontaneous fission is far less likely than α-decay, but it does occur, albeit with half-lives which are much longer than the half-life for α-decay. Spontaneous fission of $^{238}_{92}$U was first observed in 1940 by Konstantin Petrzhak and Georgy Flerov [75]. The fission half-life of $^{238}_{92}$U is around 10^{16} years – compared with the α-decay half-life of 4.5×10^9 years.

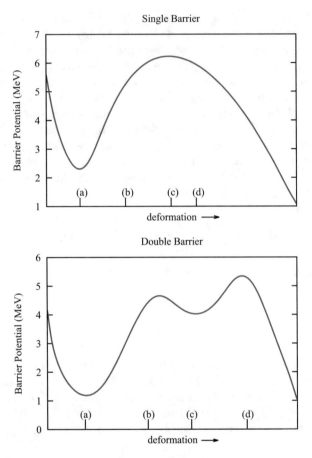

Fig. 9.2 Nuclear energy as a function of the deformation of the nucleus. This leads to a potential barrier, which the nucleus can overcome by quantum tunnelling. Tick labels (a)–(d) refer to the configurations shown in Fig. 9.1. The lower diagram shows the case for which the potential has two maxima

We can make an order-of-magnitude estimate of the height of the fission potential barrier. Suppose the two fission fragments have atomic numbers Z_1 and Z_2, respectively, and atomic mass numbers A_1 and A_2. The nuclear radii of the fragments are given by $r_1 = r_0 A_1^{1/3}$ and $r_2 = r_0 A_2^{1/3}$. The potential maximum, V_{max}, is reached just at the point of separation of two fission fragments (the scission point) and is equal to the Coulomb potential for two electrically charged spheres with charges $Z_1 e$ and $Z_2 e$ whose centres are separated by $r_1 + r_2$, as shown in Fig. 9.3.

$$V_{\text{max}} = \frac{Z_1 Z_2 \alpha \hbar c}{r_0 \left(A_1^{1/3} + A_2^{1/3} \right)}. \tag{9.1}$$

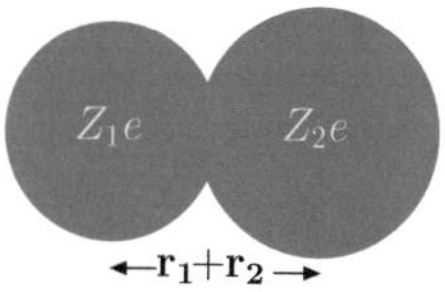

Fig. 9.3 Two charged spheres, with charges Z_1e and Z_2e and radii $r_1 = r_0A_1^{1/3}$ and $r_2 = r_0A^{1/3}$, in contact

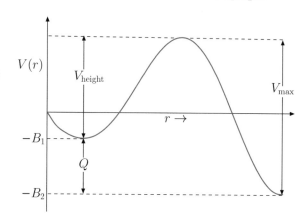

Fig. 9.4 The relation between the fission energy release, Q, the barrier height and the maximum potential energy (taken to be at scission point). B_1 is the binding energy of the parent nucleus, and B_2 is the combined binding energy of the fission fragments

The fission energy released, Q, is the difference between initial potential energy and the final potential energy (when the fission products are widely separated), which is equal to the difference between the sum of the binding energy of the fission fragments and the binding energy of the parent.

$$Q = B(A_1, Z_1) + B(A_2, Z_2) - B(A, Z_1 + Z_2). \tag{9.2}$$

The height, V_{height}, of the potential barrier is the difference between the potential energy at scission the point (9.1) and the fission energy release, as shown in Fig. 9.4

$$V_{\text{height}} = V_{\text{max}} - Q, \tag{9.3}$$

with V_{max} and Q given by (9.1) and (9.2), respectively.

The estimate obtained from (9.3) is not very good, because it involves the small difference between almost equal quantities (Q is of order 200 MeV, whereas the barrier height is of order 10 MeV) so that any fractional error in the estimate of either Q or V_{max} is amplified in the determination of the barrier height. Furthermore, (9.1) overestimates V_{max} since it assumes that when the fragments separate they can be considered as spherical charge distributions, which is not the case.

A more careful estimate of fission barrier heights [76] is shown in Fig. 9.5, which also shows corrections to these estimates from the effects of the Shell Model. Note the substantial increase in barrier heights where either the number of protons or the number of neutrons is equal to a magic number.

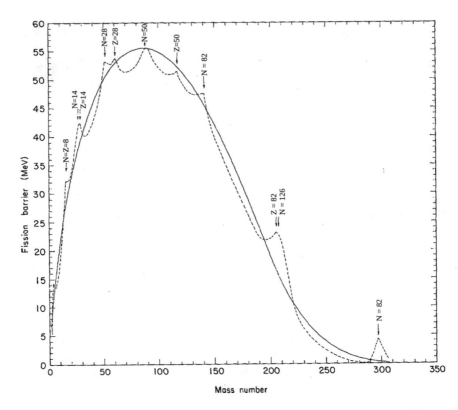

Fig. 9.5 Fission potential barrier heights calculated using the Liquid Drop Model (solid line). Corrections due to Shell-Model effects are shown by the dotted line [Reproduced from [76] by permission from Elsevier]

The tunnelling probability decreases exponentially with increasing barrier height, so that, in practice, spontaneous fission only occurs if the barrier height is less than about 10 MeV, i.e. for nuclides with atomic mass number exceeding $\sim$220.

For some nuclides, the potential barrier has two maxima with a minimum between them as shown in the lower graph of Fig. 9.2. $^{240}_{94}$Pu is an example of a nuclide with such a potential. Such nuclides have isomers with excitation energy corresponding to the metastable minimum between the two maxima. These isomers have a far smaller potential barrier (both lower and narrower) and consequently have a much larger fission decay rate than their ground states.

9.5 Fission Products

In most cases, there are two daughter nuclei (fission fragments), although in about one in 300 fission events a third nucleus is produced. This is usually a small nucleus, such as $^3_1$H (tritium) or $^4_2$He (α-particle).

The Liquid Drop Model favours splitting into two fragments of approximately equal atomic mass, Z, and atomic mass number, A. This is, however, *not* what is observed. The percentage fission yields for different atomic mass number, A, are shown in Fig. 9.6, for the case of the fission of $^{238}_{92}$U. The maximum yield occurs when the atomic mass numbers, A_+ and A_-, of the two fission fragments are in a ratio between 1.3 and 1.5. The reason for this asymmetry is not known, but a hint can be obtained from the fact that the favoured atomic mass number for the heavier nuclide is around 132 and this appears to be independent of the parent nuclide which undergoes fission. This is the atomic mass number of the doubly magic nuclide $^{132}_{50}$Sn. This *might* be the origin of the peak in the yield around $A = 132$, i.e. there is a preference for one of the fission fragments to have atomic number and atomic mass number close to that of this doubly magic nuclide.

As can be seen from Fig. 3.4, the number of neutrons per proton in stable isotopes increases with increasing atomic number. This means that the parent nuclide in a fission process always has too many neutrons for the fission fragment nuclides to be stable. Most fission processes are therefore accompanied by the emission of two or three neutrons, known as *"prompt neutrons"* as they are emitted simultaneously with the fission process.

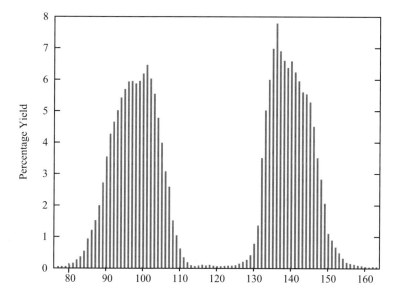

Fig. 9.6 Fission yields for $^{238}_{92}$U for different atomic mass numbers

The prompt release of energy in a fission process, Q, is the difference between the binding energy of the parent nuclide and the fission products. For example, for the fission of $^{236}_{92}U$,

$$^{236}_{92}U \rightarrow \, ^{140}_{54}Xe + ^{94}_{38}Sr + 2n, \tag{9.4}$$

the binding energies of isotopes $^{236}_{92}U$, $^{140}_{54}Xe$ and $^{94}_{38}Sr$ are 1790.4 MeV, 1160.7 MeV and 807.8 MeV, respectively, and so the energy released by this fission is 178 MeV. Most of this energy goes into the kinetic energy of the fission products (including the prompt neutrons), but in many cases the fission fragments are produced in excited states and then decay emitting γ-rays.

Notwithstanding the prompt neutron emission, the fission fragments are still neutron rich (they have too many neutrons for stability). For example, in the fission reaction of (9.4), the heaviest stable isotope of strontium has atomic mass number 88 and the heaviest stable isotope of xenon has atomic number 134. This means that there are 12 too many neutrons for stability. These neutron-rich fission products usually undergo β-decay in a fairly long decay chain until stability is reached. For example, the fission fragment $^{140}_{54}Xe$ decays in 4 stages to $^{140}_{58}Ce$ (cerium), which is stable. Several of these β-decays will be to daughter nuclides in an excited state, and so there are also γ-rays emitted in conjunction with these decays. On average there are 10 γ-rays emitted for each fission event. These secondary decays also release energy. The total energy produced in a fission process from beginning to end is usually about 10% higher than the prompt fission energy generated by the initial fission reaction.

In some of these β-decays, the daughter nuclide is produced in a sufficiently highly excited state that it can decay not by γ-ray emission, but by neutron emission. These neutrons are emitted after β-decay has occurred, and this could be as long as several minutes after the initial fission process (depending on the half-life of the β-decay). Such secondary neutrons are therefore known as "delayed neutrons". The probability for a nuclide in an excited state to emit a neutron is very small owing to the absence of Coulomb repulsion between the neutron and the daughter nucleus. Delayed neutron emission is therefore very rare and only happens in about one in a thousand fission events. They nevertheless play an important role in the design of nuclear reactors, as discussed below.

9.6   Induced Fission

Spontaneous fission occurs as a result of quantum tunnelling, whose probability is very small. Much more likely is induced fission (which is actually how fission was first observed [72]). In this case the parent nucleus, with atomic mass number A, is bombarded with a neutron. The neutron is absorbed if the binding energy of the isotope with atomic number $A + 1$ exceeds the binding energy of the isotope

Fig. 9.7 An example of fission induced by neutron bombardment of $^{235}_{92}$U

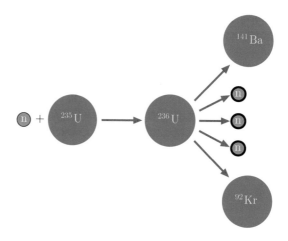

with atomic number, A, and the excess energy is released in the form of vibrational energy. If this neutron absorption energy is greater than the height of the fission potential barrier, then fission can proceed promptly – no quantum tunnelling is required.

An example of this is the process

$$n + {}^{235}_{92}\text{U} \rightarrow {}^{236}_{92}\text{U} \rightarrow {}^{141}_{56}\text{Ba} + {}^{92}_{26}\text{Kr} + 3n,$$

shown schematically in Fig. 9.7. The binding energy of $^{235}_{92}$U is 1783.9 MeV, whereas the binding energy of the more stable isotope, $^{236}_{92}$U, is 1790.4 MeV. The height of the potential barrier (for the above fission process) is 5.6 MeV, which is smaller than the binding energy of the extra neutron in the isotope $^{236}_{92}$U, so that the absorption of the neutron releases sufficient energy to overcome the potential barrier. In this case the bombarding neutrons need not be energetic. Fission can be induced by thermal neutrons which have a kinetic energy of around 0.025 eV. Nuclides for which induced fission can be accomplished using thermal neutrons are called *"fissile"*.

On the other hand, if the binding energy of the extra neutron is insufficient to overcome the potential barrier, the incident neutron needs to have sufficient kinetic energy to overcome the shortfall, thereby inducing fission. Nuclides for which fission can be induced by bombardment with neutrons of sufficient kinetic energy are called *"fissionable"*. An example of a fissionable (but not fissile) nuclide is $^{238}_{92}$U (the most abundant isotope of uranium). This has a fission barrier height of 6.3 MeV. Its binding energy is 1801.7 MeV, whereas the binding energy of $^{239}_{92}$U is 1806.5 MeV. The vibrational energy generated by the absorption of a neutron by a nucleus of $^{238}_{92}$U is therefore 4.8 MeV, which is insufficient to overcome the barrier potential. In order to induce fission in $^{238}_{92}$U, the incident neutrons need to have at least enough kinetic energy to make up the shortfall of 1.5 MeV. It is very often the case that isotopes with an odd number of neutrons are fissile, whereas isotopes with an even number of isotopes are not. This is due to the pairing term contribution to the nuclear binding energy – adding a neutron to an isotope with initially an unpaired neutron produces an isotope in which all the neutrons are paired, thereby significantly increasing its binding energy.

On the other hand, it is possible that the metastable isotope produced by bombarding a fissionable (but not fissile) isotope with neutrons with energy below the energy required to induce fission (e.g. thermal neutrons) can decay by β-decay into a fissile isotope, so that it can absorb another thermal neutron and undergo fission. $^{238}_{92}U$ is an example of this. Bombarding this with a thermal neutron can produce the reaction

$$n + {}^{238}_{92}U \;\rightarrow\; {}^{239}_{92}U \;\xrightarrow{\beta^-}\; {}^{239}_{93}Np \;\xrightarrow{\beta^-}\; {}^{239}_{94}Pu,$$

in which the nuclide $^{239}_{92}U$ decays via a two-stage β-decay to $^{239}_{94}Pu$, which is fissile. It can absorb a thermal neutron to produce $^{240}_{94}Pu$. The binding energy of the absorbed neutron is 6.5 MeV, which exceeds the fission potential barrier height. The half-lives for the above β-decays are 40 min for stage 1 and 4 days for stage 2. This should therefore be considered as a method for manufacturing a fissile material. It is used in the design of *"breeder reactors"*.

Another example, which is used in modern nuclear reactors, is $^{232}_{90}Th$. When bombarded with neutrons, this nuclide can generate the fissile isotope $^{233}_{92}U$.

$$n + {}^{232}_{90}Th \;\rightarrow\; {}^{233}_{90}Th \;\xrightarrow{\beta^-}\; {}^{233}_{91}Pa \;\xrightarrow{\beta^-}\; {}^{233}_{92}U.$$

9.7   Chain Reactions

The prompt neutrons that are emitted in a fission reaction can be absorbed by another parent nucleus, which then itself undergoes induced fission, producing yet more neutrons which can be absorbed to produce even more fission events, which in turn produce more neutrons. This leads to an avalanche of induced fission events, known as a *"chain reaction"*. The neutrons emitted in a fission reaction typically have a kinetic energy of only 1 MeV. For a non-fissile isotope, this is not usually sufficient to overcome the fission barrier and induce fission. But it is indeed possible in the case of a fissile nuclide.

The possibility of such a chain reaction was first proposed in 1939 by Herbert Anderson, Enrico Fermi and Leo Szilard [77]. In order for a chain reaction to occur, the concentration of fissile material must be above a certain value. In uranium ore,[2] the concentration of the fissile isotope, $^{235}_{92}U$, is only about 0.7% – the rest is

---

[2]Uranium ore, formally known as *"pitchblend"*, is a naturally occurring mineral, consisting mainly of uranium oxide (UO_2).

nearly all $^{238}_{92}$U. In nuclear reactors, a concentration of 3.5–4.5% is required for a chain reaction to take place. For concentrations below this, the neutrons are simply absorbed by the non-fissile isotope and the chain reaction fizzles out. The process of increasing the concentration of the fissile isotope is called *"enrichment"* and usually involves isotope separation of a gaseous compound of uranium, or other fissionable material, using a powerful centrifuge. The centrifuge rotates at a very high speed (50,000–70,000 r.p.m.), and the heavier isotope, $^{238}_{92}$U, drifts towards the wall of the centrifuge where it is extracted – leaving a higher concentration of the (lighter) fissile isotope $^{235}_{92}$U.

We define the *"neutron multiplication factor"*, k, as the number of neutrons produced at stage $n + 1$ of a chain divided by the number of neutrons produced at stage n. The number, $N_f(n)$, of fission events at stage n is given by

$$N_f(n) \propto k^n .$$

Since the number of prompt neutrons produced in each fission reaction is 2–3, k is unlikely to exceed 2. Nevertheless, this means that one fission event can trigger over 1000 fission events after 10 stages. This number, k, will depend on how many of the neutrons, produced at stage n, are absorbed by a nucleus that can undergo induced fission.

- If $k < 1$, the chain reaction will simply fizzle out and the process will halt very quickly. This is what happens in natural uranium ore, in which the concentration of $^{235}_{92}$U is so small that it is very unlikely that one of the neutrons produced in a fission event would be absorbed by another nucleus of the fissile isotope.
- If $k > 1$, then the chain reaction will grow out of control (with disastrous consequences) until all the fissile material is used up. This is achieved by enriching the natural ore so that there is a sufficiently large concentration of $^{235}_{92}$U (around 20%). Furthermore, the total mass of the fissile material must exceed a value called the *"critical mass"*. This is necessary to ensure that more than one neutron from a fission event is absorbed by the fissile material and does not simply pass through the material.

We can make a rough estimate of the critical mass [78] if we know the neutron absorption cross section σ_a. For highly enriched uranium (HEU), this is around 1.2 b, although it depends on the energy of the incident neutron. Imagine a cylinder with cross section σ_a and length l. If the volume of this cylinder is equal to the average volume occupied by one uranium atom, then l is equal to the absorption mean free path, λ_a, i.e. the average distance that a neutron will travel before absorption. The average volume occupied by one nucleus in a material is the mass of the nucleus divided by the density, ρ, of the material. For uranium, $\rho \approx 2 \times 10^4$ kg m^{-3}. Thus we have

$$\lambda_a = \frac{A m_p}{\rho \sigma_a} \approx 0.17 \, \text{m}, \tag{9.5}$$

with A set to 235. One might think that a critical sample of enriched uranium has to be this size in order for the fission neutrons to be absorbed. But this would be an overestimate. Between production and absorption, the neutron, which has a scattering cross section, σ_s, undergoes several elastic scatterings and performs a random walk in three dimensions. This means when a neutron performs a random walk of length l, its average distance from its starting point is $l/\sqrt{3\eta}$, where η is the average number of scatterings that a neutron undergoes before it is absorbed, i.e. the ratio of the scattering cross section to the absorption cross section, σ_s/σ_a. For uranium, η is equal to 3.7. This means that the critical radius, R_c, of a spherical sample of HEU is given by

$$R_c = \frac{1}{\sqrt{3\eta}} \frac{Am_p}{\rho\sigma_a} \approx 0.05\,\text{m}. \tag{9.6}$$

This corresponds to a critical mass of 10 kg. The actual value of the critical radius is around 0.085 m, giving a critical mass of around 50 kg. There is a correction to (9.6), which depends on the average number of prompt neutrons emitted per fission event [79].

The reader will be relieved to know that a sample of HEU or other fissile material whose mass exceeds the critical mass will not produce an uncontrolled chain reaction. As soon as the chain reaction starts, the energy released will cause the sample to split into smaller pieces, which are below the critical mass.

- If $k = 1$ exactly, we have a controlled reaction. This is what is needed in a nuclear reactor. The absorption is controlled by interspersing the fissionable fuel (e.g. uranium) with cadmium or boron rods that absorb most of the neutrons (cadmium and boron have a high neutron absorption cross section). The reaction is controlled automatically by moving the rods in and out so that the value of k is kept equal to one. A diagram of one type of reactor, known as a *"boiling water reactor"*, is shown in Fig. 9.8, in which the fission energy is used to boil water and the steam produced drives a turbine.

It is in this control that delayed neutrons play a crucial role, even though there are relatively few of them. The absorbers are set such that the value of k from the prompt neutrons is very slightly below one, so that the chain reaction would normally die out. However the chain reaction is revitalized by the arrival of the delayed neutrons, so that the chain reaction can be sustained. The relatively long time delay between the prompt neutrons and the delayed neutrons provides time for the mechanical adjustment of the control rods in order to keep the reactor operating safely. Several other designs of nuclear reactors are currently in use.

A new design of reactor, which uses thorium as its fuel, is currently being developed. Thorium has the advantage that it is more abundant than uranium and does not produce as much long-lived radioactive waste. Moreover, it is much safer, since the reactor can be run in sub-critical mode, $k < 1$, using fast neutrons to induce the transmutation to the fissile isotope $^{233}_{92}\text{U}$. Carlo Rubbia [80] has proposed an energy amplifier that uses fast neutrons produced by the scattering of protons from a high-energy particle accelerator against a fixed target.

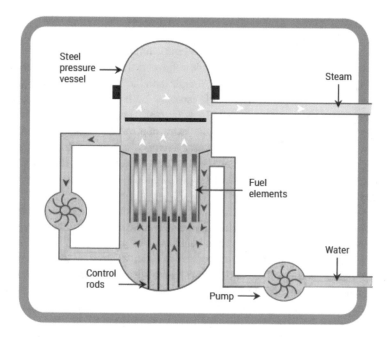

Fig. 9.8 Diagram of a boiling water nuclear reactor. The fission energy boils the water and the steam drives a turbine. The control rods are plunged into the gaps between the fuel elements to control the fission process. [*Source*: http://www.world-nuclear.org/getmedia/10b78f7b-0895-4837-9adc-f83d60d52963/boiling-water-reactor-bwr.png.aspx]

Summary

- Fission is the splitting of a heavy nucleus into two lighter fragments, whose combined binding energy exceeds the binding energy of the parent nucleus.
- The splitting occurs because the nucleus undergoes a deformation from its original, almost spherical, shape, and then develops a neck which eventually splits into two fragment nuclei.
- There are usually two fission fragments. In most cases, one of the fission fragments is more massive than the other by a factor of 1.3–1.5, with a similar asymmetry in the atomic numbers of the fragments.
- Since heavier stable nuclei have more neutrons per proton, the parent nucleus has more neutrons than the sum of the lighter fission fragments. There is an emission of prompt neutrons – about two or three per fission event.
- The fission fragments are often produced in excited states so that the fission process is accompanied by the emission of γ-rays.
- The fission fragments usually contain too many neutrons for stability, and so they decay via a chain of β-decays, releasing yet more energy, until they reach a stable isobar. These β-decays often produce daughter nuclides in excited states so that they are also accompanied by γ-ray emission.

- In some rare cases, the nuclei produced during the chain of β-decays are in a sufficiently highly excited state to be able to emit delayed neutrons.
- Fission can take place spontaneously via quantum tunnelling through the potential barrier which arises as the nucleus is deformed. The half-lives for such decays are very long and in particular much longer than the half-life for α-decay.
- Fission can be induced in fissionable isotopes by bombardment with neutrons. In most cases, the neutrons have to be sufficiently energetic so that their kinetic energy plus the binding energy of the extra neutron exceeds the height of the fission potential barrier. However, some isotopes are fissile, meaning that the absorption energy of the incident neutron alone is larger than the height of the potential barrier. For these fissile isotopes, fission can be induced by very low-energy (thermal) neutrons.
- The 2–3 prompt neutrons produced with each fission can be absorbed by other nuclei of a fissile isotope, which then undergoes fission producing yet more neutrons etc. giving rise to a chain reaction. For a sample of ore to produce such a chain reaction, it must be enriched in order to increase the concentration of the fissile isotope. The sample must also be greater than a critical mass so that the neutrons can be absorbed by another nucleus rather than escaping from the sample. The important characteristic of induced fission is the neutron multiplication factor, k. This exceeds 1 for a chain reaction. In a nuclear reactor, k is held precisely to 1 so that a fission process can be sustained without growing out of control.

Problems

Problem 9.1 Consider the induced fission process of $^{240}_{94}$Pu:

$$n + {}^{240}_{94}\text{Pu} \rightarrow {}^{134}_{52}\text{Te} + {}^{103}_{42}\text{Mo}.$$

(a) How many prompt neutrons are produced in this part of the fission process?
(b) Use the semi-empirical mass formula to estimate the total energy released in this process.
(c) $^{134}_{52}$Te decays via a β-decay chain to $^{134}_{54}$Xe, and $^{103}_{42}$Mo decays to $^{103}_{45}$Rh. How many electrons and antineutrinos are emitted in this decay chain?
(d) Find the *total* energy released by the fission process.

Problem 9.2 Using the semi-empirical mass formula, estimate which of the following isotopes of plutonium are fissile:
$^{237}_{94}$Pu, $^{238}_{94}$Pu, $^{239}_{94}$Pu, $^{240}_{94}$Pu, and $^{241}_{94}$Pu.
   (The fission barrier height for any fission process of Pu may be taken to be 6.0 MeV.)

Problem 9.3 The absorption cross section of $^{239}_{94}$Pu is 1.4 b, and the scattering cross section is 3.4 b. The density of plutonium is 2×10^4 kg m^{-3}. Estimate the critical mass of a highly enriched sample of this isotope.

Problem 9.4 What mass of uranium ore, enriched to a 4% concentration of $^{235}_{92}$U, is required to fuel a reactor, with an output of 500 MW, for a year (3.15×10^7 s)? Quote your answer to two significant figures.

(The energy release per fission may be taken to be 200 MeV.)

Chapter 10
Nuclear Fusion

If we look again at Fig. 3.2, we note that for light nuclei, the binding energy per nucleon increases with atomic mass number. This means that it is energetically favourable for two light nuclei to combine to make a heavier one. Such a process is called nuclear *"fusion"*. Somewhat less energy is usually released in a fusion reaction than in a typical fission reaction. However, since fission involves nuclei with large atomic mass numbers, the energy released *per nucleon* in a fusion reaction is much larger than in the case of fission.

Fusion was first observed by Mark Oliphant, Paul Harteck and Ernest Rutherford in 1933 [81]. They accelerated deuterons ($^2_1$H) up to an energy of 200 keV, which were then incident on a target coated with a compound such as ammonium chloride (ND_4Cl), in which the hydrogen atoms had been replaced by deuterium. This induced the reaction[1]

$$d + d \ \rightarrow \ p + t$$

with a proton, p, and tritium, t ($^3_1$H), in the final state. The energy of the fusion products was measured from the absorption depth in the target, which was calibrated using protons with known energies. The protons were emitted with an energy of 3 MeV. The initial state has negligible momentum, so the momenta of the final-state particles are equal and opposite. Tritium has three times the mass of a proton and so its energy must be 1 MeV. The energy released by this fusion reaction was therefore 4 MeV.

---

[1] The symbol "d" is used for a deuteron rather than the isotope notation, $^2_1$H, and "t" for tritium rather than $^3_1$H.

© The Author(s), under exclusive license to Springer Nature Switzerland AG 2021
A. Belyaev, D. Ross, *The Basics of Nuclear and Particle Physics*, Undergraduate Texts in Physics, https://doi.org/10.1007/978-3-030-80116-8_10

10.1   Examples of Fusion

The simplest fusion reaction would be the formation of a helium nucleus from the fusion of two protons

$$p + p \rightarrow {}^2_2\text{He} + \gamma. \tag{10.1}$$

The γ-ray is emitted in order for momentum to be conserved. However, there is no stable bound state of two protons – the isotope ${}^2_2\text{He}$ does not exist. Instead, one of the protons undergoes β^+-decay into a neutron, emitting a positron and a neutrino. This decay is not energetically possible for a free proton, but in the fusion of two protons, energy is released, which facilitates the β-decay of the proton. The fusion process is therefore

$$p + p \rightarrow d + e^+ + \nu. \tag{10.2}$$

The energy released in this reaction is

$$Q = (2m_p - m_d - m_e)c^2. \tag{10.3}$$

The mass of the deuteron is $1875.61 \, \text{MeV/c}^2$, and inserting the masses of the proton, $938.27 \, \text{MeV/c}^2$, and electron, $0.511 \, \text{MeV/c}^2$, we find that the Q-value for this reaction is $0.42 \, \text{MeV}$.

A deuteron can fuse with a proton to form ${}^3_2\text{He}$

$$d + p \rightarrow {}^3_2\text{He} + \gamma, \tag{10.4}$$

releasing energy $Q = 5.49 \, \text{MeV}$.

A deuteron can also fuse with another deuteron. We might have expected the fusion product to be the doubly magic nuclide ${}^4_2\text{He}$, i.e.

$$d + d \rightarrow {}^4_2\text{He} + \gamma. \tag{10.5}$$

Because ${}^4_2\text{He}$ has such a large binding energy, this fusion reaction would have a Q-value of $23.8 \, \text{MeV}$. However, this energy is much larger than binding energy of the fourth nucleon, so that either a proton or a neutron will immediately be emitted. The allowed fusion reactions are then

$$d + d \rightarrow t + p \tag{10.6}$$

(the reaction observed in [81]), releasing $4.0 \, \text{MeV}$, or

$$d + d \rightarrow {}^3_2\text{He} + n, \tag{10.7}$$

which releases an energy of 3.3 MeV.

Similarly, the isotope $^5_2$He is very unstable and readily emits a neutron so that the reaction that occurs in the fusion of a deuterium with tritium is

$$d + t \rightarrow {}^4_2\text{He} + n. \tag{10.8}$$

Because of the large binding energy of $^4_2$He, this reaction has a large Q-value of 17.6 MeV.

Heavier nuclei such as carbon or nitrogen can also fuse with a proton. For example,

$$^{15}_7\text{N} + p \rightarrow {}^{12}_6\text{C} + {}^4_2\text{He} . \tag{10.9}$$

The number of neutrons per proton in stable isotopes increases with the atomic number. For this reason, fusion products often possess too few neutrons for stability (this is the opposite of what happens in the case of fission). Fusion is therefore often followed by β^+-decay. For example,

$$^{14}_7\text{N} + p \rightarrow {}^{15}_8\text{O} + \gamma \rightarrow {}^{15}_7\text{N} + e^+ + \nu. \tag{10.10}$$

The isotope $^{15}_7$N is stable, whereas $^{15}_8$O is not.

Alternatively, a fusion product may undergo electron capture. For example,

$$^3_2\text{He} + {}^4_2\text{He} \rightarrow {}^7_4\text{Be} + \gamma$$
$$^7_4\text{Be} + e^- \rightarrow {}^7_3\text{Li} + \nu \tag{10.11}$$

10.2   Kinematics

In order for fusion to take place, the (centres of the) nuclei must be within the range of the strong inter-nuclear forces (a few fm). For this to happen, the (positively charged) nuclei need to overcome a Coulomb barrier, which tends to repel them. They therefore need to have some kinetic energy in order to be able to do this.

In order to be able to conserve momentum, there must be at least two final-state particles (one of which could be a photon). For two final-state particles, we have a reaction

$$1 + 2 \rightarrow 3 + 4.$$

It is most convenient to consider the kinematics in the centre-of-mass (CM) frame in which the incident nuclei have equal and opposite momentum, $\pm p$. In the non-relativistic limit, the magnitude of this momentum is related to the total kinetic energy, $T_{CM} (\equiv T_1 + T_2)$, in the CM frame, by

$$p = \sqrt{2\tilde{m}_{12}T_{CM}}, \tag{10.12}$$

where

$$\tilde{m}_{12} = \frac{m_1 m_2}{m_1 + m_2} \tag{10.13}$$

is the reduced mass of the incident particles whose masses are m_1 and m_2, respectively.

The kinetic energies, T_i $(i = 1, 2)$, of the two incident particles are

$$T_i = \frac{p^2}{2m_i} = \frac{\tilde{m}_{12}}{m_i}T_{CM}, \ (i = 1, 2). \tag{10.14}$$

The fusion process generates supplementary energy Q equal to the difference between the total rest energy of the incident nuclei and that of the fusion product or products. This means that in the CM frame, the total kinetic energy of the fusion products is $T_{CM} + Q$. In most cases, Q is much larger than T_{CM}, but this initial energy nevertheless has to be taken into account.

Now consider the two cases:

1. **Two final-state nuclei**

   In this case, it is sufficient to use non-relativistic kinematics. Since we are in the CM frame, the two nuclei in the final state again have equal and opposite momenta $\pm p'$, whose magnitude is

$$p' = \sqrt{2\tilde{m}_{34}(T_{CM} + Q)}, \tag{10.15}$$

   where $\tilde{m}_{34}$ is the reduced mass of the final-state nuclei whose masses are m_3 and m_4. The kinetic energies, T_j $(j = 3, 4)$, of the two final-state particles are

$$T_j = \frac{p'^2}{2m_j} = \frac{\tilde{m}_{34}}{m_j}(T_{CM} + Q) \ (j = 3, 4). \tag{10.16}$$

2. **One final-state nucleus**

   In this case, the second final-state particle is a photon (treated relativistically). The nucleus has mass m_3 and kinetic energy T_3, whilst the γ-ray has energy E_γ. Thus, for the total energy, we have

$$T_{CM} + Q = T_3 + E_\gamma. \tag{10.17}$$

   The magnitude of the momentum p' of outgoing γ-ray is given in terms of the γ-ray energy by

$$p' = \frac{E_\gamma}{c}. \tag{10.18}$$

In the CM frame, this is also the magnitude of momentum of the final-state nucleus so that in terms of T_3 and m_3, we have the relativistic relation

$$p' = \frac{1}{c}\sqrt{(T_3 + m_3c^2)^2 - m_3^2c^4} = \frac{1}{c}\sqrt{T_3^2 + 2m_3c^2T_3}. \qquad (10.19)$$

After some algebraic manipulation, (10.17), (10.18) and (10.19) lead to

$$E_\gamma = (T_{CM} + Q)\frac{(T_{CM} + Q + 2m_3c^2)}{2(T_{CM} + Q + m_3c^2)} \qquad (10.20)$$

and

$$T_3 = \frac{(T_{CM} + Q)^2}{2(T_{CM} + Q + m_3c^2)}. \qquad (10.21)$$

The rest energy of the nucleus, m_3c^2, is always much larger than the Q-value of the fusion reaction or the initial centre-of-mass energy so that these expressions may be approximated by

$$E_\gamma \approx (T_{CM} + Q), \qquad (10.22)$$

and

$$T_3 \approx \frac{(T_{CM} + Q)^2}{2m_3c^2}. \qquad (10.23)$$

In other words, almost all of the final-state energy is carried off by the γ-ray, with a very small fraction being carried off by the recoil of the final-state nucleus.

We may also wish to express the incoming energy in the centre-of-mass frame in terms of the incoming energy in the *"Lab frame"*, in which one of the initial nuclei (e.g. particle 2) has zero momentum, as is the case in which a stationary target is bombarded with nuclei whose kinetic energy is T_{LAB} (see Fig. 10.1).

In the CM frame, the velocities, v_1 and v_2, of the incoming particles are given by (see (10.14))

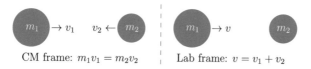

CM frame: $m_1v_1 = m_2v_2$ Lab frame: $v = v_1 + v_2$

Fig. 10.1 Incident particles with masses m_1 and m_2 viewed in the CM frame and the Lab frame (particles moving non-relativistically)

$$v_1 = \frac{\sqrt{2\tilde{m}_{12}T_{CM}}}{m_1}, \quad v_2 = \frac{\sqrt{2\tilde{m}_{12}T_{CM}}}{m_2}. \tag{10.24}$$

In the Lab frame, particle 1 has velocity[2]

$$v_{LAB} = v_1 + v_2,$$

whereas particle 2 has zero velocity. The incoming kinetic energy, T_{LAB}, is then

$$T_{LAB} = \frac{m_1}{2}(v_1 + v_2)^2 = \frac{(m_1 + m_2)^2}{m_1 m_2^2}\tilde{m}_{12}T_{CM} = \frac{(m_1 + m_2)}{m_2}T_{CM} \tag{10.25}$$

(where in the last step we have used (10.13)). Note that in the experiment of Oliphant et al. [81], the target nucleus was embedded in a compound whose mass was much greater than that of the projected deuterons, i.e. $m_2 \gg m_1$, so that the total incident kinetic energy in the CM frame and the Lab frame were almost identical.

The total (kinetic) energy of the fusion products in the Lab frame is $T_{LAB} + Q$, but the distribution of this energy between the fusion products depends on the their direction of motion relative to the direction of the incident nucleus.

10.3   Fusion Rates

The range of the strong nuclear force is a few fermi, and at this distance, the Coulomb potential is around 1 MeV. Therefore, the height of the potential barrier which the incident nucleus needs to overcome in order for fusion to occur is of the order of 1 MeV.

However, nuclei with energies less than this can tunnel through the barrier. This is the reverse process of the quantum tunnelling that occurs in α-decay. The probability for a nucleus with kinetic energy T to tunnel through the barrier is the Gamow factor discussed in Chap. 6 (see 6.18)).

$$\exp\left\{-\sqrt{\frac{E_G}{T}}\right\}.$$

The Gamow energy, E_G, is given by

$$E_G = 2\pi^2\alpha^2 Z_1^2 Z_2^2 \tilde{m}_{12}c^2, \tag{10.26}$$

---

[2]Non-relativistic kinematics is sufficient here.

where Z_1 and Z_2 are the atomic numbers of the two fusion nuclei and $\widetilde{m}_{12}$ is their reduced mass.[3]

To obtain the fusion cross section, we multiply this probability by the collision cross section. If we were to consider the nucleus as a hard sphere, we would estimate its collision cross section to be of the order of the area of a disc of radius equal to the nuclear radius. However, for microscopic objects, the "effective size" of a nucleus with momentum p is of the order of the area of a disk whose radius is equal to the its reduced de Broglie wavelength

$$\lambda = \frac{\hbar}{p}.$$

The collision cross section, σ_c, can therefore be written as

$$\sigma_c \approx \eta_{12}^2 \pi \lambda^2 = \eta_{12}^2 \pi \frac{\hbar^2}{2\widetilde{m}_{12} T_{CM,}}. \tag{10.27}$$

where η_{12} is an overall factor of order unity, which depends on the fusion reaction, and must be fit to experimental data.

The fusion cross section is then estimated to be

$$\sigma_c \approx \eta_{12}^2 \pi \frac{\hbar^2}{2\widetilde{m}_{12} T_{CM}} \exp\left\{-\sqrt{\frac{E_G}{T_{CM}}}\right\}. \tag{10.28}$$

We can see that this cross section diminishes rapidly as the kinetic energy of the incoming nuclei falls far below the energy E_G, which is typically of order 1 MeV.

Fusion can be effected by bombarding a target with accelerated nuclei. Alternatively, they may be confined in a gas at very high temperature T, such as the temperatures found in the core of stars. At these temperatures, the average energy of the fusion reactants is still considerably below the Coulomb barrier height, so that only the very small number of nuclei whose energy is in the tail of the Maxwell–Boltzmann energy distribution can overcome the potential barrier. On the other hand, at these high temperatures, the average kinetic energy of the fusion nuclei, $\sim k_B T$, is sufficient for fusion tunnelling to take place. For example, a temperature of $10^7\ K$ corresponds to an energy of about 1 keV for which the fusion cross section is a few millibarns (mb). At such high temperatures, electrons are stripped from atoms, and the gas consists of free nuclei and free electrons. This is known as a "*plasma*".

The important quantity in determining the fusion rate (the number of fusion events per unit volume per unit time) is the "*fusion reactivity*". This is defined to be the product of the fusion cross section and the relative velocity of the incident fusion

---

[3] Setting $Z_1 = 2$, $Z_2 = Z_D$ and replacing $\widetilde{m}_{12}$ by m_α and T by Q_α, we obtain the leading term in the expression for (the logarithm of) the Gamow factor for α-decay (6.18).

nuclei, $v\sigma_f$. For incident nuclei in a heat bath, this must be averaged over velocities, using the Maxwell–Boltzmann probability, $P(v)dv$, that a particle of mass m in a heat bath of temperature T has a velocity with magnitude between v and $v + dv$

$$P(v) = \sqrt{\frac{2}{\pi}} \left(\frac{m}{k_B T}\right)^{3/2} v^2 \exp\left\{-\frac{mv^2}{2k_B T}\right\}. \tag{10.29}$$

Expressing the kinetic energy in the CM frame in terms of the relative velocity v ($T_{CM} = \frac{1}{2}\widetilde{m}_{12}v^2$), the thermal averaged fusion reactivity $\langle v\sigma_f \rangle_T$ in a plasma at temperature T is

$$\langle v\sigma_f \rangle_T = (\eta_{12}\hbar)^2 \sqrt{\frac{2\pi}{(k_B)^3 \widetilde{m}_{12}}} T^{-3/2} \int_0^\infty v \exp\left\{-\sqrt{\frac{2E_G}{\widetilde{m}_{12}v^2}} - \frac{\widetilde{m}_{12}v^2}{2k_B T}\right\} dv. \tag{10.30}$$

The integral can be performed approximately in the limit $k_B T \ll E_G$ to yield

$$\langle v\sigma_f \rangle_T \approx (\eta_{12}\hbar)^2 \pi \sqrt{\frac{2}{3E_G \widetilde{m}_{12}^3}} 2^{\frac{2}{3}} \epsilon^{-\frac{2}{3}} \exp\left\{-\frac{3}{2^{2/3}}\epsilon^{-\frac{1}{3}}\right\}, \tag{10.31}$$

where ϵ is the dimensionless quantity, $\epsilon \equiv k_B T/E_G$.

This very rough estimate of the reactivity reproduces the temperature dependence fairly well for temperatures below 10^9 K, as can be seen from Fig. 10.2. The fitted

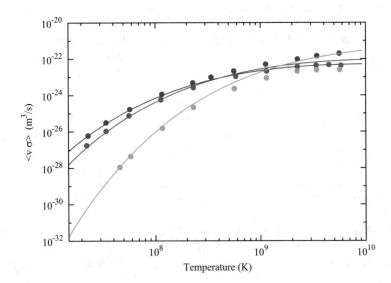

Fig. 10.2 Comparison of estimate for fusion reactivity (thermal average of $v\sigma_f$) given by (10.28) with data [82] for deuteron–deuteron fusion (blue), for tritium–$\frac{3}{2}$He fusion (red), and for tritium–tritium fusion (green)

normalization constants, η_{12}, take the values 0.08 for $d - d$ fusion, 0.5 for $t - {}_2^3\text{He}$ fusion and 0.16 for $t - t$ fusion. The agreement deteriorates at large temperatures, as expected, since the approximation used to perform the integral over velocity in (10.30) is no longer valid.

A nucleus with cross section σ and velocity v sweeps out a volume $v\sigma$ per unit time. If there are n_2 nuclei of particle 2 per unit volume, then nucleus 1 will encounter $n_2 v\sigma$ nuclei of particle 2 per unit time. If the fusion cross section is σ_f and there are n_1 nuclei of particle 1 per unit volume, then there will be $n_1 n_2 v\sigma_f$ fusion events per unit volume per unit time. After taking the thermal average, this means that in a heat bath of temperature T with concentrations n_1 and n_2 of fusion nuclei 1 and 2, respectively, the number of fusion events, N_f, per unit volume per unit time is

$$N_f = \frac{1}{(1 + \delta_{12})} n_1 n_2 \langle v\sigma_f \rangle_T , \tag{10.32}$$

where δ_{12} is zero if the fusion particles are distinct but equal to unity if the fusion particles are identical (e.g. deuteron–deuteron fusion). This factor arises because the number of ways to select a pair from n identical nuclei is $\frac{1}{2}n(n-1)$, which can be approximated by $\frac{1}{2}n^2$, compared with $n_1 n_2$ when the two fusion particles are distinct.

As an example, consider a plasma at a temperature of 10^8 K consisting of deuterons and ${}_2^3\text{He}$ each with a density of 10^{22} m^{-3}. The reactivity for the fusion process

$$d + {}_2^3\text{He} \rightarrow {}_2^4\text{He} + p$$

at $T = 10^8$ K is 2×10^{-25} m^{-3} s^{-1}. The number of fusion events per unit volume per second is

$$N_f = n_1 n_2 \langle v\sigma_f \rangle_T = 10^{22} \times 10^{22} \times \left(2 \times 10^{-25}\right) = 2 \times 10^{19} \text{ m}^{-3} \text{ s}^{-1}.$$

The binding energy of the deuteron is 2.2 MeV, that of ${}_2^3\text{He}$ is 7.7 MeV and the binding energy of ${}_2^4\text{He}$ is 28.3 MeV. The Q-value is then

$$Q = 28.3 - 2.2 - 7.7 = 18.4 \text{ MeV} \equiv 2.9 \times 10^{-12} \text{ J},$$

which is the quantity of energy produced in each fusion reaction.

The power density output, $\mathcal{P}$, of such a plasma is therefore given by

$$\mathcal{P} = 2 \times 10^{19} \times \left(2.9 \times 10^{-12}\right) \text{ W m}^{-3} = 58 \text{ MW m}^{-3}.$$

10.4   Fusion in Stars

A natural environment with temperatures of the order of those discussed above is the core of stars. These stellar cores are sufficiently hot and sufficiently dense for fusion to take place and it is this fusion process that powers the stars. This was originally suggested by Arthur Eddington [83] in 1920, who predicted the phenomenon of nuclear fusion in stars 13 years before it was first observed in a laboratory.

10.4.1   The p–p Cycle

The temperature in the core of the Sun is 1.5×10^7 K. The Sun consists mainly (91%) of hydrogen, and the main source of the Sun's energy comes from the fusion chain in which 4 protons are converted into a nucleus of $^4_2$He in a three-stage process.

$$2 \times (p + p \rightarrow d + e^+ + v) \quad [2 \times 1.42 \, \text{MeV}]$$

$$2 \times \left(d + p \rightarrow {}^3_2\text{He}\right) \quad [2 \times 5.49 \, \text{MeV}]$$

$$^3_2\text{He} + {}^3_2\text{He} \rightarrow {}^4_2\text{He} + p + p + \gamma \quad [12.89 \, \text{MeV}]. \tag{10.33}$$

We see that the total energy released in this process is 26.71 MeV.

Other fusion and β-decay processes are possible. For example, the $^3_2$He produced at stage 2 of (10.33) could fuse with a pre-existing nucleus of $^4_2$He in the reaction

$$^3_2\text{He} + {}^4_2\text{He} \rightarrow {}^7_4\text{Be} + \gamma.$$

The $^7_4$Be converts into $^7_3$Li by electron capture. The $^7_3$Li then fuses with another proton to produce a final state consisting of two nuclei of $^4_2$He. The full fusion scheme is shown in Fig. 10.3 and is called the *"p–p cycle"*, although it is actually a fusion chain rather than a cycle. We note that neutrinos are produced in some stages of this cycle. These are known as *"solar neutrinos"* and are the main source of the huge flux the neutrinos reaching the Earth ($7 \times 10^{14} \, \text{m}^{-2} \, \text{s}^{-1}$).

10.4.2   The CNO Cycle

There are also fusion reactions involving heavier nuclei. The most common of these is the *"CNO cycle"* (carbon–nitrogen–oxygen) shown in Fig. 10.4, which was independently postulated by Hans Bethe and Carl von Weizsäcker [84–86]. The cycle begins with the fusion of a proton with $^{12}_6$C to produce $^{13}_7$N. This is followed by a combination of β-decays and further fusion reactions as shown. The final state

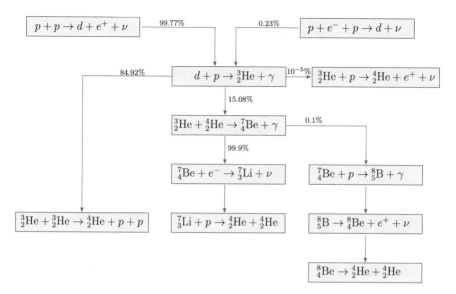

Fig. 10.3 The proton–proton fusion cycle

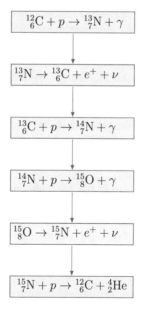

Fig. 10.4 The CNO fusion cycle

consists of the original nucleus of $^{12}_6\text{C}$ and a nucleus of ^4_2He, so that as in the case of the p–p cycle, the net reaction is the fusion of four protons into a ^4_2He nucleus. The carbon acts as a catalyst for this process and is recycled. The energy liberated in this process is 24.73 MeV.

The Coulomb barrier for the fusion of a proton and $^{12}_{6}$C is higher than the initial fusion reaction in the p–p cycle, so that the CNO cycle dominates for stars with higher core temperatures – in excess of 1.7×10^7 K. As the core temperature of the Sun is somewhat below this, the CNO cycle is subdominant and generates only about 1.7% of the total $^4_2$He produced in the Sun.

10.5   Fusion Reactors

For over 60 years, scientists and engineers have been trying to harness nuclear fusion reactions in order to provide a controlled fusion reactor, which can be used to supply energy, in analogy with fission reactors. Such reactors would have huge advantages:

- The fuel would be mainly seawater. Seawater has 1.5×10^{-4} deuterons for each normal hydrogen atom. This must be increased to around 20% in order to be used in a fusion reactor. There are three different ways of enhancing the concentration of heavy water (D_2O).

  - **Electrolysis:** Hydrogen is ionized more easily than deuterium so that during the process of electrolysis more hydrogen nuclei than deuterons drift to the cathode, leaving a solution with enhanced deuterium.
  - **Distillation:** The boiling point of heavy water is slightly higher than that of ordinary water, so that with careful distillation the ordinary water can be boiled off.
  - **Chemical Separation:** The chemical potential of deuterium in a compound differs from that of hydrogen. The difference depends on the compound and the temperature. This can be used to exchange deuterium between one compound and another. The method that is most commonly used is the *"Girdler sulphide process"* in which hydrogen sulphide (H_2S) is passed through water which is kept in hot and cold tanks. As the hydrogen sulphide mixes with water in the hot tank, deuterons from the water replace the hydrogen atoms in the hydrogen sulphide, thereby depleting the heavy water concentration, whereas in the cold tank deuterons from the hydrogen sulphide replace the hydrogen atoms in the water, thereby enriching the heavy water concentration.

  These enrichment processes are not expensive, and there would be sufficient supply for millions of years.
- The process produces very little radioactive waste, although some radioactive nuclides are generated as a result of the interactions of the neutrinos produced in the fusion process with the walls of the reactor.
- There is no carbon emission.

For reasons that are not understood, $d - t$ fusion (10.8) has a reactivity which is about two orders of magnitude larger than $d - d$ fusion. This makes it a far more suitable process for a nuclear reactor, as the plasma does not need to be maintained at such a high temperature. Unlike deuterium, the concentration of tritium in seawater is negligible so that it has to be manufactured. This is achieved by bombarding ^7_3Li with neutrons.

$$^7_3\text{Li} + n \rightarrow {}^4_2\text{He} + t + n. \tag{10.34}$$

The neutrons are produced in the fusion reaction and escape from the plasma. The plasma is therefore enclosed in a blanket of lithium.[4]

The main difficulties in constructing a fusion reactor are maintaining the plasma at high temperature, T, for a sufficiently long period, τ, and maintaining the high particle densities, n_1 and n_2, of the reactants. In addition to the fusion reactants, there will be $Z_1 n_1 + Z_2 n_2$ electrons per unit volume, where Z_1 and Z_2 are the atomic numbers of the reactants, so that the total number of particles per unit volume, n, is

$$n = (Z_1 + 1)n_1 + (Z_2 + 1)n_2.$$

By the principle of equipartition of energy, the energy per unit volume required to heat up a such a plasma to temperature T is

$$E_{\text{in}} = \frac{3}{2} n k_B T. \tag{10.35}$$

If the plasma can be maintained for a period τ, the energy per unit volume produced is

$$E_{\text{out}} = n_1 n_2 \tau \langle v \sigma_f \rangle_T Q, \tag{10.36}$$

where Q is the Q-value of the fusion reaction. For the reactor to be *"exothermic"* i.e. to produce more energy than it consumes, we require $E_{\text{out}} > E_{\text{in}}$ so that

$$n\tau > 12 \xi_{12} \frac{k_B T}{Q \langle v \sigma_f \rangle_T}, \tag{10.37}$$

where

$$\xi_{12} = \frac{4 n_1 n_2}{(Z_1 + 1)n_1 + (Z_2 + 1)n_2}.$$

For the $d - t$ fusion process with equal densities of deuterium and tritium, $\xi_{12} = 1$. The criterion (10.37) is called the *"Lawson criterion"*.

---

[4]Lithium is in plentiful supply and can also be extracted from seawater.

There are two methods which are used for maintaining the necessary high temperature and high density plasma required to satisfy the Lawson criterion:

- **Magnetic Confinement:** The charged particles are confined by means of a magnetic field. The plasma either has the shape of a cylinder with pinched ends with the magnetic field in the axial direction or a toroidal shape using both an axial and a toroidal magnetic field. The latter configuration was invented in the 1950s by Andrei Sakharov and Igor Tamm and is called a *"tokamak"*.
- **Inertial Confinement:** Pellets of fuel are heated up by simultaneous bombardment with laser beams from all directions.

Magnetically confined plasmas can be maintained for up to several seconds, compared with less than a microsecond for inertial confinement. On the other hand, far higher densities can be achieved with inertial confinement.

There are many fusion reactor projects worldwide. There has recently been a substantial increase in investment in fusion reactor research. So far, no fusion reactor has been able to sustain an exothermic process. The ITER[5] project in Bouches, France, predicts that it will achieve a sustained fusion plasma by the end of 2025. Meanwhile, the private company TAE Technology has recently announced that it has been able to sustain a plasma at a temperature of 6×10^7 K and that it will be able to provide commercial fusion power by 2030. The reader should be aware, however, that the promise of fusion power within 20 years has been with us since the 1950s.

Summary

- Two light nuclei can fuse to form a heavier nucleus with a larger binding energy. The momentum is carried off either by a γ-ray or a second outgoing nucleus.
- In order for fusion to occur, the fusion nuclei must tunnel through the potential barrier provided by the Coulomb repulsion between the nuclei.
- The tunnelling probability is given (approximately) by the Gamow factor, as in the case of α-decay.
- The collision cross section between two nuclei is approximately given by the area of a disc whose radius is equal to the reduced de Broglie wavelength. The fusion cross section is the product of the collision cross section and the tunnelling probability.
- Fusion can be effected either by colliding the incident nuclei into each other, or by confining them to a very high density and high temperature plasma (a gas of nuclei and free electrons).

---

[5]International Thermonuclear Experimental Reactor – also Latin for "the path".

- The core of stars is of sufficiently high temperature and sufficiently dense for fusion to take place. This is the process responsible for the energy radiated from stars.
- The dominant process in the Sun is the fusing of 4 protons to $^4_2$He, which proceeds through a chain of fusion and β-decay reactions. The β-decays are the source of a large number of neutrinos emitted from stars. This is called the p–p cycle. A fusion processes involving heavier nuclei, the CNO cycle, which begins with the fusion of a proton and $^{12}_6$C, also occurs in the Sun, but it is subdominant. Fusion processes involving heavier nuclei only dominate in stars with higher core temperature.
- Extracting energy from a controlled fusion reactor would provide a cheap and plentiful source of energy, which is carbon-free and produces very little radioactive waste. In order to achieve an exothermic process, the plasma density, multiplied by the time interval over which the plasma must be maintained, has to exceed a certain value (the Lawson criterion). Deuterium–tritium fusion has the largest reactivity. The tritium is manufactured by bombarding $^7_3$Li with neutrons.
- Plasmas can be confined using a magnetic field. Such magnetically confined plasmas can be maintained for several seconds but can only achieve relatively low densities.
- Inertial confinement plasmas in which pellets of nuclear fuel are heated by lasers can achieve very high densities but can only be maintained for the order of a microsecond.

Problems

Problem 10.1 A proton with kinetic energy 0.87 MeV is incident on a nucleus of $^{12}_6$C at rest. What is the total kinetic energy in the centre-of-mass frame? The proton and $^{12}_6$C fuse into $^{13}_7$N and a γ-ray. The binding energy of $^{12}_6$C is 92.16 MeV and that of $^{13}_7$N is 94.11 MeV. Calculate the kinetic energy of the nitrogen nucleus and the energy of the γ-ray in the centre-of-mass frame.

Problem 10.2 For the fusion of deuterium and tritium, estimate the ratio of the reactivity at a temperature of 10 keV to the reactivity at a temperature of 100 keV.

Problem 10.3 One hydrogen atom in 7000 is deuterium. Calculate the total mass of seawater required to fuel a 500 MW fusion reactor for 1 year (3.15×10^7 s). The reaction is deuterium-tritium fusion, which has a Q-value of 17.6 MeV.
(The mass of a molecule of water is 2.99×10^{-26} kg.)

Chapter 11
Charge Independence and Isospin

11.1 Mirror Nuclides and Charge Independence

Two nuclides that are related by the interchange of protons and neutrons are called *"mirror nuclides"* (or sometimes *"mirror nuclei"*). If we examine two mirror nuclides, we find that their binding energies are almost the same.

In fact, the only term in the semi-empirical mass formula that is *not* invariant under $Z \leftrightarrow (A - Z)$ is the Coulomb term in (3.7) discussed in Chap. 3. The difference between the binding energies of two mirror nuclides is the difference in the Coulomb term, given by

$$\Delta B_{mirror} \equiv B(A, Z) - B(A, A - Z) = -a_C A^{2/3}(2Z - A). \qquad (11.1)$$

This difference is very small in comparison with the total value of the binding energy since inside a nucleus electromagnetic forces are much weaker than the strong inter-nucleon forces (strong interactions). Therefore, mirror nuclides have very similar binding energies despite the extra Coulomb energy for nuclides with more protons.

Not only are the binding energies (and therefore the ground-state energies) very similar for mirror nuclides, but so too are the energies of their excited states. As an example, let us look at Fig. 11.1 that presents the energy levels for mirror nuclides $^{7}_{3}\text{Li}$ and $^{7}_{4}\text{Be}$, where we see that for all the states the energies are very close, with the $^{7}_{4}\text{Be}$ states being slightly higher because it has one more proton than $^{7}_{3}\text{Li}$.

All this suggests that whereas the electromagnetic interactions clearly distinguish between protons and neutrons, the strong interactions, responsible for nuclear binding, are charge independent.

Now let us look at a pair of mirror nuclides whose proton number and neutron number differ by two, together with the isobar between them. The example we take is $^{6}_{2}\text{He}$ and $^{6}_{4}\text{Be}$, which are mirror nuclides. Each of these has a closed shell of two protons and a closed shell of two neutrons. The unclosed shell consists of

© The Author(s), under exclusive license to Springer Nature Switzerland AG 2021
A. Belyaev, D. Ross, *The Basics of Nuclear and Particle Physics*, Undergraduate
Texts in Physics, https://doi.org/10.1007/978-3-030-80116-8_11

Fig. 11.1 Energy levels for
the mirror nuclides $^7_3$Li and
$^7_4$Be

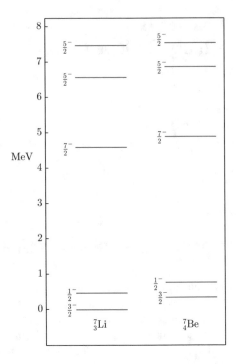

two neutrons for $^6_2$He and two protons for $^6_4$Be. The nuclide "between" is $^6_3$Li, which has one proton and one neutron in the outer shell.

Using only the principle of charge independence of the strong interactions, we would have expected all three nuclides to display the same energy-level structure. We see from Fig. 11.2 that although there are states in $^6_3$Li that are close in energy to the states of the mirror nuclides $^6_2$He and $^4_2$Be, there are also states in $^6_3$Li that have no equivalent in the two mirror nuclides. This difference can be understood from the Pauli exclusion principle. In the case of $^6_2$He or $^4_2$Be two protons or alternatively two neutrons in the ground state in the outer shell cannot have the same spin state. The only possible spin state for them is the configuration with antiparallel spins, whereas for $^6_3$Li, in which the nucleons in the outer shell are not identical, the Pauli principle does not apply and there are extra states in which the neutron and proton are in the same spin state, i.e. the configuration with parallel spins.

Further exploration allows us to conclude that nuclear forces are charge independent between any two nucleons (*pp*, *nn* or *np*) that are in the same spin and parity state. In other words, the neutron and proton are identical particles from the "point of view" of nuclear force, provided the Pauli exclusion principle is respected, as required.

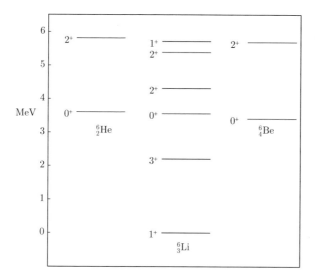

Fig. 11.2 Energy levels for $^6_2$He, $^7_3$Li and $^6_4$Be. The levels in blue are the $^7_3$Li levels, which are absent for $^6_2$He or $^6_4$Be, because they would require wavefunction that is symmetric under the interchange of two identical nucleons

11.2 Isospin

This strong-interaction feature of the proton and neutron can be described with a formal (mathematical), but useful quantum property. This property is the *"isotopic spin vector"* and the corresponding quantum number is *"isotopic spin"* or *"isospin"*. As we will see later, the concept of isospin plays an important role not only in Nuclear Physics but also in Particle Physics.

Let us consider the analogy between spin and isospin. If we have two electrons with z-component of spin set to $s_z = +\frac{1}{2}$ and $s_z = -\frac{1}{2}$ (in units of $\hbar$), then we can distinguish them by applying a (non-uniform) magnetic field in the z-direction – the electrons will move in opposite directions.[1] But in the absence of this external field these two spin orientations cannot be distinguished and we are used to thinking of these as two states of the same particle.

Similarly, if we could "switch off" electromagnetic interactions, we would not be able to distinguish between a proton and a neutron. As far as the strong interactions are concerned these are just two states of the same particle – a nucleon. One can therefore think of an imagined (internal) space in which the nucleon has an isospin, which is mathematically analogous to spin. The isospin vector is a vector in this internal space, known as *"isospace"*. This vector is an abstract object, in

---

[1] This was proved in the Stern–Gerlach experiment, which demonstrated that the spatial orientation of angular momentum is quantized [87, 88].

contrast with angular momentum in real space. Isospin has a quantum description that is mathematically identical to the description of angular momentum, except that isospin is a dimensionless quantity, which is not associated with any type of angular momentum. The proton and neutron are now considered to be the same particle – i.e. a nucleon with different values of the third component of isospin.

Since this third component can take two possible values, we assign $I_3 = +\frac{1}{2}$ for the proton and $I_3 = -\frac{1}{2}$ for the neutron. The nucleon therefore has isospin $I = \frac{1}{2}$, in the same way that the electron has spin $s = \frac{1}{2}$, with two possible values of the third component. As far as the strong interactions are concerned this just represents two possible quantum states of the same particle. If there were no electromagnetic interactions, these particles would be totally indistinguishable in all their properties – mass, spins, *etc.*

In analogy with conservation of angular momentum, isospin is conserved in any transition mediated by the strong interactions. This is an example of the symmetry that is called "isotopic invariance" or just "isospin invariance". Inside the nucleus the strong forces between nucleons do not distinguish between particles with different third components of isospin and would lead to identical energy levels, but there are electromagnetic interactions that break this symmetry and lead to small differences in the energy levels of mirror nuclides. In this respect isotopic invariance is an approximate symmetry.

The electromagnetic interactions are determined by the electric charge,[2] Q, of the particles, and in the case of nucleons this electric charge is related to the third component of isospin by

$$Q = I_3 + \frac{1}{2}. \tag{11.2}$$

Other particles can also be classified as isospin multiplets. For example there are three pions, $\pi^+ \pi^0$, π^- (discussed in Chap. 12), which have almost the same mass and zero spin *etc.* There are three of them with different charges but they behave similarly under the influence of the strong interactions. Therefore, three pions form an isospin multiplet with $I = 1$ and three possible third components, namely $+1$, 0, -1. In the case of pions the electric charges are equal to I_3. This comes from more general formula for the charge Q:

$$Q = I_3 + \frac{Y}{2}, \tag{11.3}$$

where Y is a property of particles called *"hypercharge"*, which will be defined in Chap. 15. For now we need to know that nucleons have $Y = 1$ and pions have

---

[2]In Nuclear and Particle Physics, electric charge is usually quoted in nuclear units for which the charge of the proton is unity.

$Y = 0$. The concept of isospin is also applied to weak interactions and for this purpose leptons are assigned a hypercharge $Y = -1$.

Particles that are members of an isospin multiplet have the same properties, with the exception of their electric charge, i.e. they have the same spin and almost the same mass (the small mass differences being due to the electromagnetic interactions that are not isospin invariant). We will see later that particles can have other properties (called "*strangeness*", "*charm*", etc.) and members of an isospin multiplet will have the same values of these properties as well.

In the same way that two electrons can have a total spin $S = 0$ or $S = 1$, two nucleons can have a total isospin $I = 0$ or $I = 1$. The concept of isospin can be generalized for nuclides (A, Z) as follows:

$$I_3 = \frac{Z - N}{2} = \frac{2Z - A}{2} \tag{11.4}$$

$$I \leq \frac{Z + N}{2} , \tag{11.5}$$

where N is the number of neutrons in the nuclide. In analogy with angular momentum, the allowed values of I_3 for a multiplet with isospin I are given by

$$-I \leq I_3 \leq I \text{ (in integer steps).}$$

For two electrons we may write the total wavefunction as

$$\Psi_{12} = \Psi(\mathbf{r}_1, \mathbf{r}_2)\chi(s_1, s_2), \tag{11.6}$$

where $\chi(s_1, s_2)$ is the spin part of the wavefunction.
For $S = 1$ the spin part of the wavefunction is

$$\chi(s_1, s_2) = |\uparrow\uparrow\rangle, \quad S_z = +1$$
$$\chi(s_1, s_2) = \frac{1}{\sqrt{2}}(|\uparrow\downarrow\rangle + |\downarrow\uparrow\rangle), \quad S_z = 0$$
$$\chi(s_1, s_2) = |\downarrow\downarrow\rangle, \quad S_z = -1, \tag{11.7}$$

which is symmetric under interchange of the two spins. Since by Fermi–Dirac statistics the total wavefunction must be antisymmetric under exchange of the two electrons, the spatial part of the wavefunction must then be antisymmetric under the interchange of the positions of the electrons, i.e.

$$\Psi(\mathbf{r}_1, \mathbf{r}_2) = -\Psi(\mathbf{r}_2, \mathbf{r}_1). \tag{11.8}$$

For the case of $S = 0$,

$$\chi(s_1, s_2) = \frac{1}{\sqrt{2}} (|\uparrow\downarrow\rangle - |\downarrow\uparrow\rangle), \tag{11.9}$$

which is antisymmetric under interchange of spins so that it must be accompanied by a symmetric spatial part of the wavefunction

$$\Psi(r_1, r_2) = +\Psi(r_2, r_1). \tag{11.10}$$

In the case of nucleons the complete wavefunction also includes its isospin part, $\chi_I(I_1, I_2)$, and takes the form:

$$\Psi_{12} = \Psi(r_1, r_2)\chi_S(s_1, s_2)\chi_I(I_1, I_2). \tag{11.11}$$

For isospin of the two-particle system $I = 1$ we have

$$\chi_I(I_1, I_2) = |p\,p\rangle, \quad I_3 = +1$$
$$\chi_I(I_1, I_2) = \frac{1}{\sqrt{2}} (|p\,n\rangle + |n\,p\rangle), \quad I_3 = 0$$
$$\chi_I(I_1, I_2) = |n\,n\rangle, \quad I_3 = -1, \tag{11.12}$$

which is symmetric under the interchange of the isospins of the two nucleons, so that it must be accompanied by a combined spatial and spin wavefunction that must be antisymmetric under simultaneous interchange of the two positions *and* the two spins.

On the other hand, the state with $I=0$,

$$\chi_I(I_1, I_2) = \frac{1}{\sqrt{2}} (|p\,n\rangle - |n\,p\rangle), \tag{11.13}$$

is antisymmetric under the interchange of the two isospins. Therefore, when the nucleons are combined in this isospin state, they must be accompanied by a combined spatial and spin wavefunction that is *symmetric* under simultaneous interchange of the two positions *and* the two spins.

Now we can apply this concept to the three nuclei $^6_2$He and $^6_4$Be and $^6_3$Li that we discussed previously. Their closed shells of neutrons and protons have a total isospin zero so we do not need to consider these in determining the isospin of the nuclei. We note that $^6_2$He has two neutrons in the outer shell so its isospin must be $I = 1$, with $I_3 = -1$, whereas the $^6_4$Be has two protons in the outer shell so its isospin must be $I = 1$, with $I_3 = +1$. Since both these nuclides are in an $I = 1$ state, i.e. the isospin part of their wavefunction is symmetric, the remaining part of the wavefunction (spatial and spin parts) must be antisymmetric under simultaneous interchange of the two positions and the two spins. On the other hand, the nucleus $^6_3$Li has one proton and one neutron in the outer shell and can therefore be *either* in an $I = 1$ state like the other two nuclei *or* in an $I = 0$ state, which is not possible for the other two. The strong interactions will give rise to different energy levels depending on the total isospin of the nucleons in the outer shell (in the same way that atomic energy levels depend on the total angular momentum J). Thus we see that two of the states shown for ${}^6_3\text{Li}$ can be identified as $I = 1$ states and they approximately match states for the other two nuclei. These states provide the $I_3 = 0$ component of the $I = 1$ multiplets. The other states are $I = 0$ states and have wavefunctions that are symmetric under the simultaneous interchange of the positions and the spins of the two nucleons in the outer shell that have no counterpart in ${}^6_2\text{He}$ or ${}^6_4\text{Be}$.

The fact that the ground states of ${}^6_2\text{He}$ and ${}^6_4\text{Be}$ have spin zero and the ground state of ${}^6_3\text{Li}$ has spin one can be deduced from the isospin of these ground states. For ground-state wavefunctions the orbital angular momentum, ℓ, is zero, and since the symmetry of the spatial part of the wavefunction is given by $(-1)^\ell$, this means that the spatial part of the wavefunction is symmetric under the interchange of the positions of the two nucleons in the outer shell. We know that the overall wavefunction for the two nucleons in the outer shell must be antisymmetric under interchange because the nucleons are fermions. It therefore follows that the isospin part and the spin part of the wavefunction must have opposite symmetry. Thus for the ground states of ${}^6_2\text{He}$ and ${}^6_4\text{Be}$, which are in $I = 1$ (symmetric) isospin states, the spin part of the wavefunction must be antisymmetric, and therefore, the spins of the two outer shell nucleons must combine to give spin $S = 0$. On the other hand, for the ground state of ${}^6_3\text{Li}$, which[3] is in an $I = 0$ (antisymmetric) isospin state, the spin part of the wavefunction must be symmetric and therefore the spins of the two outer shell nucleons must combine to give spin $S = 1$.

Summary

- The similarity of binding energies and energy levels with the same spin and parity of mirror nuclides leads to the conclusion that the strong interactions that bind nucleons in a nucleus do not distinguish between protons and neutrons.
- This strong interaction of protons and/or neutrons can be described with isotopic spin vector ($\boldsymbol{I}$) and the corresponding quantum number – "isotopic spin" or "isospin". According to this concept, the proton and neutron are considered to be the same particle (a nucleon) with different values of the third component of isospin, I_3: $I_3 = +\frac{1}{2}$ for the proton and $I_3 = -\frac{1}{2}$ for the neutron.
- The idea of isospin is important for both Nuclear and Particle Physics and will be used in the following chapters.

---

[3]The isospin of the ground state of ${}^6_3\text{Li}$ could be either $I = 0$ or $I = 1$. It is known from experiment that in the ground state has isospin. $I = 0$ – there are no corresponding states for ${}^6_2\text{He}$ or ${}^6_4\text{Be}$ with approximately the same energy.

- In analogy with angular momentum, nuclear states or elementary particles can be classified in multiplets with isospin I whose third component I_3 ranges from $-I$ to $+I$. The electric charge of a member of such a multiplet increases with an increasing value of I_3.
- The properties of states of nuclides can be deduced by treating the nucleons as identical particles and incorporating a factor into the wavefunction, which accounts for the isospin of the states. For example, for two nucleons in an $I = 1$ (isotriplet) state, the isospin part of the wavefunction is symmetric, so that the overall antisymmetry of the wavefunction implies that the remaining part of the wavefunction must be antisymmetric, whereas for the $I = 0$ (isosinglet) state the isospin part of the wavefunction is antisymmetric, implying that the remaining part of the wavefunction must be symmetric.

Problems

Problem 11.1 What are the allowed isospins of a state consisting of two pions?

Problem 11.2 There are no known bound states of two protons or of two neutrons, but deuterium – a bound state of a proton and a neutron – has a binding energy of 2.2 MeV. What is the isospin of deuterium? What is the spin of a deuteron in its ground state (the proton and neutron are in the $1s$ state)?

Problem 11.3 What are the third components of isospin of the nuclides

$$\substack{6\\2}\text{He}, \quad \substack{6\\3}\text{Li}, \quad \substack{6\\4}\text{Be ?}$$

Problem 11.4 Find the spin, parity and third component of isospin for $^{13}_{7}\text{N}$, $^{13}_{6}\text{C}$ and $^{13}_{5}\text{B}$ in the ground, using Shell Model predictions.

Part II
Particle Physics

Chapter 12
The Forces of Nature and Particle Classification

12.1 What Is Particle Physics?

Particle Physics explores the properties of elementary particles and their interactions with each other through the fundamental forces of Nature. The only known stable particles are the proton, neutron, electron and neutrino. Many other particles exist, but they decay very rapidly to the stable particles (possibly in stages). Such decays are energetically allowed since the particles are more massive than the stable particles. Because of their instability, these particles are not found naturally but can be produced in experiments using incident particles at sufficiently high energies, i.e. energies exceeding the rest energy of these massive particles. For this reason, Particle Physics is also called *"High Energy Physics"* (HEP).

The most massive particles that have been discovered so far are the top-quark with a mass of $173 \, \text{GeV}/c^2$, the Higgs boson with a mass of $125 \, \text{GeV}/c^2$, the Z-boson with a mass of $91.2 \, \text{GeV}/c^2$ and the W-boson with a mass of $80.4 \, \text{GeV}/c^2$. The masses of these particles are measured with accuracy better than 1%. All these particles are around 100 times heavier than the proton. So we need really high energies in order to produce them.

Another way of seeing that we need high energies is to note that we need to probe very short distances in order to explore the properties and possible substructure of known particles – to find out if they are truly elementary, i.e. do not consist of smaller constituents.

At the very least we want to probe distances that are small compared with a typical nuclear radius, i.e.

$$x \ll 1 \, \text{fm} = 10^{-15} \, \text{m}.$$

In order to do this the uncertainty in the position, Δx, must be much smaller than 1 fm, and by Heisenberg's uncertainty principle

© The Author(s), under exclusive license to Springer Nature Switzerland AG 2021
A. Belyaev, D. Ross, *The Basics of Nuclear and Particle Physics*, Undergraduate
Texts in Physics, https://doi.org/10.1007/978-3-030-80116-8_12

$$\Delta x \Delta p \geq \hbar/2 \tag{12.1}$$

the uncertainty in momentum Δp must obey the inequality

$$\Delta p \gg \frac{\hbar}{1\,\text{fm}} = 197\,\text{MeV/c}. \tag{12.2}$$

This in turn means that the momenta (and consequently the energy) of the particles used as a probe must be much larger than this value.

In fact, the weak interactions have a range that is about three orders of magnitude shorter than this and so particles used to investigate the mechanism of weak interactions have to have energies of at least 100 GeV.

12.2   Forces of Nature

There are four forces of Nature:

- **Strong interactions:** These are strong, but short-range interactions, which bind nucleons together in a nucleus and generate interactions between nuclei in various nuclear reactions.[1] The quantum theory of strong interactions is called *"Quantum Chromodynamics (QCD)"* and will be discussed in Chap. 20.
- **Electromagnetic interactions:** These are interactions between charged particles, which couple to electric fields. Neutral particles can also participate in electromagnetic interactions through their magnetic dipole moments that couple to magnetic fields. The classical theory of Electromagnetism is Maxwell's theory of electromagnetic interactions. The corresponding quantum theory is called *"Quantum Electrodynamics (QED)"* and was developed in the late 1940s by Richard Feynman [89], Julian Schwinger [90] and Schin'ichirō Tomonaga [91].
- **Weak interactions:** These are very short-range interactions that are responsible for β-decay. The quantum theory that describes the weak interactions is called the *"electroweak theory"* or *"Quantum Flavourdynamics"*, which is a theory describing both the weak and electromagnetic interactions. The electroweak theory was developed in the 1960s by Sheldon Glashow [92], Abdus Salam [93], and Steven Weinberg [94].
- **Gravitational interactions:** Gravity affects all particles. However, it is the weakest of the fundamental forces. At the classical level, gravity is explained by General Relativity. Although much progress has been made in the last 35 years, mainly by the development of String Theory, a complete quantum theory of gravity has yet to be fully formulated.

---

[1] The strong force also binds quarks, the constituents of nucleons, and mediate interactions between other strongly interacting particles.

12.3 Particle Classification

Apart from the force carriers associated with these forces of Nature (discussed below), there are two categories of directly observed particles:

12.3.1 Leptons

Derived from the Greek word *leptos* meaning small, these are particles that only participate in weak and possibly electromagnetic interactions. The (negatively charged) electron is an example of a lepton and also the electrically neutral (anti)neutrino, which is created in conjunction with the electron in weak-interaction processes. Both of these particles are fermions and have intrinsic spin $\frac{1}{2}$.

In addition to the electron and its neutrino, Nature has provided us with two more pairs of leptons. These have the same properties as the electron and neutrino, but the charged lepton has a much larger mass. The second charged lepton is the muon (μ), whose mass is $106\,\text{MeV/c}^2$ (207 times that of the electron). The third charged lepton is the tau-lepton (τ), whose mass is $1.78\,\text{GeV/c}^2$ (3484 times that of the electron). Each charged lepton has associated with its own distinct neutrino with negligible mass. For this reason neutrinos are labelled ν_e, ν_μ, ν_τ, to indicate which charged lepton they are associated with. We refer to these three different pairs of leptons as leptons with a different *"lepton flavour"*. Until now it was assumed that apart from the masses these three charged leptons behaved in an identical way. However, recent data [95] on the decay of a meson known as a *"B-meson"* show a discrepancy between the decay rates into a final state containing an electron–positron pair and the decay rate into a similar final state but containing a $\mu^+\mu^-$ pair.

The heavier charged leptons can decay into lighter leptons, via the weak interactions, in such a way that only leptons appear in the final state. In order for (spin) angular momentum to be conserved, there must be an odd number of leptons in the final state. However, lepton number is conserved,[2] meaning that one of the particles in the final state must be an antilepton. Moreover, it turns out that lepton flavour is also conserved, so that when a heavy charged lepton decays, the final state contains a neutrino of that particular lepton flavour.

The heavy lepton flavour decay modes are therefore

$$\mu^- \to e^- + \nu_\mu + \bar{\nu}_e$$

$$\tau^- \to e^- + \nu_\tau + \bar{\nu}_e \tag{12.3}$$

or

$$\mu^- + \nu_\tau + \bar{\nu}_\mu.$$

---

[2] So far, there are no data that contradict this conservation law.

If lepton flavours were not conserved, then decays with a photon instead of the neutrino and antineutrino such as

$$\mu^- \; \rightarrow \; e^- + \gamma,$$

would be possible. This decay mode has been extensively searched, but so far no such decays have been found.

12.3.2   Hadrons

Non-elementary particles[3] that interact strongly are called *"hadrons"*. These are further divided into two sub-categories:

- *"Mesons"*: These are bosons with integer spin. The lightest of these are pions (π), which come with one of three charges: $\pi^\pm$ and π^0. The mass of the π^0 is 135 MeV/c^2, which is slightly less than the mass of the $\pi^\pm$, whose mass is 140 MeV/c^2.

  The $\pi^\pm$ decays via the weak interactions into leptons. The decay mode is nearly always

$$\pi^\pm \; \rightarrow \; \mu^\pm + \nu_\mu(\bar{\nu}_\mu), \tag{12.4}$$

with a mean lifetime of 8.5×10^{-8} s. This is regarded as a "long-lived" particle.[4]

The π^0 usually decays by the electromagnetic interactions into two photons. As the electromagnetic interactions extend over much longer distance than the weak interactions, electromagnetic reaction rates are much greater so that the π^0 has a much shorter lifetime than the $\pi^\pm$. The lifetime of the π^0 is only 8.5×10^{-17} s.

Since the discovery of the pion in 1947 [96], more than 200 additional mesons have been identified, with masses considerably larger than the pion. Often, these mesons decay to pions or other lighter mesons rather than directly to leptons. Lepton number has to be conserved, so that whenever there is a lepton in the final state, there is also an antilepton. A decay into leptons only is called a *"leptonic decay"*, whereas a decay into mesons plus leptons is called a *"semi-leptonic decay"*, and a decay into mesons only is called a *"hadronic decay"*. For example, the meson K^+ can decay to one of several different final states (known as *"decay channels"*) amongst which are

---

[3]Composite particles constructed out of quarks, discussed in Chap. 15.

[4]Particles that live long enough to produce a track of macroscopic length are conventionally called "long-lived".

Table 12.1 Classification of particles

Particle type		Strong interaction	Weak interaction	Electromagnetic interaction	Spin
Leptons		No	Yes	Some	$\frac{1}{2}$
Hadrons	Mesons	Yes	Yes	Yes	Integer
	Baryons	Yes	Yes	Yes	Half odd-integer

$$K^+ \rightarrow \mu^+ + \nu_\mu \text{ (leptonic)}$$
$$\rightarrow \pi^0 + e^+ + \nu_e \text{ (semi-leptonic)}$$
$$\rightarrow \pi^+ + \pi^0 \text{ (hadronic)}. \qquad (12.5)$$

• *"Baryons"*: The name baryon comes from the Greek *barys* meaning heavy. Protons and neutrons are examples of baryons. Originally, the known baryons were heavier than the known mesons. The lightest baryons (the proton and neutron) have a mass more than six times greater than the lightest mesons – pions. The distinctions between baryons and mesons are:

  – Baryons have half odd-integer spin. Most known baryons have spin $\frac{1}{2}$ (such as protons and neutrons) or spin $\frac{3}{2}$, but some baryons have been identified with larger spins – always half odd-integer. Baryons are fermions, whereas mesons are bosons.
  – Unlike mesons, the number of baryons (baryon number) is conserved.[5] A baryon can therefore decay into a lighter baryon plus one or more mesons and possibly leptons, but there must be a baryon in the final state.

The classification of particles is summarized in Table 12.1.

12.3.3 Antiparticles

All particles have antiparticles, with exactly the same mass, intrinsic spin and lifetimes but the opposite values of other properties (quantum numbers) such as electric charge. For example, the positron has exactly the same mass and spin as the electron but with opposite charge and lepton number. Some neutral mesons, such as the π^0, are their own antiparticle, whereas the antiparticle of the π^+ is the π^- (note that π^+ and π^- have the same mass and lifetime).

---

[5]There exist theories in which this conservation law is violated, but so far there is no experimental evidence for any such theories.

12.4   Particles and Fields

In order to perform calculations for the interactions of high energy particles we need to use Quantum Field Theory, which synthesizes Quantum Physics and Special Relativity. This is way beyond the scope of this book, but it is helpful to have a qualitative understanding of the basic idea behind that theory.

Let us consider the photon. This is a quantum of an electromagnetic wave – a travelling oscillation of an electric and magnetic field. Therefore, we can view a photon as a quantum excitation of an electromagnetic field. The energy of the photon is the energy of the excitation of the electromagnetic field. The energy density at a particular point in space at a particular time is proportional to the square of the amplitude of the wave motion of the electric or magnetic field.[6] In terms of a particle with a given energy, whose exact location is uncertain, this means that the square of the amplitude at a given point is proportional to the probability that the photon is located at that point. Conversely, the electromagnetic field has a particle (the photon) associated with it. These particles have energy equal to the quantized excitation energies of the electromagnetic field.

For a state with no photons, the energy of the electromagnetic field is zero, implying that the magnitude of the electric or magnetic field is zero. With increasing number of photons with a given frequency, the amplitude of the electric or magnetic field wave with that frequency increases, such that the energy stored in the electromagnetic field is equal to the total energy of the photons.

All this also applies to particles other than the photon. In the same way that a photon has a field associated with it, every particle has its own field associated with it. A state with a particle of a given momentum implies an excitation of a wave motion of the corresponding field with a wavelength equal to the de Broglie wavelength of the particle. Conversely, these fields have particles associated with them. The energies of these associated particles are equal to the quantized excitation energies of the field. Particles cannot exist without fields associated with them and fields cannot exist without particles associated with them.

The reason why we are not familiar with the fields associated with massive particles, whereas we are familiar with the electromagnetic field, is that the field associated with a massless particle is long range, so that it can be observed directly in a laboratory. The electric field, say, between the plates of a capacitor can be viewed as the exchange between the capacitor plates of a large number of photons with wavelengths that are of the order of or larger than the separation of the plates. The field associated with a particle of mass m is attenuated over a range, r, which is of the order of the Compton wavelength of the particle, $r \approx 2\pi\hbar/mc$. This is approximately 400 fm in the case of an electron and smaller for more massive particles. Fields with such a small attenuation range cannot be observed directly.

---

[6]The amplitudes of the electric and magnetic fields are related by Maxwell's equations.

The wave motion of a field associated with a given particle is closely related (not quite identical) to the de Broglie wave (or "matter wave"). This means that the square of the amplitude of field's wave motion at any given point is proportional to the probability that the particle is located at that point. So if the uncertainty in the position of a particle is Δx, the wave is a wavepacket of width Δx and the probability of finding the particle at a point $\mathbf{x}$ is proportional to the square of the amplitude of the wave at $\mathbf{x}$.

A state in which a given particle is absent implies that the amplitude of oscillation of the field corresponding to that particle is zero[7] – i.e. the magnitude of the field is zero. The one exception is the Higgs field associated with the Higgs boson, which we will discuss in detail in Chap. 18.

12.5   Force Carriers

A force has a field associated with it and, as discussed above, every field has a particle associated with it. Thus we see that each force has a particle associated with it, which is a *"force carrier"*. With the exception of gravity, these force carriers have intrinsic spin one (gravity is mediated by gravitons that have spin two). For reasons that we will not go into, they are more normally called *"gauge bosons"*.

The potential, $V_f(r)$, of a force of type f decreases with distance r from the source of the force as

$$V_f(r) \propto \frac{e^{-M_f c/r\hbar}}{r}, \tag{12.6}$$

where M_f is the mass of the corresponding force carrier. For forces whose gauge bosons have a non-zero mass, the potential has an exponential attenuation in addition to the relatively mild $1/r$ dependence. The range of the force is equal to the Compton wavelength of the gauge boson.

- **Electromagnetic force carriers:** The gauge boson of electromagnetic interactions is the photon. Since the photon is massless (it travels with the speed of light), electromagnetic interactions are long range and there is no exponential attenuation of the potential.
- **Weak-interaction force carriers:** Weak interactions can change the electric charge of the particle undergoing the weak interaction. For example the conversion of a neutron into a proton in a β-decay, generated by the weak interactions. In order to conserve electric charge, the gauge bosons themselves carry electric charge $\pm e$. They are called "$W^\pm$-bosons". The W-boson mass is $80.4\,\text{GeV/c}^2$.

---

[7]What is really meant here is that all the wave oscillations of the field are in their ground state, so that there are no excitations. It is the quantum expectation value of the field that is zero for states with no particles.

These interactions are extremely short range – of the order of the Compton wavelength of the W-boson, which is about 2.5×10^{-3} fm.

In addition to weak-interaction processes in which the interacting particle changes electric charge, there are weak interactions in which there is no change in electric charge. Such neutral weak interactions are mediated by a neutral gauge boson called the "Z-boson", whose mass, $91.2\,\mathrm{GeV}/c^2$, is a little greater than that of the charged W-boson.

• **Strong interaction force carriers:** The gauge bosons of the strong interaction are called "*gluons*" and will be discussed in more detail in Chap. 20. These gluons have a peculiar property, namely that their mass is zero, but they are "confined", meaning that they can only propagate over short distances (a few fm – related to the Compton wavelength of pions, which are the lightest bound states of the strong interactions). This means that the strong interactions are short range, despite the gluons having zero mass. The exact mechanism that leads to this property ("*confinement*") is not fully understood, but it is associated with the fact that the interactions mediated by gluons become stronger as the interacting particles move further apart.

• **Gravity force carriers:** Finally, there is the "*graviton*". Gravitational interactions are long range (galaxies are known to interact with each other through gravity), so the graviton is massless like the photon. The graviton is distinguished from the other gauge bosons in that it carries intrinsic spin two. The recent discovery of gravitational waves [97] has confirmed the existence of gravitons and has also verified Einstein's General Theory of Relativity.

There is one further particle, which does not fit into any of these classes. This is the spin zero Higgs boson. This particle, whose properties will be discussed in detail in Chap. 18, is responsible for generating the masses of all particles with which it interacts.

12.6 Particle Decays

As mentioned above, all of these particles with the exception of the proton, the neutron, the electron and the neutrinos are unstable. They are massive particles that have sufficient rest energy to decay into two or more stable light particles, and sometimes also photons.

12.6.1 Long-Lived Particles

Some of these particles are "long-lived", meaning that their lifetimes[8] exceed 10^{-13} s. Particles with such "long" lifetimes can leave a visible track in a type of detector called a *"tracker"*, described in Chap. 14. The particles typically have energies much larger than their rest energy, meaning that they are moving with a velocity very close to the speed of light. For example, if they have an energy 100 times larger than their rest energy, then

$$\gamma \equiv \frac{1}{\sqrt{1 - v^2/c^2}} \sim 100.$$

A particle with a lifetime of $\tau = 10^{-13}$ s, moving at such a high speed leaves a track of length l, where

$$l = \gamma c \tau \approx 3 \, \text{mm},$$

which can easily be detected and measured in modern tracking devices.

12.6.2 Resonances

On the other hand, particles with lifetimes much shorter than 10^{-13} s ("short-lived" particles) decay far too fast to leave an identifiable track in a detector. Instead, their lifetime has to be deduced from their *"decay width"*, Γ.

   If an unstable particle of mass M is produced in a scattering process in which the incident particles annihilate each other, and then decays almost immediately, the cross section has a peak at a CM energy $E = Mc^2$, where all of the incoming energy goes into producing that particle at rest. This peak is known as a *"resonance"*. However, if the particle only lives for a very short time, τ, the peak is *not* a perfectly sharp peak but is spread out over an energy range Γ, reflecting the fact that from Heisenberg's uncertainty principle (in terms of time and energy) a process that occurs over a short period, τ, leads to an energy uncertainty, Γ, where

$$\Gamma = \frac{\hbar}{\tau} \equiv \hbar \lambda, \tag{12.7}$$

λ being the decay rate. More precisely, in the region $E \sim Mc^2$, the cross section $\sigma(E)$, as a function of CM energy, E, is proportional to the *"Breit–Wigner distribution"*, $f(E)$:[9]

---

[8]In Particle Physics one usually talks about a lifetime, which is the inverse of the decay rate, rather than a half-life. This is the time at which the probability that the particle has *not* decayed is $1/e$.
[9]The motivation for this energy dependence will be discussed in Chap. 16

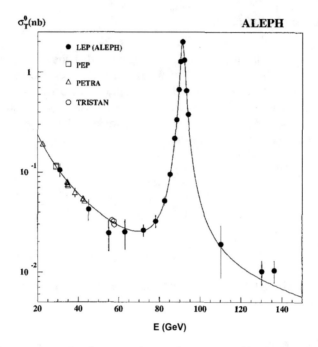

Fig. 12.1 Cross section at for e^+e^- scattering to $\mu^+\mu^-$, measured by the ALEPH collaboration at LEP. E is the total energy of the incident particles in the CM frame. [Reproduced from [98] by permission from Elsevier]

$$f(E) = \frac{1}{\left(E - Mc^2\right)^2 + \frac{1}{4}\Gamma^2}. \qquad (12.8)$$

We see from (12.8) that this distribution has a peak at $E = Mc^2$ and falls to one-half of this value at $E = Mc^2 \pm \frac{1}{2}\Gamma$. The energy interval between the two values of E for which the cross section is one-half of its peak value is called the *"full width at half maximum (FWHM)"* and is equal to Γ.

As an example, consider the cross section for the process

$$e^+ + e^- \rightarrow \mu^+ + \mu^-,$$

measured by the Aleph collaboration at the Large Electron–Positron Collider (LEP), shown in Fig. 12.1. The cross section has a peak at $E = 91.2\,\text{GeV}$. This is due to the production and almost immediate decay of the Z-boson. The two-stage process is

$$e^+ + e^- \rightarrow Z \rightarrow \mu^+ + \mu^-.$$

The value of the cross section at the Z-peak is 2.00 nb, so that in the region $E \sim$ 91.2 GeV, the cross section may be written as

$$\sigma(E) = \frac{2.00 \, [\text{nb}] \times \frac{1}{4} \Gamma_Z^2}{(E - 91.2 \, [\text{GeV}])^2 + \frac{1}{4} \Gamma_Z^2}. \tag{12.9}$$

Inserting the measured values of the cross section and the CM energies of the experimental points adjacent to the peak, $\sigma(92.05) = 1.37$ nb and/or $\sigma(90.21) = 1.28$ nb into (12.9) and solving for Γ we get

$$\Gamma_Z = 2.5 \, \text{GeV}.$$

Alternatively, we can draw a smooth curve which passes through the points surrounding the maximum, takes the value $\frac{1}{2} \times 2.00$ nb at $E = 90.0$ GeV and 92.5 GeV, giving a width of $(92.5\text{–}90.0)$ GeV.

From (12.7), the lifetime of the Z-boson is then

$$\tau_Z = \frac{\hbar}{\Gamma_Z} = \frac{1.05 \times 10^{-34} [\text{J s}]}{(2.5 \times 10^9 [\text{eV}]) \times (1.6 \times 10^{-19} [\text{C}])} = 2.6 \times 10^{-25} \, \text{s}.$$

A resonance does *not* have to occur as a function of the CM energy of the incident scattering particles. Very often, a scattering process produces more than two particles in the final state and one can look for a resonance in the CM energy of any two of the final-state particles. As an example, consider the scattering process

$$p + p \rightarrow p + p + e^- + e^+.$$

One can plot the cross section as a function of E_{ee}, where E_{ee} is the energy of the outgoing electron–positron pair in *their* CM frame, i.e. the frame in which the final-state electron–positron pair has zero total momentum. E_{ee} is called the CM energy of the $e^+ e^-$ channel. The cross section could have a resonance peak when E_{ee} is equal to the rest energy, Mc^2, of an unstable particle of mass M, which can decay into an electron–positron pair. This particle can be produced together with other particles in the scattering process, without displaying a resonance in the incident CM energy. It then decays almost immediately into an electron–positron pair, giving rise to a resonance in the variable E_{ee}. In the case where the intermediate particle is a Z-boson, the two-stage process would be

$$p + p \rightarrow p + p + Z \rightarrow p + p + e^+ + e^-.$$

The intermediate state consists of three particles whose total energy is equal to the incident energy of the scattering protons. The total cross section does *not* display a resonance in the incident CM energy. However, since the intermediate Z-boson can decay into an electron–positron pair, there will be a resonance in the differential cross section as a function of E_{ee}, the CM energy of the final-state electron–positron pair, with a peak at $E_{ee} = M_Z c^2$.

12.6.3 Branching Ratios and Partial Widths

Most particles can decay into more than one channel, f. If the decay rate into that channel is λ_f, and the total decay rate is λ, then the branching ratio, BR_f, is defined as (in analogy with the branching ratio for multi-modal radioactive decays discussed in Chap. 5)

$$\mathrm{BR}_f = \frac{\lambda_f}{\lambda}. \tag{12.10}$$

The sum of all the branching ratios is one,

$$\sum_f \mathrm{BR}_f = 1.$$

We also define the *"partial width"*, Γ_f as[10]

$$\Gamma_f = \mathrm{BR}_f \Gamma. \tag{12.11}$$

For example, the ω-meson can decay into the following channels:

$$\omega \to \pi^+ + \pi^- + \pi^0 \; (\mathrm{BR} = 89.2\%)$$
$$\to \pi^0 + \gamma \; (\mathrm{BR} = 8.3\%)$$
$$\to \pi^+ + \pi^- \; (\mathrm{BR} = 1.5\%) \tag{12.12}$$

(with a branching ratio of 1% into miscellaneous other channels). The total width of the ω-meson is 8.5 MeV, so the partial widths are

$$\Gamma_{\omega \to \pi^+ \pi^- \pi^0} = 0.892 \times 8.5 = 7.6\,\mathrm{MeV},$$

$$\Gamma_{\omega \to \pi^0 \gamma} = 0.083 \times 8.5 = 0.71\,\mathrm{MeV},$$

$$\Gamma_{\omega \to \pi^+ \pi^-} = 0.013 \times 8.5 = 0.13\,\mathrm{MeV}.$$

---

[10] Γ is often called the *"total width"* to distinguish it from the partial width.

Summary

- There are four fundamental forces of Nature:

  - Strong interactions (short range)
  - Electromagnetic interactions (long range)
  - Weak interactions (very short range)
  - Gravitational interactions (long range)

- There are two types of particles:

  - Leptons, which only partake in weak and electromagnetic interactions.
  - Hadrons, which also interact via the strong interactions. Hadrons are further subdivided into mesons that have integer spin and baryons that have half odd-integer spin.

- All particles have fields associated with them. The corresponding particles are quantum excitations of wave motions of the fields.
- The particles associated with the fundamental force fields are called force carriers or gauge bosons. The interactions can be viewed as the exchange of these gauge bosons.
- Long-lived particles with a lifetime exceeding 10^{-13} s can leave a detectable track in a tracking detector. The lifetimes of short-lived particles can be inferred from the width of their resonance peak in a scattering experiment in which the particles are produced.

Problems

Problem 12.1 From the following decay modes, indicate whether the decaying particle is a lepton or a hadron and in the case of hadrons indicate whether it is a baryon or a meson.

(a) $\Sigma^+ \rightarrow p + \pi^0$
(b) $\tau^+ \rightarrow \mu^+ + \nu_\mu + \bar{\nu}_\tau$
(c) $D^+ \rightarrow \pi^+ + \mu^+ + \mu^-$
(d) $\eta \rightarrow \gamma + \gamma$
(e) $B^- \rightarrow \tau^- + \bar{\nu}_\tau$
(f) $\Lambda^0 \rightarrow p + e^- + \bar{\nu}_e$

Problem 12.2 Before the development of QCD, the strong interactions were described by the Yukawa theory, which consisted of the exchange of π-mesons. What is the range of the interactions in this theory?

Problem 12.3 In the scattering process

$$\pi^+ + p \rightarrow \pi^+ + p,$$

the cross section displays a resonance at (CM) energy 1232 MeV, due to the production of an intermediate Δ^{++}-baryon. The value of the cross section at this peak is 198 mb. At an energy of 1198 MeV the cross section is 145 mb. Calculate the lifetime of the Δ^{++}-baryon.

Problem 12.4 Show that close to the resonance of a particle with mass M and width Γ, the Breit–Wigner distribution can be rewritten in the relativistic form

$$f(s) \propto \frac{1}{\left(s - M^2 c^4\right)^2 + M^2 c^4 \Gamma^2} + \cdots ,$$

where $\cdots$ indicate corrections that vanish at $s = M^2 c^4$. The quantity s is given by

$$s = (E_1 + E_2)^2 - \left(p_1 + p_2\right)^2 c^2,$$

where E_1 and E_2 are the energies of the incident particles and p_1 and p_2 are their momenta (in *any* reference frame).

By applying a Lorentz transformation to the energy and momentum of the incident particles, show that s is a Lorentz invariant quantity.

Chapter 13
Particle Accelerators

In order to achieve the very high energies required to produce and study particles and their properties, the particles need to be accelerated in *"accelerators"*. Incident particles are accelerated to these high energies and scattered against another particle. Typically, there is enough energy to smash the initial particles up and produce (possibly many) different particles in the final state, some of them with considerably higher masses than the incident particles. In such cases the scattering is called *"inelastic scattering"*. Conversely, a scattering event in which the final-state particles are the same as the initial particles is called *"elastic scattering"*. Rutherford scattering or Mott scattering are examples of elastic scattering. The word "elastic" here means that none of the incoming energy is used up in the production of other particles.

Our Universe is the natural accelerator of particles, providing them at very high energies known as a *"cosmic rays"*. Cosmic rays mainly consist of high energy protons (90%). The studies of cosmic rays interacting with the Earth's atmosphere led to discovery of particles beyond those observed in atoms, such as the positron [99], muon [100], pion [96], kaon [5] and several more.

The era of consistent exploration of new particles started with high energy accelerators built on Earth. The first high energy accelerator was the Cosmotron with an energy of 3.3 GeV at BNL[1] that started operating in 1953. This was soon superseded by the 6.2 GeV Bevatron at LBL[2] in 1954, followed by the Synchrophasotron built at JINR[3] in 1957 with an energy of 10 GeV and then by the 20 GeV synchrotron at CERN in 1959. Since then many much more powerful accelerators have been built.

For example, at the LEP accelerator (which operated from 1989 until 2000 and collided electrons against positrons) at CERN, one possible process was

---

[1] Brookhaven National Laboratory.

[2] Lawrence Berkeley Laboratory.

[3] Joint Institute for Nuclear Research.

© The Author(s), under exclusive license to Springer Nature Switzerland AG 2021   201
A. Belyaev, D. Ross, *The Basics of Nuclear and Particle Physics*, Undergraduate Texts in Physics, https://doi.org/10.1007/978-3-030-80116-8_13

$$e^+ e^- \rightarrow W^+ W^-, \tag{13.1}$$

in which the electron and positron annihilate each other and produce two W-bosons instead. To produce a pair of W-bosons with $m_W = 80.4\,\text{GeV}/c^2$, a total centre-of-mass energy greater than $2 \times m_W c^2$ is required. The cross section for the process $e^+ e^- \rightarrow W^+ W^-$ is the total number of events in which two W-bosons are produced per unit incident flux (i.e. the number of W-boson pairs produced divided by the number of particle scatterings per unit area).

It is now believed that there might exist particles with masses that are an order of magnitude larger than the mass of weak bosons that can be probed at the LHC. The LHC is presently colliding protons at 13 TeV (1.3×10^{13} eV) CM energy, which will possibly be increased to 14 TeV in the future. This new energy frontier with their correspondingly small distances that can be probed at the LHC, and at even more powerful accelerators in the future, could shed a light on physics beyond the Standard Model[4] (BSM) that predicts the existence of new particles.

13.1 Fixed Target Experiments vs. Colliding Beams

In fixed target experiments, a target particle at rest is bombarded with an accelerated projectile. In a colliding beam experiment *both* the incident particles are accelerated and the two beams are brought together at an intersection point where they collide into each other. Colliding beam accelerators are therefore also known as *"colliders"*.

The total energy of a projectile particle plus the target particle depends on the reference frame. The frame that is relevant for the production of high mass particles is the centre-of-mass (CM) frame for which the projectile and target have equal and opposite momentum of magnitude p. For simplicity, let us suppose that the projectile and target particles are the same, or possibly particle–antiparticle (e.g. proton–proton, proton–antiproton or electron–positron) so that their masses, m, are the same. This means that in this frame both the particles have the same energy, E_{CM}. (Since we are usually dealing with relativistic particles, E_{CM} refers to the sum of the kinetic energy, T_{CM}, *plus* rest energy, mc^2.)

Consider the quantity, s, introduced by Stanley Mandelstam in 1958 [101]

$$s = \left(\sum_{i=1,2} E_i \right)^2 - \left(\sum_{i=1,2} \boldsymbol{p}_i \right)^2 c^2. \tag{13.2}$$

In the centre-of-mass frame, where the momenta are equal and opposite, the second term vanishes and we have

---

[4]The *"Standard Model"* is a theory of the interactions of all the particles required to describe the strong, weak and electromagnetic interactions. It is discussed in more detail in subsequent chapters.

$$s = 4E_{\text{CM}}^2 , \tag{13.3}$$

i.e. s is the square of the total incoming energy in the CM frame – this is a quantity that appears often in Particle Physics and the notation s is always used. For one particle we know that $E^2 - p^2 c^2$ is equal to $m^2 c^4$ and is therefore the same in any frame of reference even though the quantities E and p will be different in different frames. Likewise the above quantity, s, is Lorentz invariant – it is the same in any frame of reference.

Let us consider now a fixed target collision in which a particle of mass m_1 is incident on a particle of mass m_2 at rest. The target particle at rest (the Lab frame) has energy $m_2 c^2$ and zero momentum, whereas the projectile has energy E_{LAB} and momentum p_{LAB}, so that we have

$$\begin{aligned} s &= \left(E_{\text{LAB}} + m_2 c^2 \right)^2 - p_{\text{LAB}}^2 c^2 \\ &= E_{\text{LAB}}^2 + m_2^2 c^4 + 2 m_2 c^2 E_{\text{LAB}} - p_{\text{LAB}}^2 c^2 \\ &= m_1^2 c^4 + m_2^2 c^4 + 2 m_2 c^2 E_{\text{LAB}}, \end{aligned} \tag{13.4}$$

where in the last step we have used the relativity relation

$$E_{\text{LAB}}^2 - p_{\text{LAB}}^2 c^2 = m_1^2 c^4 . \tag{13.5}$$

Equating the two expressions for s, (13.3) and (13.4), and taking a square root, we obtain the relation

$$E_{\text{CM}} = \frac{1}{2} \sqrt{ m_1^2 c^4 + m_2^2 c^4 + 2 m_2 c^2 E_{\text{LAB}} } . \tag{13.6}$$

For example, taking the proton mass to be approximately $1 \text{ GeV}/c^2$, if we have an accelerator that can accelerate protons up to an energy of 100 GeV, the total centre-of-mass energy achieved is only about 14 GeV – far less than the energy required to produce a particle of mass $100 \text{ GeV}/c^2$.

The solution to this problem is to use colliding beams of particles. In these experiments both beams of particles involved in the scattering move in the opposite directions with their high energies maintained by means of a magnetic field. At various points the beams intersect and scattering takes place. If the (equal mass) incident particles have the same energy and therefore equal but opposite momentum, then the experiment is taking place in the centre-of-mass frame and the full energy delivered by the accelerator, $\sqrt{s} = 2E_{\text{CM}}$, can be used to produce high mass particles.

13.2   Luminosity

The luminosity, $\mathcal{L}$, is the number of particle collisions per unit area per second, usually quoted in $cm^{-2}\ s^{-1}$. This is a very important characteristic of particle accelerators. The number of events of type f that occur per second in the scattering of particles, i, is the cross section $\sigma_{i \to f}$ multiplied by the luminosity. For example, for the production of two W-bosons at LEP, the cross section, $\sigma(e^+ e^- \to W^+ W^-)$, is about 15 pb ($15 \times 10^{-12}$ b) for $\sqrt{s} = 180$ GeV and the luminosity of LEP was $10^{32} cm^{-2}\ s^{-1}$. The number of these pairs of W-bosons produced per second is given by [5]

$$\frac{dN_{W^+ W^-}}{dt} = (15 \times 10^{-12} \times 10^{-24})[cm^2] \times (10^{32})[cm^{-2} s^{-1}] = 1.5 \times 10^{-3} s^{-1},$$

where the first term in parenthesis is the cross section in units of cm^2 and the second is the luminosity in units of $cm^{-2}\ s^{-1}$. So, the general formula for rate of the reaction i to f, $R_f = dN_f/dt$ is

$$R_f = \sigma_{i \to f} \times \mathcal{L}. \tag{13.7}$$

The cross section $\sigma_{i \to f}$ is therefore the probability per unit incident flux per unit time that the initial particles i will scatter to produce the state f.

For integrated luminosity over time $L = \int \mathcal{L} dt$ the number of observed events, N_f in which the state f is produced, is given by

$$N_f = \sigma_{i \to f} \times L. \tag{13.8}$$

Integrated luminosity is often quoted in inverse femtobarns, fb^{-1}, so that the number of events observed is the cross section in fb (femto $\equiv 10^{-15}$) multiplied by the integrated luminosity in these units.

For colliding beams, luminosity $\mathcal{L}_c$ is proportional to the number "bunches" of particles in each beam, n (typically 5–100), the revolution frequency, f (usually between 10^3 and 10^6 Hz), $\mathcal{N}_1, \mathcal{N}_2$ – the number of particles in each bunch (typically $\simeq 10^{10} - 10^{11}$) and inversely proportional to the beam cross section, A (which is of the order of 10^{-5} cm^2):

$$\mathcal{L}_c = \frac{nf \mathcal{N}_1 \mathcal{N}_2}{A}. \tag{13.9}$$

As the cross section is a probability, the expected number of events N_f given by (13.8) is an average value. As in the case of radioactivity, the actual number of events

---

[5]Usually in this book we use either SI or nuclear units. However, particle experimentalists use the CGS system (cm,g), which is what we use in this chapter.

observed is a random distribution centred on N_f but with a standard deviation $\sqrt{N_f}$, meaning that there is a 68% probability that the number of events observed will lie within the range $N_f - \sqrt{N_f}$ to $N_f + \sqrt{N_f}$. Therefore, in order to measure the above cross section at LEP to an accuracy of 1% it was necessary to collect 10000 such W-pairs, which, at a rate of 1.5×10^{-3} per second, took about 3 months.

In the case of a fixed target experiment the luminosity $\mathcal{L}_{ft}$ depends on the incident beam flux, F, in particles per second, on the target thickness, δ, and density of protons (or other scattering particles) ρ_p:

$$\mathcal{L}_{ft} = F\delta\rho_p , \tag{13.10}$$

which can be expressed through fixed target material density ρ as

$$\mathcal{L}_{ft} = F\delta\frac{\rho}{m_p}, \tag{13.11}$$

where m_p is the mass of the proton. For example for proton–proton scattering with a flux of 10^{12} protons per second and a 10 cm thick target with density $\rho = 1\,\mathrm{g\,cm}^{-3}$ one has

$$\mathcal{L}_{ft} = 10^{12}[\mathrm{s}^{-1}]10[\mathrm{cm}]\frac{1[\mathrm{g\,cm}^{-3}]}{1.67 \times 10^{-24}[\mathrm{g}]} = 6.0 \times 10^{36}\,\mathrm{cm}^{-2}\mathrm{s}^{-1}.$$

There is a price to pay for colliding beam experiments in terms of luminosity. In colliding beams it is necessary to focus the incident beams as tightly as possible using magnetic fields, in order to maximize the luminosity. This is very difficult task. Luminosities of 10^{34} cm^{-2}s^{-1} have been achieved so far at the LHC, which means that, at present, the reaction rate is down by almost three orders of magnitude compared with a fixed target experiment. After an upgrade to the high-luminosity LHC, scheduled for 2027, the luminosity is expected to reach 10^{35} cm^{-2}s^{-1}.

13.3 Types of Accelerators

As we have discussed, the general aim of accelerators is to collide two particles at (very) high energy in order to create new particles from the combined energy and quantum numbers, or to probe inside one of the particles to see what it is made of.

Only stable charged particles can be accelerated: namely electrons, positrons, protons, antiprotons and some ions. The muon, which is long-lived (lifetime, $\tau \simeq 2 \times 10^{-6}$s) has been considered for use in a potential future muon collider.

Single DC stage accelerators such as the Van de Graaff generator can accelerate electrons and protons up to about 20 MeV.

There are two general types of modern accelerators – circular (cyclic) and linear.

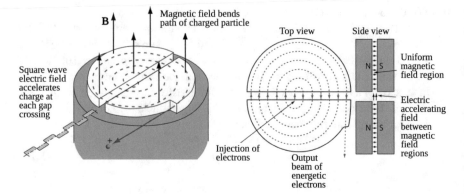

Fig. 13.1 The schematic diagram of a cyclotron. [*Source*: http://hyperphysics.phy-astr.gsu.edu/ hbase/magnetic/cyclot.html (edited)]

13.3.1   Cyclotrons

The prototype design for all circular accelerators is the cyclotron. In this device, which is shown schematically in Fig. 13.1, the (charged) particles to be accelerated move in two hollow metallic semi-disks (D's) with a large magnetic field, B, applied normal to the plane of the D's. The particles move in a spiral from the centre and an alternating electric field is applied between the D's. The frequency of alternation is exactly equal to the frequency of rotation of the charged particles, such that when the particles cross from one of the D's to the other, the electric field always acts in the direction that accelerates the particles.

A charged particle with charge e moving with velocity $\mathbf{v}$ in a magnetic field $\mathbf{B}$ experiences a force $\mathbf{F}$, where

$$\mathbf{F} = q\mathbf{v} \times \mathbf{B}. \tag{13.12}$$

When the magnetic field is perpendicular to the plane of motion of the charged particle, this force is always towards the centre and gives rise to centripetal acceleration, so that at the moment when the particles are moving in a circle of radius r. The magnitude of the force, F, is given by

$$F = Bev = m\frac{v^2}{r}. \tag{13.13}$$

Dividing both sides of (13.13) by v, we see immediately that the angular velocity $\omega = v/r$ is constant, so that the frequency of the alternating electric field remains constant. The maximum energy that the particles can acquire depends on the radius, R, for which the velocity has its maximum value v_{max},

$$v_{max} = \frac{BeR}{m}, \tag{13.14}$$

leading to a maximum kinetic energy

$$T_{max} = \frac{1}{2}mv_{max}^2 = \frac{B^2e^2R^2}{2m}. \tag{13.15}$$

This works fine if the energy of the particle remains non-relativistic. However, in high energy accelerators the particles are accelerated to energies that are extremely relativistic – the particles are travelling very nearly with the velocity of light (e.g. at the LHC v/c is about $1 - 10^{-8}$!). Taking relativistic effects into account, the angular velocity is now

$$\omega = \sqrt{1 - v^2/c^2}\frac{Be}{m}. \tag{13.16}$$

This means that as the particles accelerate, either the frequency of the applied electric field must vary – such machines are called *"synchrocyclotrons"* – or the applied magnetic field must be varied (or both) – such machines are called *"synchrotrons"*.

In a synchrotron, dipole magnets keep the particles in a circular orbit, R, which is determined by the particle momentum, p, and the value of the magnetic field, B, as

$$p[\text{GeV } c^{-1}] = 0.3 \times B[\text{Tesla}] \times R[m], \tag{13.17}$$

which follows from (13.14). Quadrupole magnets are used to focus the beam. A schematic picture of the cross section of the dipole/quadrupole magnet and a segment of the magnet ring of a synchrotron are shown in Fig. 13.2.

Since the bending field B is limited, the maximum energy is limited by the size of the ring. The Super Proton Synchrotron (SPS) at CERN has a radius R = 1.1 km and accelerates protons up to an energy of 450 GeV. Particles are accelerated by resonators (RF Cavities). The bending field, B, is increased with time as the

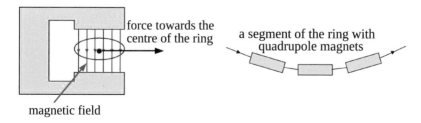

Fig. 13.2 The cross section of the dipole/quadrupole magnet (left) and the segment of the magnet ring (right) of a synchrotron

momentum increases so as to keep R constant. Electron synchrotrons are similar to proton synchrotrons except that the energy losses are much greater.

One of the main limiting factors of synchrotron accelerators is the synchrotron radiation. A charged particle moving in a circular orbit is accelerating (even if the speed is constant) and therefore radiates. The energy radiated per turn per particle is

$$\Delta E = \frac{4\pi e^2 v^2 \gamma^4}{3Rc^2},$$ (13.18)

where e is the charge, and $\gamma = 1/\sqrt{1 - v^2/c^2} = E/m$, from which follows that

$$\Delta E \propto 1/m^4.$$ (13.19)

For relativistic electrons and protons of the same momentum the ratio of energy losses is very large for electrons versus protons:

$$\frac{\Delta E_e}{\Delta E_p} = \left(\frac{m_p}{m_e}\right)^4 \simeq 10^{13} \ .$$ (13.20)

13.3.2   Linear Accelerators

The energy loss due to synchrotron radiation can be avoided in a linear accelerator. In such a machine the particles are accelerated by means of a radio-frequency (RF) oscillating applied electric field along a long tube.

Proton linear accelerators (Linacs) use a succession of drift tubes ("*klystrons*") of increasing length (to compensate for increasing velocity) as shown in the schematic diagram in Fig. 13.3. Particles travel in a vacuum. At radio frequencies ($< 10^9$ Hz), there is no field inside the drift tubes. The oscillating voltages applied to adjacent cylindrical electrodes have opposite polarity, so that there is an accelerating electric field in the gap between each pair of electrodes. The frequency of oscillation of the voltage is such that by the time the particle passes to the next gap between electrodes, the voltage is reversed so that the electric field is always accelerating. Particles are therefore accelerated down the tube in bunches, which are synchronized with the oscillating voltage. Proton linacs of $\sim$10–70 m give energies of 30–200 MeV. These linacs are usually used as injectors for higher energy machines.

The largest linear collider was the Stanford Linear Collider (SLC) at SLAC[6] in California. A schematic map of SLAC is presented in Fig. 13.4. The SLC was 3 km

---

[6]Stanford Linear Accelerator Center.

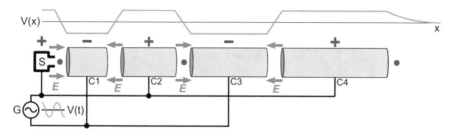

Fig. 13.3 A schematic diagram of a linac. [*Source*: https://en.wikipedia.org/wiki/File: Linear_accelerator_animation_16frames_1.6sec.gif]

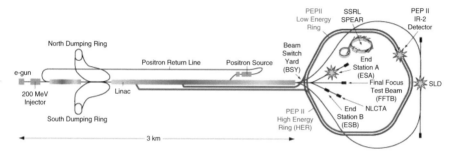

Fig. 13.4 Schematic map of SLAC

long accelerating both electrons and positrons up to energies of 50 GeV, so it had a total energy of 100 GeV in the CM frame.

The frequency of the oscillating voltage was in the microwave region. At such frequencies, the klystrons, whose transverse dimensions were larger than the microwave wavelengths, acted as waveguides. Unlike the case of previously designed linacs, there was an oscillating electric field in the longitudinal direction, inside the klystrons. In regions where this longitudinal electric field was positive, positively charged particles were accelerated along the tube, and in regions where it was negative, negatively charged particles were accelerated along the tube. In this way *both* electrons and positrons could be injected into the tube at a separation of precisely half the wavelength of the microwave and were accelerated along the tube simultaneously, as shown schematically in Fig. 13.5. Above a few MeVs the electrons and positrons were moving with almost the speed of light, so that the klystrons were uniform in length.

The oscillating electric field also served to re-focus the bunches, which travelled in the second or fourth quarter of the wave, as shown in Fig. 13.5. The particles that had fallen behind the centre of the bunch experienced a stronger field and were accelerated more. Likewise particles that had got ahead of the centre of the bunch experienced a weaker field and were accelerated less, so that the net effect was to bunch the particles closer together.

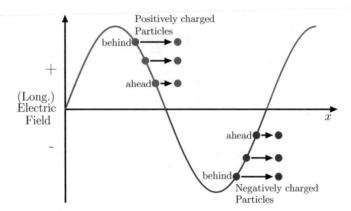

Fig. 13.5 Snapshot of the oscillating (longitudinal) electric field inside the klystrons at the SLC. *x* refers to the distance along the tube

At the end of the accelerator, the electron and positron beams were directed by a magnetic field into two opposing arcs and collided at the intersection point (SLD) shown in Fig. 13.4.

13.4 Particle Accelerators: Past, Present and Future

Two known stable particles – electron and proton (and their antiparticles) – gave rise to three pairs of scattering particles used in the most powerful recent and present colliders:

1. e^+e^- SLC (Stanford, 1989–1998) and LEP (CERN, 1989–2000) colliders
2. ep HERA (DESY, 1992–2007) collider
3. $p\bar{p}$ Tevatron (Fermilab, 1992–2011) and pp LHC (CERN, 2008–present) colliders

We briefly discuss each of these below.

13.4.1 LEP

LEP was the most powerful electron–positron (e^+e^-) synchrotron collider to date with a circumference of 27 km. It was built at CERN, in Geneva, and operated from 1989 until 2000. At the end of 2000 it was discontinued and dismantled to make way for the LHC that re-used the LEP tunnel located roughly 100 m underground.

The initial energies of the electron and positron beams were about 45.5 GeV each, making a total CM energy of 91 GeV, to enable production and study of the Z-boson. Electrons and positrons were accelerated and collided in the same vacuum

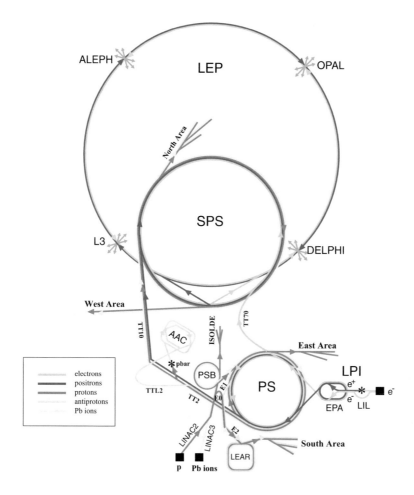

Fig. 13.6 Schematic view of LEP collider facilities. [*Source*: Rudolf LEY, PS Division, CERN, 02.09.96]

LEP: Large Electron Positron collider
SPS: Super Proton Synchrotron
AAC: Antiproton Accumulator Complex
ISOLDE: Isotope Separator OnLine DEvice
PSB: Proton Synchrotron Booster
PS: Proton Synchrotron

LPI: Lep Pre-Injector
EPA: Electron Positron Accumulator
LIL: Lep Injector Linac
LINAC: LINear ACcelerator
LEAR: Low Energy Antiproton Ring

ring at LEP. It had four collision points with each with a large detector: ALEPH, DELPHI, L3 and OPAL (see Fig. 13.6) that analyzed events from collisions.

In 1995 LEP was upgraded for a second operation phase that allowed the collider CM energy to reach 209 GeV. This was very close to the 216 GeV threshold needed for the associated production of Z- and Higgs bosons, which will be the dominant production mode of Higgs bosons at future e^+e^- colliders with energy not far above the threshold.

The high luminosity, $\mathcal{L}$, at LEP reached $10^{32}\,\text{cm}^{-2}\text{s}^{-1}$. Apart from the Higgs boson, the high energy physics (HEP) community was also hoping to discover the so-called supersymmetric partners[7] of the Standard Model at LEP, but unfortunately none were found. In spite of the lack of Higgs boson or new physics discovery, LEP's precision measurements of the Standard Model particles, particularly measurements of the properties of Z- and W-bosons, remain very valuable (and some of the measurements are still the most accurate to date notwithstanding new data from the LHC). In addition, some lower limits on the masses of some new physics particles at LEP remain the best so far.

13.4.2   HERA

The Hadron–Electron Ring Accelerator (HERA), at DESY[8] in Hamburg, collided 27.5 GeV electrons (positrons) and 920 GeV protons. The corresponding CM energy was $\sqrt{s} = 318\,\text{GeV}$. HERA consisted of two accelerator rings, an electron (or positron) ring and a proton ring, which intersected at four collision points, where four experiments: H1, ZEUS, HERMES and HERA-B were situated. The circumference of each accelerator ring was 6.3 km. The HERA collider layout is depicted in Fig. 13.7. In the PETRA ring, electrons and protons were pre-accelerated to energies of 14 GeV and 50 GeV, respectively, and were then accelerated by the HERA rings up to their final energies. HERA operated from 1992 to 2007. During this period HERA delivered an integrated luminosity of about 500 pb^{-1} for each collider experiment. HERA was the only electron–proton (ep) collider (so far). The data analyzed from the experiments have provided important new information on the proton structure and the interactions of high energy photons with matter. HERA has also provided important lower limits on masses of new particles predicted by BSM physics.

13.4.3   Tevatron

The Tevatron was a circular proton–antiproton collider located at the FNAL,[9] near Chicago and operated from 1983 until 2011. The circumference of the accelerator

---

[7]Supersymmetry (SUSY) is the theory that predicts an existence of the partners of the Standard Model particles. SUSY particles have the same quantum numbers as Standard Model but intrinsic spin differing by $\frac{1}{2}$.

[8]Deutsches Elektronen-Synchrotron.

[9]Fermi National Accelerator Laboratory, also known as "Fermilab".

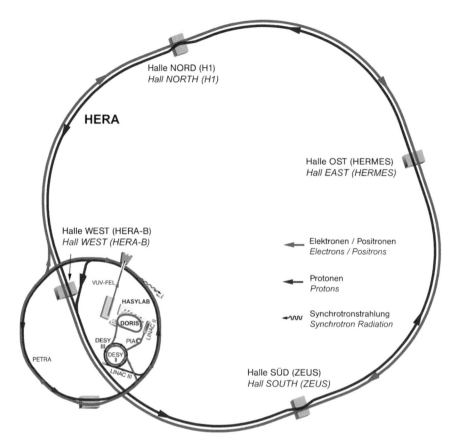

Fig. 13.7   Schematic view of the HERA accelerator complex. [*Source*: https://media.desy. de/DESYmediabank/image/previews/3.12534.1115.3526970954356.HERA-Ring_Grafik.jpg? collection=SearchResult]

ring was 6.3 km, accelerating protons and antiprotons up to an energy of about 1 TeV (hence its name). The layout of the Tevatron is shown in Fig. 13.8. This was a synchrotron type accelerator in which very high magnetic fields were achieved using superconducting (electro-)magnets, capable of maintaining very large currents thereby producing large magnetic fields. The Tevatron had two collision points and correspondingly two detectors – DØ and CDF, each of which recorded about 0.1 fb^{-1} of data in Run I ($\sqrt{s} = 1.8$ TeV) and about 10 fb^{-1} of data in Run II ($\sqrt{s} = 1.96$ TeV). The highest luminosity was 4×10^{32} cm^{-2} s^{-1} recorded at the end of Run II. The main achievement of the Tevatron was the discovery of the top-quark in 1995, and the exploration of its properties. This was the last quark predicted by the Standard Model to be discovered. It turned out that this elementary particle is also the most massive particle of the Standard Model, being more than 170 times more massive than the proton!

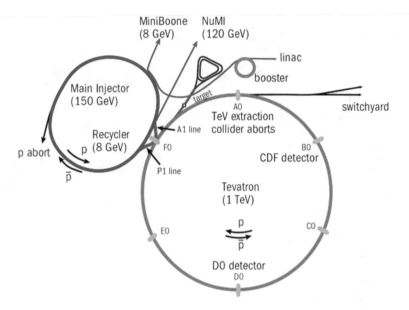

Fig. 13.8 Layout of the Tevatron. [*Source*: http://images.iop.org/objects/ccr/cern/51/9/22/CCtev2_09_11.jpg]

13.4.4 LHC

The LHC is the most powerful proton–proton collider so far. The LHC project started in September 2008. After an accident in October 2008, which caused it to be shut down, it resumed its operation in November 2009 and is now colliding protons against protons with energies $6.5 \oplus 6.5$ TeV resulting in a total CM energy of 13 TeV. It uses the 27 km LEP tunnel. In contrast with LEP or the Tevatron, which collided particles and antiparticles that had been accelerated in the same vacuum beam pipes, at the LHC beams of protons travelling in opposite directions are accelerated in separate vacuum beam pipes with the help of a specially designed magnetic field configuration. Beams collide at four intersection points where detectors ATLAS, CMS, ALICE and LHCb are situated. The schematic map of the LHC collider facilities is presented in Fig. 13.9.

The highest luminosity recorded so far was 2×10^{34} cm^{-2} s^{-1} and the total integrated luminosity for both ATLAS and CMS is about 150 fb^{-1} in Run 2.[10] The main achievement of the LHC so far is the well-known discovery of the Higgs boson in 2012, the last particle predicted by the Standard Model to be discovered. Since then, LHC experimental groups have explored the properties of the Higgs boson and

---

[10]Run 2 took place between 2015 and 2018 at $\sqrt{s} = 13$ TeV after Run 1 between 2009 and 2013 at $\sqrt{s} = 7$–8 TeV.

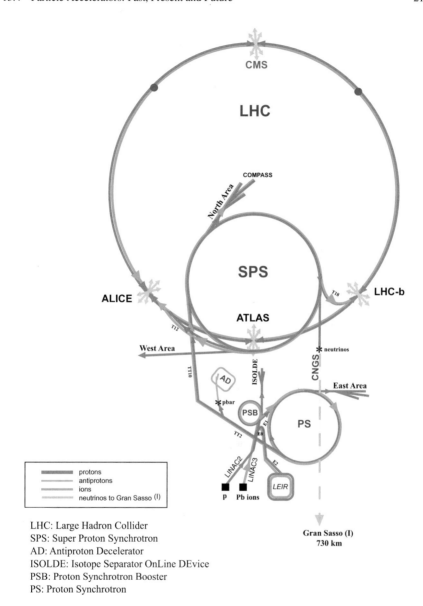

LHC: Large Hadron Collider
SPS: Super Proton Synchrotron
AD: Antiproton Decelerator
ISOLDE: Isotope Separator OnLine DEvice
PSB: Proton Synchrotron Booster
PS: Proton Synchrotron

Fig. 13.9   Schematic map of the LHC collider facilities. [*Source*: https://ps-div.web.cern.ch/
PSComplex/accelerators.pdf (edited)]

Table 13.1 Summary of the colliders discussed above

Collider	Particles	Shape and circumference	Energy	Location	Years
SLC	e^+e^-	Linear, 3 km	2×50 GeV	Stanford USA	1989–1998
LEP	e^+e^-	Circular, 27 km	up to 2×104.5 GeV	CERN	1989–2000
HERA	e^-p e^+p	Circular, 6.3 km	27.5 GeV ($e^\pm$), 920 GeV (p)	DESY Germany	1992–2007
Tevatron	$p\bar{p}$	Circular, 6.3 km	up to 2×980 GeV	Fermilab USA	1992–2011
LHC	pp	Circular, 27 km	up to 2×7 TeV	CERN	2008–present

the top-quark in detail. They have also produced numerous results on precision tests of the Standard Model as well as carrying out searches for particles predicted by BSM physics. No new physics has been found so far, but it is still possible that it is just around the corner. The next LHC run will be Run 3, followed by another high-luminosity LHC Run that will continue to search for new physics and could possibly discover some of the particles that various BSM theories predict. In Table 13.1 we present a short summary of the colliders discussed above.

13.4.5 Hadron vs. e^+e^- Colliders, and Their Future

It is worth considering the advantages and drawbacks of hadron colliders compared with e^+e^- accelerators. Let us note that the Tevatron $p\bar{p}$ collider with $\sqrt{s} = 1.96$ TeV had enough energy to *produce* a Higgs boson and these were actually produced at this collider, but in insufficient quantities for an unambiguous discovery. At the same time if LEP had just enough energy to produce Higgs boson (in association with Z-boson as discussed earlier), LEP would definitely have discovered it. This means that merely going to high energy may be not sufficient to discover a particle that we are looking for. In particular, an e^+e^- collider is quite a "clean" machine in the sense that besides e^+e^- no other particles are involved in the collision. Also the CM energy of e^+e^- is fixed and known. On the other hand, in case of high energy $p\bar{p}$ or pp collisions only pairs of constituents of the proton, i.e. quarks, antiquarks or gluons (known as *"partons"* – to be discussed in Chap. 20), are collided and the rest of the protons do not participate in the interaction. The CM energy of the interacting partons is not fixed and its value in a given event is not known. Furthermore, the *"background events"* from strongly interacting partons at hadron colliders have much higher rate than background events from e^+e^- that are determined by the electroweak interactions. Background events are events coming from the processes different from the signal process under investigation but nevertheless contributing to the same observable[11]. This is especially important when the signal process has low cross section such as the associated production of the Higgs boson with Z. In this case the signal to background ratio will be higher at e^+e^- in comparison with a hadron collider, making the e^+e^- collider the better tool for the precision study of the processes as soon as the collider energy is high enough to produce the particles under study. In particular, the precision measurement of the properties of the Higgs boson is one of the main objectives of the program for a future e^+e^- International Linear collider (ILC) with the CM energy 250–500 GeV. Apart from the ILC, the following future e^+e^- colliders are under discussion: 3 TeV CLIC (Compact Linear Collider) and 91–365 GeV future circular e^+e^- collider (FCC-ee).

On the other hand, synchrotron radiation losses in a storage ring of a given circumference are much larger for e^+e^- than pp or $p\overline{p}$. For this reason hadron colliders can reach much higher energy as can be seen from a comparison of the energies of LEP versus LHC, which use the same tunnel. The ability to reach higher energy can be crucial even for a "messy" hadron collider environment. This is especially important in searches for new heavy particles produced by the strong interactions. That is why the Tevatron was able to discover the top-quark in Run 1, which had comparatively low luminosity. From this point of view hadron colliders are great for the discovery of heavy strongly interacting particles. Therefore, amongst the planned future colliders is a 100 TeV future circular pp collider (FCC-hh) with a 100 km ring.

One should also mention the plasma acceleration technique for accelerating charged particles. This uses the electric field associated with an electron plasma wave or other high-gradient plasma structures (known as a "wakefield"), created by laser pulses or energetic particle beams. This technique has the potential to build ultra-high energy particle accelerators of much smaller size than modern accelerators. Devices under development have accelerating gradients that are several orders of magnitude larger than current particle accelerators over very short distances and about one order of magnitude larger at a distance of around 1 m.

Summary

- Accelerators and detectors are the main ingredients of the modern experimental high energy physics .
- An important characteristic of an accelerator is the Lorentz invariant quantity s, which is the square of the total incident energy CM frame.

---

[11]For example, in the search for the decay of a Higgs particle into two photons, there are other processes that can lead to the observation of two photons but which are not generated by the decay of a Higgs, so that some of the events in which a pair of photons is observed are background events.

- Luminosity is another very important characteristic of the accelerator. The relation of the total number of events, integrated luminosity and cross section is given by

$$N = \sigma \times L .$$

- There are two types of accelerators:
  - Fixed target accelerators in which a target particle at rest is bombarded with an accelerated projectile
  - Colliders, in which both the incident particles are accelerated (often with equal and opposite momentum so that scattering takes place in the CM frame) and stored in storage rings, where they collide with each other at various intersection points

- Much larger CM energies can be achieved with colliders than in fixed target experiments, but with lower luminosity than in fixed target experiments.
- In some colliders, the incident particles are accelerated in the storage rings themselves (circular accelerators), where particles of opposite electric charge move in opposite directions under the influence of a magnetic field. With others, the particles are injected into the storage rings having been accelerated in a linear accelerator (linac) by an oscillating electric field. In a linac, bunch particles and antiparticles can be accelerated by opposite phases of the same oscillating electric field.
- The energy of the circular accelerators is limited by the synchrotron radiation, whilst a linear collider is free from it.
- Accelerators using different pairs of incident particles, e^+e^-, $e^\pm p$, $p\bar{p}$ and pp, complement to each other. Hadron colliders can be considered as discovery machines, whilst e^+e^- colliders are optimal for the precision measurements of the properties of the discovered particles.

Problems

Problem 13.1 At the planned electron–proton collider eLHC, 7 TeV protons will be scattered against electrons with an energy of 67 GeV. What is the centre-of-mass energy of the proton–electron system? [Note: Protons and electrons with these energies are both extremely relativistic so that their masses can be neglected when determining their momenta.]

Problem 13.2 The cross section for the production of the Higgs boson production and decay to two photons at the 13 TeV LHC is 110 fb. If the collider luminosity is 10^{34} cm^{-2} s^{-1}, how accurately will one be able to determine this cross section after 3 months running (hypothetically neglecting background processes and assuming only statistical errors)?

Problem 13.3 The LHC has a circumference of 27 km and the protons move around at the speed of light. In the 2017 run, there were 2556 bunches per ring each containing 1.15×10^{11} protons. The luminosity achieved was $1.58 \times 10^{34}\,\mathrm{cm^{-2}s^{-1}}$. Estimate the area of the beams at the points of intersection.

Chapter 14
Particle Detectors

In order to investigate the properties of subatomic particles by examining how they behave when they undergo inelastic scattering at very high energy, we need to be able to identify the particles which appear in the final state, and, wherever possible, to determine their energy. Sophisticatedly developed particle detectors are designed to achieve this goal.

High energy particles interact with the medium through which they are passing. They scatter and lose their energy, initiating different processes which are used in various particle detectors. Amongst these processes are nuclear elastic and inelastic scattering, *"Cherenkov radiation"* and *"bremsstrahlung"* (radiation from deceleration of charged particles), excitation and ionization processes.

The natural "detector" for cosmic rays is the Earth's atmosphere. The interaction of cosmic rays with the atmosphere gives rise to particle showers and the production of new particles. The size of the detector can be considerably reduced in comparison with the depth of the atmosphere as the density of the detector body is correspondingly increased. The relative contributions from different processes depend on the type of particle, its energy, and the type of material the particle is passing through. These processes are used in various particle detectors discussed below.

14.1 Particle Interaction with Matter

Energy Loss by Ionization Relativistic charged particles and photons interact with electrons in the medium, leading either to ionization or atomic excitation. The ionization energy loss depends only on the charge and velocity of the projectile, but not the mass. The net effect of multiple collisions leads to an energy loss, the rate of which is called *"stopping power"*.

© The Author(s), under exclusive license to Springer Nature Switzerland AG 2021
A. Belyaev, D. Ross, *The Basics of Nuclear and Particle Physics*, Undergraduate
Texts in Physics, https://doi.org/10.1007/978-3-030-80116-8_14

Fig. 14.1 Mean energy loss rate in liquid hydrogen, gaseous helium, carbon, aluminium, iron, tin and lead, as a function of the velocity-dependent factor $\beta\gamma$. [*Source*: https://pdg.lbl.gov/ 2019/reviews/rpp2019-rev-passage-particles-matter.pdf (Fig. 33.2)]

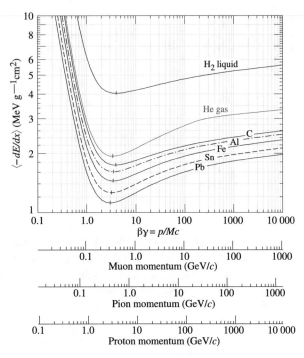

$$\frac{dE}{dx} \equiv \frac{dE}{d(\rho z)} = -\frac{E}{X_0}. \qquad (14.1)$$

Here $x = \rho z$ is the product of the linear depth, z, of the particle in the stopping medium, with its density ρ, so that dE/dx is in MeV g^{-1} cm^2. X_0 is called the "*radiation length*", although it is not actually a length, the units of X_0 are gm cm^{-2}. It is a measure of the mass per unit area of absorber material that the particle must penetrate for its energy to be attenuated to $1/e$ of its original value.

The dependence of energy loss on $\beta\gamma = p/(Mc)$ is presented in Fig. 14.1. The quantity X_0 is of the same order for different materials owing to the fact that the definition of radiation length is normalized to the material density, ρ. One can see that typical values of dE/dx are about 1 MeV g^{-1} cm^2. For values of $\beta \ll 1$, $dE/dx \propto 1/\beta^2$ which is essential for the stopping of slow particles like α-particles discussed earlier.

The ionization path from charged particles in the medium can be used to determine particle properties. This was used in the "*bubble chamber*" invented in 1952 by Donald Glaser [102].

In a bubble chamber a supersaturated liquid in a metastable phase detects charged particles which create an ionization track, around which the liquid vaporizes, forming microscopic bubbles. The density of the bubbles around a track is proportional to the density of ionized atoms and is related to the particle's energy loss. The bubbles

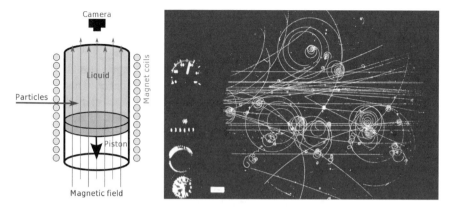

Fig. 14.2 Left: a schematic picture of bubble chamber. [*Source*: https://upload.wikimedia.org/wikipedia/commons/thumb/b/b8/Bubble-chamber.svg/800px-Bubble-chamber.svg.png] Right: the image of real particle tracks in CERN's first liquid hydrogen bubble chamber from 1960 [*Source*: https://cds.cern.ch/record/39474/files/11465.jpeg?subformat=icon-1440]

grow in size as the chamber is expanded by withdrawing a piston embedded in the chamber, and at some point they are large enough to be seen or photographed by cameras mounted around the chamber. The schematic picture of a bubble chamber is shown in Fig. 14.2(left) whilst Fig. 14.2(right) shows a bubble chamber event recorded at CERN, demonstrating multiple tracks from charged particles. The constant magnetic field in the chamber causes the charged particle to travel in a helical path, the radius of which is determined by the charge-to-mass ratio and the velocity. Since the charge of long-lived particles is known to be $\pm e$, we can deduce the particle's momentum from the radius of the curvature of the track. Moreover the lengths of the tracks and kinks in a bubble chamber allows the observation of the decay of longed-lived particles and hence the measurement of their lifetime.

Bubble chambers are no longer in use, but have been replaced by *"wire chambers"*. This is a gas filled chamber containing a network of wires held at a high potential. A charged particle ionizes the gas and the ions drift on to a wire, causing it to discharge. By tracking which wires "fire" in this way, one can reproduce the track of the charged particle. A *"drift chamber"* is a particularly accurate type of wire chamber since it takes into account the time taken for an ion to drift from the point of ionization to the nearest wire. Another type of modern detector is a scintillation counter, similar to that used in the detection of radioactivity, as discussed in Chap. 5. The signal from the wire chamber or scintillation counter is recorded directly by a computer. The interface between the detector and computer is called the *"readout"*. Very often, the readout software incorporates a *"trigger"* which only records an event if a certain criterion is satisfied. An example of such a trigger might be the requirement that at least one particle had a *"transverse momentum"* (i.e. the component of its momentum perpendicular to the direction of the incident scattering particles) larger than a certain specified value. A trigger substantially reduces the number of recorded events, avoiding overload of the data acquisition system. This is in stark contrast with the era of bubble chambers, during which each bubble-chamber photograph had to be reviewed by a team of scanners, who selected those few photographs which may contain an event of interest.

Energy Loss by Radiation (bremsstrahlung) When a charged particle is accel-erated or decelerated, it emits radiation and thus loses energy. This is called deceleration radiation (bremsstrahlung). It occurs when charged particles enter a material and are accelerated or decelerated by the electric field of the material's atomic nuclei and electrons. The intensity of bremsstrahlung radiation is inversely proportional to the square of the particle mass. In the case of electrons, the energy loss by bremsstrahlung dominates the energy loss by ionization, whereas for heavier particles such as the muon, with a mass of $208\,m_e$, or a charged pion with mass $280\,m_e$, the ionization energy loss is far greater than the bremsstrahlung loss. Thus, energy loss is determined by bremsstrahlung in the case of electrons but by ionization for other charged particles.

Cherenkov and Transition Radiation If the velocity of a charged particle is larger than the speed of light in the material, Cherenkov radiation is emitted.[1] The emission angle, θ_c, of Cherenkov photons with respect to the particle direction is given by

$$\cos\theta_c = \frac{1}{\beta n}\,, \tag{14.2}$$

where βc is the velocity of the particle, and n is the refractive index of the medium in which it is travelling. Note from (14.2) that the emission angle, θ_c only exists if $\beta n > 1$, for which the particle is travelling faster than the speed of light in the medium. When a charged particle crosses the boundary between two media with different refractive indices, additional radiation known as *"transition radiation"* takes place. The energy loss due to the emission of Cherenkov or transition radiation is small compared with the energy loss due to ionization or bremsstrahlung but both are used in high energy and cosmic-ray physics detectors.

14.2   Modern Detectors and Their Components

Since the bubble chambers era, the contemporary approach has significantly evolved to make use of the range of various detectors and technologies used at TeV colliders. The very big modern detector complex such as ATLAS (45 m long and 25 m in diameter) or CMS (21.6 m long and 14.6 m in diameter) at the LHC consist of several modules aimed at detecting different particles. In Fig. 14.3 the schematic

---

[1]This is the optical analogue of the "sonic boom" observed when an aircraft travels faster than the speed of sound.

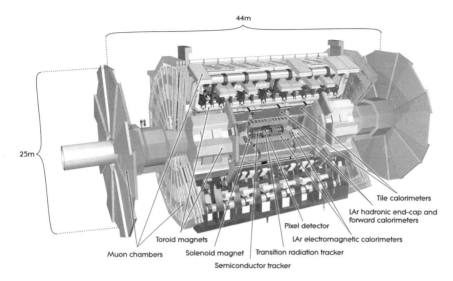

Fig. 14.3   Schematic   view   of   ATLAS   detector   [*Source*:   https://mediaarchive.cern.ch/
MediaArchive/Photo/Public/2008/0803012/0803012_01/0803012_01-A4-at-144-dpi.jpg]

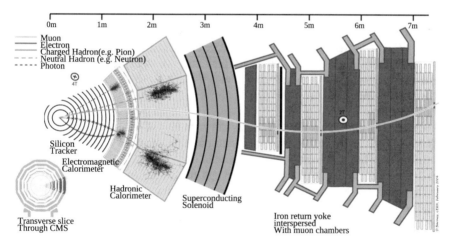

Fig. 14.4   Schematic section of the CMS detector [*Source*: https://cds.cern.ch/record/2205172/
files/CMS%20Slice.gif]

view of ATLAS detector is presented, whilst in Fig. 14.4 the schematic section of
CMS is shown.

ATLAS and CMS both have cylindrical structures around the vacuum pipe with
the centre at the collision points, and aim to cover the full 4π solid angle around the
collision events. The principle structure of ATLAS and CMS is similar, but details
of the detectors are quite different.

Fig. 14.5 The schematic diagram of the primary vertex and secondary vertex [103]. "IP" is the impact parameter, of one of the displaced tracks

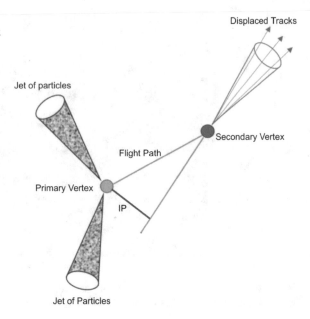

Magnets Modern detectors need a very strong magnetic field in order to bend high energy charged particles sufficiently to be able to determine the charge-to-mass ratio of the outgoing particles from the curved path they follow in the magnetic field. For example, the strengths of the magnetic fields in the ATLAS and CMS detectors are 2 and 4 T, respectively.

Tracker The first part of the detector as shown in Fig. 14.4 is the *"silicon tracker"*. This is the closest detector to the collision point and consists of parallel strips of millions of closely spaced diodes. It records the location, magnitude and timing of the signals from particles passing through the diode strip. This is the modern analogue of a bubble chamber but it works much more effectively and does not require human visual analysis of the images. It provides a very fast automatic readout of particle tracks and allows the determination of the transverse momentum of the particles to within 1%, provided its value is not too high (below a few tens of GeV). For higher values of momentum, the accuracy is diminished owing to the reduced precision of the determination of the track curvature in the magnetic field. In addition, the inner part of the tracker allows one to identify the so-called *secondary vertices* (or *"displaced vertices"*) from long-lived heavy flavour baryons, as shown qualitatively in Fig. 14.5. The typical physical scale of this picture can range from 1 to 10 mm, depending on the distance between the *"primary vertex"* which is at the beam collision point and the secondary vertex at which the massive long-lived particle decays. The charged tracks from this decay will be displaced by the distance from the interaction point by a perpendicular distance equal to the impact parameter (defined in Chap. 1), as shown in Fig. 14.5. The value of the impact parameter is used to tag long-lived particles.

ECAL The next detector, located further from the beam axis, is the *"electro-magnetic calorimeter"* (ECAL). In general, calorimeters measure particle energies with high accuracy and, unlike trackers, they are designed to stop the particles. In particular, the ECAL is sufficiently large and dense to stop electrons (positrons) and photons. The CMS ECAL is constructed from extremely dense but optically clear crystals of scintillating lead tungstate which emit light when particles go through them. The ATLAS ECAL uses liquid argon as its active medium. The ATLAS and CMS ECALs have quantities of scintillating material corresponding to about 25 radiation lengths. This is sufficient to absorb almost all of the energy of electrons and photons with initial energies up to 1 TeV. As with scintillation counters designed to detect radioactivity, the intensity of the light emitted by a scintillator crystal is proportional to the energy of the incident particle. An important characteristic for the calorimeters is the energy resolution, ΔE, which depends on the particle energy, E, and is parametrized as

$$\frac{\Delta E}{E} = \frac{a}{\sqrt{E}} \oplus \frac{b}{E} \oplus c \equiv \sqrt{\frac{a^2}{E} + \frac{b^2}{E^2} + c^2}, \qquad (14.3)$$

where a, b and c (often expressed as percentages) are characteristic parameters of the detector. The first term is the stochastic term related to the fluctuations in the number of signal generating processes, the second term is the noise term caused by other particles from other collision events and the third term is the calorimeter specific term (related to imperfections *etc.*) and is particularly important at high energies. The "$\oplus$" symbol means that terms entering the energy resolution should be added in quadrature as indicated in (14.3). For ATLAS these parameters are [104]: $a = 0.1\,\mathrm{GeV}^{1/2}$, $b = 0.17\,\mathrm{GeV}$, and $c = 0.007$ whereas for CMS they are $a = 0.028\,\mathrm{GeV}^{1/2}$, $b = 0.12\,\mathrm{GeV}$, and $c = 0.003$.

HCAL The hadronic calorimeter (HCAL) finds the position, energy, and arrival time of hadrons which are involved in the strong interactions. The HCAL of CMS contains layers of brass and steel absorber alternating with fluorescent plastic scintillators. The ATLAS HCAL consists of steel and scintillating tiles. Optical fibres feed the light pulses to photodetectors which amplify the signal. The total light energy captured over a given region provides a measure of a particle's energy. The HCAL provides wide angle coverage and records the energy and continuous position of individual hadrons. The total depth of the calorimeter system is about ten interaction lengths. Although hadrons pass through the ECAL without being stopped, they interact with the medium of the ECAL losing some of their energy. For this reason the effective energy resolution of the HCAL is actually the combination (again in quadrature) of the energy resolution of the ECAL and the energy resolution of the HCAL. The HCAL + ECAL combined energy resolution for hadrons is given by Cavallari [104]: $a = 0.53\,\mathrm{GeV}^{-1/2}$, $b = 0$, $c = 0.057$ for ATLAS and $a = 0.85\,\mathrm{GeV}^{-1/2}$, $b = 0$, $c = 0.074$ for CMS.

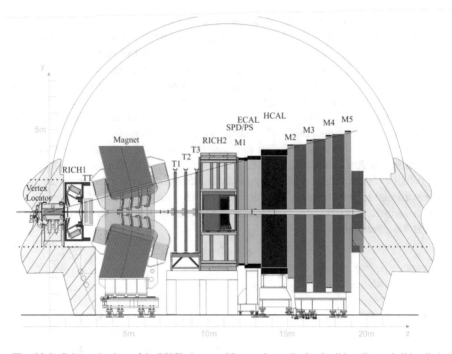

Fig. 14.6 Schematic view of the LHCb detector [*Source*: https://upload.wikimedia.org/wikipedia/commons/3/36/Lhcbview.jpg]

Muon Detectors Since the energy loss for a muon arises mainly by ionization (in contrast with the electron) and the muon does not interact strongly with the medium, muons can travel in the medium further than any other charged particles of the Standard Model. Therefore, in general purpose detectors, the outermost layers are usually designed to capture deeply penetrating muons, which can travel through several metres of iron and are not stopped by the calorimeters. The muon calorimeters are situated at the outer part of the cylindrical structure and consist of various types of detectors, which operate in an analogous way to a wire chamber or drift chamber.

Cherenkov Detectors Cherenkov radiation can be effectively used to identify charged particles created in the collisions. In particular, in the LHCb detector, a schematic view of which is shown in Fig. 14.6, charged hadron identification is effected by two Ring Imaging Cherenkov (RICH) detectors respectively denoted in the figure as RICH1 and RICH2.

In contrast with ATLAS and CMS, LHCb only has half of the full 4π coverage, as one can see from the Fig. 14.6. The RICH detectors allow the reconstruction of Cherenkov angles from the recorded patterns, thereby providing information about the particle velocity. Combining this with the momentum, determined from magnetic bending, one can deduce the particle mass, thereby identifying the particle.

Summary

- Particle detection is based on the effects of the interactions of particles with the medium through which they travel. These interactions include such effects as:
  - Ionization,
  - Atomic excitation,
  - Bremsstrahlung,
  - Cherenkov radiation,
  - Nuclear interactions.

- The following known particles can be seen in different parts of multi-purpose general detector:
  - Electrons – in the tracking chamber and the electromagnetic calorimeter (ECAL),
  - Photons – in the ECAL,
  - Neutral hadrons – in the Hadron calorimeter (HCAL),
  - Charged hadrons – in the tracking chamber, the ECAL and the HCAL,
  - Muons – in the tracking chamber, the ECAL, the HCAL and the muon chamber.

Problems

Problem 14.1 Calculate how thick the plate of steel should be in order to have the stopping power for high energy charged particles identical to the one of the Earth's atmosphere [use the same energy loss dE/dx for the atmosphere and the steel]. [The density of steel is $8 \times 10^3 \, kg \, m^{-3}$ and the atmospheric pressure is 1.01×10^5 Pa, and the acceleration due to gravity is $9.8 \, m \, s^{-2}$.]

Problem 14.2 The tracks left by particles in a bubble chamber form spirals. These are paths of a particle moving in a circle of decreasing radius.

(a) Explain why the radius of the circle decreases as the particle moves though the medium of the bubble chamber.
(b) Explain why some of the tracks are clockwise spirals whereas others are anticlockwise.

Problem 14.3 The Higgs boson has a mass of $125 \, GeV/c^2$. It can decay into two photons or into two hadrons of equal mass, such as B-mesons. What is the energy of each photon or hadron from the decay of a Higgs particle at rest?

Problem 14.4 For the ECAL and ECAL + HCAL energy resolution parameters given in Sect. 14.2 for ATLAS and CMS, calculate:

(a) The mass resolution for the Higgs boson when it decays (from rest) into two photons for ATLAS and CMS ECAL;

(b) The mass resolution for the Higgs boson when it decays into two B-mesons for ATLAS and CMS ECAL + HCAL.

[Note that in order to find the energy resolution of the decaying particle, the energy resolutions of the decay products must be combined in quadrature.]

Chapter 15
Constituent Quarks

15.1 Quarks

In 1950, the only particles that had been discovered were the stable particles – electron, proton (and their antiparticles) and neutron – and a few mesons, namely the charged pions ($\pi^{\pm}$) and charged kaons ($K^{\pm}$). In fact, the existence of pions had been predicted in 1935 by Hideki Yukawa [105], who proposed a model of strong interactions in which pions acted as their force carriers. At that time, these particles were considered to be elementary, i.e. could not be broken down into constituents.

With the rapid development of particle accelerators and the greatly increased effort in Particle Physics experiments, by the early 1960s, a large number of massive mesons and baryons had been observed. In 1964, this led Murray Gell-Mann [106] and George Zweig [107] (independently) to propose that hadrons were in fact bound states of elementary particles called *"quarks"*.[1] Baryons are bound states of three quarks, whereas mesons are bound states of a quark and an antiquark.

Quarks are spin-$\frac{1}{2}$ particles and have electric charge either $+\frac{2}{3}$ or $-\frac{1}{3}$, in atomic units. So, for example, a proton is made up of two quarks with electric charge $+\frac{2}{3}$ and one quark with electric charge $-\frac{1}{3}$, whereas a neutron consists of one quark with electric charge $+\frac{2}{3}$ and two quarks with electric charge $-\frac{1}{3}$.

Quarks also have another property called *"flavour"*. For example, the quark with flavour "strange" (s-quark) has a quantum number called strangeness associated with it. Its *"strangeness"*, S, is assigned to be $S = -1$ (the negative sign arises because the K^+, which actually contains an $\bar{s}$-antiquark, was assigned strangeness $S = +1$). This means that any hadron that contains one s-quark has strangeness -1, a hadron that contains two s-quarks has strangeness -2 *etc.* Antiparticles have the opposite quantum numbers from their corresponding particles, so that

---

[1] The name "quark" was coined by Gell-Mann, who came across the word in "Finnegan's Wake" by James Joyce.

© The Author(s), under exclusive license to Springer Nature Switzerland AG 2021
A. Belyaev, D. Ross, *The Basics of Nuclear and Particle Physics*, Undergraduate
Texts in Physics, https://doi.org/10.1007/978-3-030-80116-8_15

mesons can have positive or negative strangeness as they can contain s-quarks or $\bar{s}$-antiquarks (or both). Strangeness is conserved in strong interactions, but not in weak interactions. This means that although the decay of a K^+ ($S = 1$) into a pion plus other particles with zero strangeness is energetically possible, it requires a violation of the conservation of strangeness. Such a decay can therefore only proceed via the weak interactions, and consequently it takes place relatively slowly, which is why kaons live sufficiently long to leave discernible tracks in a detector. This conservation, exhibited by the strong interactions but not by the weak interactions, applies to *all* the flavours of quarks.

There are six known flavours of quarks. Their properties are summarized in Table 15.1. All of these quarks have antiparticles (antiquarks), which have opposite flavours – including opposite electric charge and opposite strangeness, but the same spin and mass.

The "masses" indicated in the last column of Table 15.1 are approximate. The reason for this is that in order to measure the mass of a particle, one needs to be able to create the free particle in a laboratory. It turns out that like the gluons, these quarks are confined to within a hadron or distances of the order of 1 fm, so that precise measurement of their masses is not possible. The approximate masses quoted are deduced from the masses of mesons or baryons constructed out of them.[2] For example, the J/Ψ, which is the lowest mass bound state of a c-quark and a $\bar{c}$-antiquark, has a mass of 3.1 GeV/c^2, from which we deduce that the c-quark and its antiquark have a mass of around 1.5 GeV/c^2. Likewise a proton, whose mass is close to 1 GeV/c^2, consists of two u-quarks and a d-quark, whereas a neutron, whose mass is almost the same, consists of two d-quarks and a u-quark. From this we deduce that the u-quark and d-quark both have a mass of approximately 0.33 GeV/c^2.

Quarks also possess another quantum number. They come in three types known as *"colours"* (not literally colour – the word is just used as a label). The three quarks that make up a baryon have different colours, whereas the quark and antiquark that make up a meson are of opposite colour.

Table 15.1 The six flavours of quarks

Symbol	Flavour	Electric charge (e)	Isospin	I_3	Mass **GeV**/c^2
u	Up	+2/3	1/2	+1/3	≈ 0.33
d	Down	−1/3	1/2	−1/2	≈ 0.33
c	Charm	+2/3	0	0	≈ 1.5
s	Strange	−1/3	0	0	≈ 0.5
t	Top	+2/3	0	0	≈ 173
b	Bottom (beauty)	−1/3	0	0	≈ 4.5

---

[2]This is not a unique way of defining quark masses. Another method introduces the concept of a *"bare mass"*, which, in the case of the lighter quarks, is considerably lower than the values quoted in Table 15.1. The exact definition of this bare mass is outside the scope of this book.

The u- and d-quarks form a doublet of *strong*[3] isospin with third component, I_3, equal to $\pm 1/2$, respectively. They transform into each other under isospin transformations (rotations in isospace). The other quarks are isospin singlets (they do not change under isospin transformations). Strong interactions are invariant under strong isospin transformations. As explained in Chap. 11, this means that, as far as strong interactions are concerned, the u- and d-quarks can be considered to be the same particle, but in different quantum states, denoted by the quantum number, I_3. The weak and electromagnetic interactions are *not* invariant under this isospin transformation. Hadrons, which differ only by the replacement of a u-quark by a d-quark or vice versa (i.e. by changes only in the value of I_3), have almost the same mass – the small mass difference being due to isospin violating electromagnetic interactions. A particle containing of N_u, u-quarks and N_d, d-quarks and possibly also $N_{\bar{u}}$, $\bar{u}$-antiquarks and $N_{\bar{d}}$, $\bar{d}$-antiquarks is labelled by its value of I_3, where

$$I_3 \; = \; \frac{1}{2}\left(N_u - N_d - N_{\bar{u}} + N_{\bar{d}}\right), \tag{15.1}$$

and by its isospin I. For a given value of isospin, I, there are $(2I+1)$ possible values of I_3. A multiplet of particles with isospin I is sometimes referred to as a "$2I + 1$"-plet. For example, a particle with isospin zero is called a "singlet", whereas the three particles with isospin one are called a "triplet".

Rotations, whether in ordinary space or in (the internal) isospace, are described mathematically by a symmetry group[4] called "*SU(2)*", whose fundamental building block has two components. In the case of transformations in ordinary space (which generate rotations), these two components are the two values of the z-component of spin for a spin-$\frac{1}{2}$ particle. States with larger angular momentum can be built out of several of these two-component blocks. For example, two spin-$\frac{1}{2}$ particles can combine to give a spin-one state with three possible values of the z-component of spin or a singlet state with spin zero. For isospin transformations, the two components of the fundamental building block are the u-quark and d-quark. States consisting of several such quarks can have higher values of isospin.

Isospin transformations are generated by three basic operations:

1. replacement of a d-quark by a u-quark or $\bar{u}$-antiquark by $\bar{d}$-antiquark, thereby increasing I_3 by one,
2. replacement of a u-quark by a d-quark or $\bar{d}$-antiquark by $\bar{u}$-antiquark, thereby decreasing I_3 by one, and
3. reversal of the relative sign of the wavefunction of a d-quark (or antiquark) relative to the wavefunction of a u-quark (or antiquark). Reversal of the

---

[3]This isospin was discussed in Chap. 11. It is sometimes called "*strong isospin*" to distinguish from "*weak isospin*" under which *all* quarks are arranged in doublets. This will be discussed in Chap. 17.

[4]We use the term "*symmetry group*" to mean a set of transformations, under which certain interactions are invariant.

wavefunction for *both* the u-quark and d-quark simply reverses the sign of the entire wavefunction, which has no physical effect.

15.2   Hadrons from u- and d-Quarks and Antiquarks

15.2.1   Baryons

The lightest baryons constructed out of only u- and d-quarks are shown in Table 15.2.

- Three spin-$\frac{1}{2}$ quarks can give a total spin of either $\frac{1}{2}$ or $\frac{3}{2}$ and these are the spins of the baryons. For these "low-mass" particles, the orbital angular momentum of the quarks is zero so that the spin of the baryon is the vector sum of the spins of the three quarks. On the other hand, excited bound states of quarks (heavier baryons) with non-zero orbital angular momenta are also possible, and in these cases, the determination of the spins of the baryons is more complicated.
- Three isospin-$\frac{1}{2}$ quarks produce states which either have $I = \frac{1}{2}$ (the lightest being nucleons) or $I = \frac{3}{2}$. The $I = \frac{3}{2}$ states are the Δ baryons and have four possible values of I_3 corresponding to electric charges which range between -1 and $+2$. The electric charge, Q, of these baryons is related to I_3 by

$$Q = I_3 + \frac{1}{2}\,. \tag{15.2}$$

- The masses of particles with the same isospin (but different I_3) are almost the same – the differences being due to the electromagnetic interactions which distinguish members of the isospin multiplet with different electric charges. If it were possible to "switch off" the electromagnetic interactions, these masses would be exactly equal.
- The baryons that consist of three u-quarks or three d-quarks only occur for spin $\frac{3}{2}$ (we return to this later).

Table 15.2 Baryons constructed out of u- and d-quarks

Baryon	Quark content	Spin	Isospin	I_3	Mass (MeV/c^2)
p	uud	1/2	1/2	+1/2	938
n	udd	1/2	1/2	−1/2	940
Δ^{++}	uuu	3/2	3/2	+3/2	1232
Δ^{+}	uud	3/2	3/2	+1/2	1232
Δ^{0}	udd	3/2	3/2	−1/2	1232
Δ^{-}	ddd	3/2	3/2	−3/2	1232

15.2.2 Mesons

The lightest mesons formed from u- and d-quarks and their antiquarks are shown in Table 15.3.

- A spin-$\frac{1}{2}$ quark and an antiquark with the same spin in a bound state with zero orbital angular momentum produce mesons of spin zero or spin one.
- Two isospin doublets can combine to form an $I = 1$ triplet – the π-mesons (pions) or ρ-mesons – or neutral $I = 0$ singlets. The electric charge, Q, of these mesons is simple given by

$$Q = I_3 . \tag{15.3}$$

- The neutral mesons have $I_3 = 0$. The π^0 and ρ^0 are in an isospin $I = 1$ state and have masses that are similar to those of their charged counterparts. The ω, whose mass is a little larger than that of the ρ-mesons, has isospin $I = 0$. In general, the neutral mesons are not pure $u\bar{u}$, or $d\bar{d}$ states, but quantum superpositions of these, as shown in Table 15.3.

The strong interactions conserve flavour. In a scattering process mediated by the strong interactions, a quark can annihilate against an antiquark of the same flavour, releasing energy that can then be converted into mass and used to create a more massive particle. An example of this is

$$
\begin{array}{cccc}
\pi^- & + & p & \rightarrow & \Delta^0 \\
(d\bar{u}) & & (uud) & & (udd)
\end{array} . \tag{15.4}
$$

In this process, a u-quark from the proton and a $\bar{u}$-antiquark from the pion annihilate and the extra energy goes into the extra mass of the Δ^0, as shown in Fig. 15.1. The Δ^0 is very short-lived and rapidly decays either back into p, π^- or to n, π^0. For the n, π^0 decay channel, there are two possible mechanisms. A d-quark, $\bar{d}$-antiquark pair can be created and the $\bar{d}$-quark binds with one of the d-quarks in the decaying Δ^0 to make the π^0 as shown in Fig. 15.2a, or a u-quark, $\bar{u}$-antiquark

Table 15.3 Lightest mesons consisting of a u-quark or d quark and a $\bar{u}$- or $\bar{d}$-antiquark

Meson	Quark content	Spin	Isospin	I_3	Mass (MeV/c^2)
π^+	$u\bar{d}$	0	1	$+1$	140
π^0	$\frac{1}{\sqrt{2}}(u\bar{u} - d\bar{d})$	0	1	0	135
π^-	$d\bar{u}$	0	1	-1	140
ρ^+	$u\bar{d}$	1	1	$+1$	775
ρ^0	$\frac{1}{\sqrt{2}}(u\bar{u} - d\bar{d})$	1	1	0	775
ρ^-	$d\bar{u}$	1	1	-1	775
ω	$\frac{1}{\sqrt{2}}(u\bar{u} + d\bar{d})$	1	0	0	782

Fig. 15.1 The quark picture of the production of a Δ^0 in $\pi^- - p$ scattering. An antiquark is denoted by a line with the direction of the arrow reversed

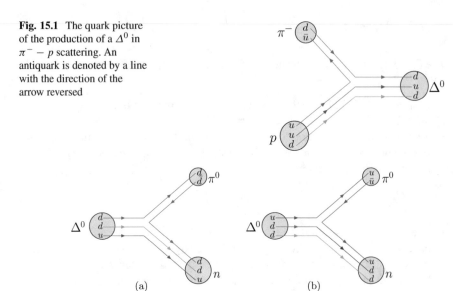

Fig. 15.2 The quark picture of the decay of a Δ^0 into a π^0 and a neutron, either by creation of a d-quark, $\bar{d}$-antiquark pair (**a**) or a u-quark, $\bar{u}$-antiquark pair (**b**)

pair can be created and the $\bar{u}$-antiquark binds with the u-quark in the decaying Δ^0 to make the π^0 (which is a superposition and $u\bar{u}$ and $d\bar{d}$), as shown in Fig. 15.2b. The short-lived intermediate Δ^0 appears as a resonance in the scattering cross section.

Quark–antiquark pair creation is possible in any particle–particle scattering process provided there is sufficient energy to create the final-state particles. Thus, for example, it is possible to have the inelastic process

$$p + p \rightarrow n + n + \pi^+ + \pi^+$$
$$(uud) \quad (uud) \quad (udd) \quad (udd) \quad (u\bar{d}) \quad (u\bar{d}). \tag{15.5}$$

In this process, two pairs of d-quarks and $\bar{d}$-antiquarks are created. The d-quarks bind with the u- and d-quarks from the incoming protons to form neutrons, whereas the $\bar{d}$-antiquarks bind with the remaining u-quarks in the incoming protons to form pions. In the CM frame, in which the total momentum is zero, the lowest energy that the incoming protons require for this process to occur is equal to the sum of the rest energies of two neutrons and two pions. At this minimum energy, known as the *"threshold energy"* for the process (15.5), all the particles in the final state are at rest – they all have zero momentum. The threshold value for the invariant quantity s (see (13.3)) is

$$s_{\text{threshold}} = \left(2m_n c^2 + 2m_\pi c^2\right)^2. \tag{15.6}$$

15.3   Hadrons with s-Quarks (or $\bar{s}$-Antiquarks)

15.3.1   Hyperons

Baryons that contain one or more s-quarks are known as *"hyperons"*.[5] The lightest
hyperons and their properties are listed in Table 15.4.

- As in the case of non-strange baryons, these can have spin $\frac{3}{2}$ or spin $\frac{1}{2}$.
- – Hyperons with one s-quark have strangeness $S = -1$ and isospin $I = 0$ or
  $I = 1$.
  – Hyperons with two s-quarks have strangeness $S = -2$ and isospin $I = \frac{1}{2}$.
  – Hyperons with three s-quarks have strangeness $S = -3$ and isospin $I = 0$.
  The electric charge, Q, of hyperons is related to the strangeness, S, and I_3 by

$$Q = I_3 + \frac{1}{2} + \frac{S}{2}. \tag{15.7}$$

The Ω^- ($S = -3$) had not yet been discovered when the Quark Model
was invented. Its existence was a prediction of the model. Furthermore, its mass
was predicted from the observation that the masses of the spin-$\frac{3}{2}$ hyperons were
approximately a linear function of strangeness, namely

$$M_{\Sigma^*} - M_\Lambda \approx M_{\Xi^*} - M_{\Sigma^*} \approx 150\,\text{MeV/c}^2,$$

giving a predicted value for M_Ω of

$$M_\Omega = M_{\Xi^*} + 150 = 1680\,\text{MeV/c}^2.$$

Table 15.4 The lightest hyperons

Baryon	Quark content	Strangeness	Spin	Isospin	I_3	Mass (**MeV/c**^{2})
Σ^+	uus	-1	$1/2$	1	$+1$	1189
Σ^0	uds	-1	$1/2$	1	0	1193
Σ^-	dds	-1	$1/2$	1	-1	1197
Ξ^0	uss	-2	$1/2$	$1/2$	$+1/2$	1315
Ξ^-	dss	-2	$1/2$	$1/2$	$-1/2$	1321
Λ	uds	-1	$1/2$	0	0	1116
Σ^{*+}	uus	-1	$3/2$	1	$+1$	1383
Σ^{*0}	uds	-1	$3/2$	1	0	1384
Σ^{*-}	dds	-1	$3/2$	1	-1	1387
Ξ^{*0}	uss	-2	$3/2$	$1/2$	$+1/2$	1531
Ξ^{*-}	dss	-2	$3/2$	$1/2$	$-1/2$	1535
Ω^-	sss	-3	$3/2$	0	0	1672

---

[5]Hyperons are constructed out of $u-$, $d-$ and s-quarks only.

The original measured value of the Ω^- mass [108] was $1686 \pm 12\,\text{MeV/c}^2$ (its current experimental value is $1672.5 \pm 0.3\,\text{MeV/c}^2$).

15.3.2 Strange Mesons

The lightest mesons containing an s-quark and/or an $\bar{s}$-antiquark are shown in Table 15.5.

- The mesons can have either spin zero or spin one.
- Mesons containing an s-quark have strangeness $S = -1$ and mesons containing an $\bar{s}$-antiquark have strangeness $S = +1$. Note that the η is a superposition of bound states of $u\bar{u}$, $d\bar{d}$ and $s\bar{s}$. The electric charge, Q, of strange mesons is related to their strangeness, S, and I_3 by

$$Q = I_3 + \frac{S}{2}. \tag{15.8}$$

The various expressions, (15.2), (15.3), (15.7) and (15.8), for the electric charges can be combined into a single expression:

$$Q = I_3 + \frac{Y}{2}, \tag{15.9}$$

where Y is the hypercharge, defined to be

$$Y = B + S, \tag{15.10}$$

where B is the baryon number. For baryons $B = 1$, and for mesons $B = 0$.

Table 15.5 Mesons containing an s-quark and/or an $\bar{s}$-antiquark

Meson	Quark content	Strangeness	Spin	Isospin	I_3	Mass (MeV/c^2)
K^+	$u\bar{s}$	+1	0	1/2	+1/2	494
K^0	$d\bar{s}$	+1	0	1/2	−1/2	498
$\overline{K^0}$	$s\bar{d}$	−1	0	1/2	+1/2	494
K^-	$s\bar{u}$	−1	0	1/2	−1/2	495
η	$(u\bar{u}, d\bar{d}, s\bar{s})$	0	0	0	0	548
K^{*+}	$u\bar{s}$	+1	1	1/2	+1/2	892
K^{*0}	$d\bar{s}$	+1	1	1/2	−1/2	896
$\overline{K^{*0}}$	$s\bar{d}$	−1	1	1/2	+1/2	896
K^{*-}	$s\bar{u}$	−1	1	1/2	−1/2	892
ϕ	$s\bar{s}$	0	1	0	0	1019

15.4 The Eightfold Way

The *"eightfold way"* is a method of classifying hadrons made from $u-$, $d-$ and s-quarks and their antiquarks, devised independently by Murray Gell-Mann [109] and Yuval Ne'eman [110]. It extends the concept of isospin invariance, which is invariance under the interchange of u- and d-quarks, to an invariance under interchange of any of the u-, d- or s-quarks. Whereas isospin invariance is described mathematically by the group SU(2), whose fundamental building block has two components – the u- and d-quarks – the eightfold way is described mathematically by a symmetry group called *"SU(3)"*, which has a three-component fundamental building block – the $u-$, $d-$ and s-quarks. Extending the three basic SU(2) operations, SU(3) transformations are generated by eight basic operations, namely the six operations consisting of the replacement of any of the three quarks by one of the two other quarks and the two possible sign reversals of the wavefunction of the d-quark or the s-quark, relative to that of the u-quark. This total of eight operations is the origin of the name "eightfold way".[6]

Clearly, the strong interactions are not exactly invariant under such transformations. For example, the masses of strange hadrons are considerably different from those which do not contain s-quarks. Nevertheless, it is helpful to assume that strong interactions may be partitioned into an SU(3) invariant part, which is indeed invariant under the interchange of any of these three quarks, plus a correction which is the contribution to the strong interactions that does *not* respect SU(3) invariance. This SU(3) non-invariant part is responsible for the differences in mass of hadrons with different strangeness.

Particles with the same spin, and approximately the same mass, can be displayed as sites on a plot of strangeness against I_3. Such a diagram is called a *"weight diagram"*.

15.4.1 Weight Diagrams for Baryons

The weight diagrams for the multiplet of eight spin-$\frac{1}{2}$ baryons (octet) and the multiplet of ten spin-$\frac{3}{2}$ baryons (decuplet) are shown in Figs. 15.3 and 15.4, respectively.

The rows contain the isospin multiplets. However, in the case of the octet, there are *two* particles (Σ^0 and Λ), both with $S = -1$ and $I_3 = 0$ but with different isospin, I. The $I_3 = 0$ and $S = 0$ site in the weight diagram for an octet has multiple occupancy.

The masses of the particles in one multiplet are very roughly the same (to within about 30%). For example, the mass of the Σ particles is about $1190 \, \text{MeV}/c^2$,

---

[6]The name is also an allusion to the "eightfold path" of Buddhism.

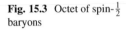

Fig. 15.3 Octet of spin-$\frac{1}{2}$ baryons

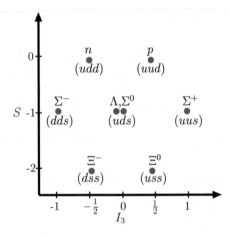

Fig. 15.4 Decuplet of spin-$\frac{3}{2}$ baryons

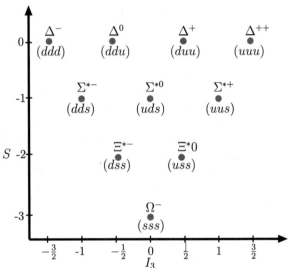

compared with the mass of a nucleon of about 940 MeV/c^2. The masses of particles with the same I_3 and strangeness but different isospin, I, are also somewhat different. For example, the mass of the Λ, which has isospin zero, is 1116 MeV/c^2, although it has the same strangeness and I_3 as the Σ_0, which is part of an $I = 1$ triplet, and has a mass of 1193 MeV/c^2. If strong interactions were invariant under SU(3) transformations, in the same way that they are invariant under SU(2) (isospin) transformations, i.e. if we could somehow switch off the part of the strong interactions which was *not* invariant under SU(3) transformations, the masses of all the particles in one multiplet would be almost the same, with small differences arising only from electromagnetic interactions.

15.4.2 Weight Diagrams for Mesons

The weight diagrams for the lightest spin-zero mesons and the lightest spin-one mesons are shown in Figs. 15.5 and 15.6, respectively. The spin-zero mesons form an octet in a similar way to the spin-$\frac{1}{2}$ baryons, whereas the spin-one mesons are a nonet – the site with $S = 0$ and $I_3 = 0$ has three entries (ρ^0, ϕ and ω), all of which are superpositions of $u\bar{u}$, $d\bar{d}$ and $s\bar{s}$.

These meson multiplets contain mesons and their antiparticles (obtained by replacing each quark by its antiquark and vice versa). Conversely, the baryon multiplets have separate antiparticle multiplets, which are bound states of three antiquarks. Note that besides having opposite electric charge, particles and their antiparticles have opposite strangeness (e.g. the K^+ with strangeness $S = +1$ is the antiparticle of the K^- whose strangeness is $S = -1$). Mesons, such as π^0, ρ^0, η, which are bound states of a quark and an antiquark of the same flavour, are their

Fig. 15.5 Weight diagram for the lightest spin-zero mesons

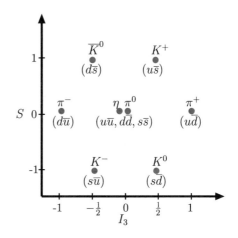

Fig. 15.6 Weight diagram for the lightest spin-one mesons

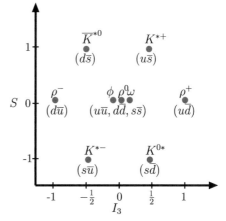

own antiparticles. Pions are anomalously light so that the mesons in the spin-zero octet have masses that vary by a factor of three. On the other hand, the mesons in the spin-one nonet have masses which are within 25% of each other.

15.4.3   Associated Production and Decay

In strong-interaction processes, quark flavour is conserved. s-quarks cannot be converted into other quarks by the strong interactions. An s-quark can, however, be created in conjunction with an $\bar{s}$-antiquark in a strong-interaction process. This means that in a scattering experiment (e.g. proton–proton or pion–proton scattering), the strong interactions can only create a particle containing a strange quark if at the same time there is a particle containing an $\bar{s}$-antiquark, so that the total strangeness is conserved. An example of such a process is

$$\begin{array}{ccccc}
\pi^- & + & p & \to & \Lambda & + & K^0 \\
(d\bar{u}) & & (duu) & & (dus) & & (\bar{s}d)
\end{array} \tag{15.11}$$

What happens is that a u-quark annihilates against a $\bar{u}$-antiquark and an s-quark, $\bar{s}$-antiquark pair is created, as shown in Fig. 15.7.

This reaction is only possible above a threshold energy, i.e. the total energy of the pion and proton in the CM frame must exceed the combined rest energy of the Λ and K^0.

By contrast, the process

$$\begin{array}{ccccc}
\pi^- & + & p & \to & \Lambda & + & \pi^0 \\
(d\bar{u}) & & (duu) & & (dus) & & (\bar{d}d, \bar{u}u)
\end{array} \tag{15.12}$$

is forbidden in the strong interactions because the strangeness of the initial and final states is not the same. Process (15.12) *can* proceed by the weak interactions, but the cross section for such a process is much smaller than the cross section for process (15.11).

Fig. 15.7 Quark picture of the associated production of particles with opposite strangeness

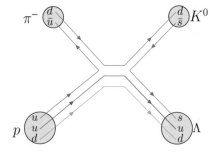

It is, however, possible to scatter charged kaons against nucleons. The K^- contains an s-quark, and it is therefore possible to produce strange baryons in this process, such as

$$K^- + n \rightarrow \Lambda + \pi^-$$
$$(s\bar{u}) \quad (udd) \quad (dus) \quad (\bar{u}d). \tag{15.13}$$

All flavours have been conserved in this reaction. Conversely, $K^+ - n$ scattering will *not* produce a strange baryon because a strange baryon contains s-quarks, but no $\bar{s}$-antiquarks, whereas the K^+ contains an $\bar{s}$-antiquark, but no s-quark.

Similarly, in the decays of more massive particles containing s-quarks (or $\bar{s}$-antiquarks), the decay can proceed via the strong interactions into lighter particles with strangeness. This strangeness conserving decay process is very rapid – leading to a resonance with a large width, but it is only possible if the decay products have a total strangeness which is equal to the strangeness of the decaying particle. For such a process to occur, the mass of the decaying particle must be larger than the combined mass of the decay products.

An example is

$$K^{*+} \rightarrow K^0 + \pi^+$$
$$(u\bar{s}) \quad (d\bar{s}) \quad (u\bar{d}). \tag{15.14}$$

A d-quark and $\bar{d}$-antiquark pair is created as shown in Fig. 15.8. The initial and final states both contain an $\bar{s}$-antiquark, so that flavour is conserved. The masses are $m_{K^*} = 892\,\text{MeV/c}^2$, $m_{K^0} = 494\,\text{MeV/c}^2$ and $m_{\pi^+} = 140\,\text{MeV/c}^2$. The mass of the K^* exceeds the sum of the masses of the kaon and pion, so that there is enough energy for the decay to proceed. Since this is a strong-interaction decay, the K^* has a very short lifetime. It is only seen as a resonance in the CM energy of the final-state $K - \pi$ system (i.e. the total energy of the final-state K^0 and π^+ in the frame in which the sum of their momenta is zero).

Likewise, the Ξ^* has enough mass to decay into a Ξ^7 plus a pion – the initial and final states both having strangeness, $S = -2$. Similarly, the Σ^* can decay into a

Fig. 15.8 Quark picture of the strong-interaction decay of a K^{*+} meson into K^0 and π^+

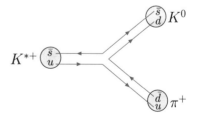

---

[7] Ξ is pronounced /ksai/, but it is also known as the "cascade" particle.

Σ plus a pion or into a Λ plus a pion, conserving strangeness. Again, these strong-interaction decays proceed very rapidly.

Most of the lighter strange particles do not have enough energy to decay into other strange particles. They have to decay through the weak interactions and therefore have a much longer lifetime, leaving a track in a detector.

Combining associated production and decay, we get a two-stage process such as

$$\pi^+ + n \;\rightarrow\; K^{*+} + \Lambda \;\rightarrow\; K^0 + \pi^+ + \Lambda. \tag{15.15}$$

The observed particles are the K^0, π^+ and Λ. The K^{*+} is only observed as a resonance in the final-state $K - \pi$ channel. It is identified by measuring the energies, E_K and E_π and momenta $\boldsymbol{p}_K$ and $\boldsymbol{p}_\pi$ of the final-state K^0 and π^+. From these, we can construct the relativistically invariant quantity

$$s_{K\pi} \equiv (E_K + E_\pi)^2 - c^2 \left(\boldsymbol{p}_K + \boldsymbol{p}_\pi\right)^2.$$

This is the square of the combined energy, $E_{K\pi}$, of the K and π in their CM frame (for which $\left(\boldsymbol{p}_K + \boldsymbol{p}_\pi\right) = \boldsymbol{0}$). The cross section, when plotted against the energy $E_{K\pi}$, displays a resonance at $E_{K\pi} = 892\,\text{MeV}$, which is the rest energy of the K^{*+}. The width of the resonance is $50\,\text{MeV}$.

15.5   Heavy Flavours

When the quark model was invented, only u-, d- and s-quarks were postulated and all known hadrons could be built out of these three quarks and their antiquarks.

A fourth quark – the c (charm)-quark – was predicted from studies of the weak interactions [111]. In 1974, a very narrow resonance in the $e^+ e^-$ channel at a CM energy of 3.1 GeV was observed simultaneously at SLAC [112] and at BNL [113]. This is the J/Ψ particle.[8] This resonance was too narrow (i.e. its lifetime was too long) to be a meson formed out of $u-$, $d-$ or s-quarks, which would have decayed rapidly via the strong interactions into lighter mesons. This was a spin-one bound state of a c-quark and a $\bar{c}$-antiquark. It has a mass roughly twice the mass of the c-quark. Shortly afterwards, similar particles were observed, both with spin one and spin zero and in the following years several mesons and baryons containing a c-quark (or $\bar{c}$-antiquark) were found. Mesons containing a c-quark are called "D-mesons", whereas a neutral baryon containing one c-quark is denoted by Λ_c.

In 1977, a similar resonance, Υ,[9] was observed at FNAL in the $\mu^+\mu^-$ channel at a CM energy of 9.5 GeV [114]. This is a bound state of a b (bottom)-quark and

---

[8]It was named Ψ by the SLAC collaboration and J by the BNL collaboration.

[9]Pronounced /upsilon/.

a $\bar{b}$-antiquark. Several mesons and baryons containing a b-quark (or $\bar{b}$-antiquark) have since been found.

For hadrons containing c-quarks and/or b-quarks, the hypercharge, Y of (15.10), is amended to

$$Y = \mathcal{B} + S + C - B, \tag{15.16}$$

where the charm number, C, is the number of c-quarks minus the number of $\bar{c}$-antiquarks in the hadron and B is the number of b-quarks minus the number of $\bar{b}$-antiquarks in the hadron.

The last quark – the t (top)-quark – was discovered at FNAL in 1995 [115]. Although this quark decays via the weak interactions, it has such a short lifetime ($\approx 5 \times 10^{-25}$ s) that it does not live long enough to form a bound state with any other quark. It was identified from its decay into a W^+ gauge boson and a b-quark. The t-quark appeared as a resonance in the W^+b-quark system with an energy of 173 GeV. Why this quark has a mass which is so much larger than any other quark remains a mystery.

15.6 Quark Colour

There is a difficulty within the quark model when applied to baryons. This can be seen by looking at the Δ^{++} or Δ^- or Ω^-, which are bound states of three quarks, all of the same flavour. For these low-mass states, the orbital angular momentum is zero, and so the spatial parts of the wavefunctions for these baryons are symmetric under interchange of the position of two of these (identical flavour) quarks.

We know that the total wavefunction for a baryon must be antisymmetric as baryons have half-odd-integer spin and therefore obey Fermi–Dirac statistics. The spin part of the wavefunction of baryons constructed from three identical flavour quarks should therefore be antisymmetric. However, these baryons have spin-$\frac{3}{2}$, which means that the spin part of the wavefunction is symmetric (for example, the $S_z = +\frac{3}{2}$ state is the state in which all three quarks have $s_z = +\frac{1}{2}$ and this is clearly symmetric under the interchange of two spins).

This puzzle is solved by postulating that quarks come in three possible quantum states, denoted by their "colour": r, g or b. The antisymmetry of the baryon wavefunction is restored by requiring that the baryon wavefunction be antisymmetric under the interchange of two colours. If a baryon is composed of three quarks with flavours f_1, f_2 and f_3, then these should also have a colour index, e.g. f_1^r, f_1^g or f_1^b. The colour antisymmetric wavefunction is written as

$$\Psi_{\text{baryon}} = \frac{1}{\sqrt{6}} \left(|f_1^r f_2^g f_3^b\rangle + |f_1^b f_2^r f_3^g\rangle + |f_1^g f_2^b f_3^r\rangle \right.$$

$$\left. - |f_1^b f_2^g f_3^r\rangle - |f_1^r f_2^b f_3^g\rangle - |f_1^g f_2^r f_3^b\rangle \right). \tag{15.17}$$

We can see that this changes sign if we interchange any two colours. This means that in order to have a totally antisymmetric wavefunction (including the colour part), the spin and spatial parts must be *symmetric* so that a particle in which all three quarks have the same flavour (and zero orbital angular momentum) must be symmetric under the interchange of any two of the spins – and this is the spin-$\frac{3}{2}$ state.

A state of three different colours, which is antisymmetric under the interchange of any two of the colours, is called a "colour singlet" state – we can think of it as a colourless state. The quarks themselves are a colour triplet – meaning that they can be in any one of three colour states.

It is assumed that all physically observable particles (i.e. all hadrons) are colour singlets (colourless particles). This means that it is not possible to isolate individual quarks and observe them. Indeed no individual quark has ever been observed. This is called *"quark confinement"* and it provides the reason as to why the strong interactions are short range, despite the fact that the strong-interaction force carriers – the gluons – are massless. It means that you cannot pull a quark too far away from the other quarks or antiquarks in the hadron to which it is bound.

For mesons, we also require that the quarks and antiquarks bind in such a way that the meson is also in a colour singlet state. In the case of a quark and antiquark bound state, this means that the wavefunction is a superposition of a quark with colour r bound to an antiquark with colour $\bar{r}$, a quark with colour g bound to an antiquark with colour $\bar{g}$ and a quark with colour b bound to an antiquark with colour $\bar{b}$. Thus, for example, the wavefunction for the π^+ is written as

$$|\pi^+\rangle = \frac{1}{\sqrt{3}}\left(|u^r\bar{d}^{\bar{r}}\rangle + |u^g\bar{d}^{\bar{g}}\rangle + |u^b\bar{d}^{\bar{b}}\rangle\right). \tag{15.18}$$

The colourless property is achieved by requiring that a quark of a given colour binds with an antiquark of opposite colour. For example, the $\bar{u}^{\bar{r}}$-antiquark is the antiparticle of the u^r-quark. Then we have to "average" over all the colours by taking a superposition of all three possible colour pairs, as shown in (15.18).

15.7   "Exotic" Hadrons

Nearly all baryons discovered so far conform to the Quark Model picture of a bound state of three quarks. However, it is possible to have a bound state of four quarks and an antiquark. Such baryons are called *"pentaquarks"*. In 2015, the LHCb collaboration [116], whilst investigating the weak-interaction decay of Λ_b^0 (a b–u–d-quark bound state), which decays as

$$\Lambda_b^0 \rightarrow K^- + J/\Psi + p,$$

observed two resonances in the the J/Ψ, p channel, i.e. in the CM energy of the final-state J/Ψ and p. The resonances were found at CM energies of J/Ψ and p of 4380 MeV and 4450 MeV. This was consistent with the production and decay of a pentaquark consisting of 2 u-quarks, a d-quark, a c-quark and a $\bar{c}$-antiquark. Further similar states were observed by LHCb in 2019 [117].

Likewise, it is possible for a meson to be a bound state of two quarks and two antiquarks, called "tetraquarks". Several candidates for such a particle have been reported, but the first confirmed tetraquark is the $Z_c^{\pm}$ with a mass of 3900 MeV/c^2. It has been observed by three independent collaborations [118–120]. It is seen to decay into a $\pi^{\pm}$ and a J/Ψ, from which we can deduce that its quark content is c-quark, a $u(d)$-quark, a $\bar{c}$-antiquark and a $\bar{d}(\bar{u})$-antiquark. In 2020, the LHCb collaboration announced the observation of a resonance at an energy of 6900 MeV [121], which decays into two J/Ψ's. This is a tetraquark consisting of two c-quarks and two $\bar{c}$-antiquarks. In the same year, the LHCb collaboration observed the first tetraquark with non-zero (net) charm [122]. This is a charm meson consisting of a u- and d-quarks and a $\bar{c}$- and $\bar{s}$-antiquarks and was seen as a resonance in the D^-, K^+ channel at a CM energy of 2900 MeV.

Summary

- Quarks are the fundamental building blocks from which all hadrons are built.
- Quarks have spin-$\frac{1}{2}$ and carry electric charge $\frac{2}{3}e$ or $-\frac{1}{3}e$. They come in six possible flavours, u, d, s, c, b and t.
- Baryons are built from three quarks and mesons are built out of a quark and an antiquark.
- The lighter baryons and mesons constructed out of u-, d- and s-quarks and their antiquarks, with the same spin, can be arranged in multiplets of the symmetry SU(3). These multiplets consist of hadrons with the same spin and similar masses, but different charge and strangeness.
- The strong interactions conserve flavour. Pairs of particles with opposite flavour (e.g. strangeness) can be produced from the scattering of particles which do not possess that flavour.
- Quarks come in three "colours". All hadrons are colourless. For mesons this means that a meson is a bound state of a quark and antiquark of opposite colour, whereas for baryons it means that the colour part of the wavefunction is antisymmetric under exchange of the colours of two of the quarks. Since the baryons are fermions, the remaining part of the wavefunction must be symmetric. This means that baryons constructed out of three quarks of the same flavour have spin-$\frac{3}{2}$, since the spin part of the wavefunction must also be symmetric.
- Pentaquarks, which are baryons consisting of four quarks and one antiquark, as well as tetraquarks, which are mesons consisting of two quarks and two antiquarks, have also been shown to exist.

Problems

Problem 15.1 Write down the quark content of all baryons containing one c-quark. In each case, state the isospin, the third component of isospin, the strangeness and the electric charge (in atomic units).

Problem 15.2 Consider $\pi^- + p \rightarrow K^+ + \pi^0 + \Sigma^-$ process.

(a) What is the minimum CM energy for this process to occur?
(b) What is the minimum CM energy for this process to display a resonance in the $K\pi$ channel due to an intermediate K^{*0}?

$(m_{K^{*0}} = 892 \,\text{MeV/c}^2, \ m_{K^+} = 494 \,\text{MeV/c}^2, \ m_{\pi^0} = 135 \,\text{MeV/c}^2, \ m_{\Sigma^-} = 1197 \,\text{MeV/c}^2.)$

Problem 15.3 In a scattering event, the energies and momenta of two final-state particles a and b are measured to be

$$E_a = 1632 \,\text{MeV}, \quad \boldsymbol{p}_a = (858, -402, -941) \,\text{MeV/c}$$
$$E_b = 754 \,\text{MeV}, \quad \boldsymbol{p}_b = (390, 109, 401) \,\text{MeV/c}.$$

(a) What are the masses of particles a and b?
(b) What is the total energy in the CM frame of a and b?

Problem 15.4 Draw the quark diagram for the strong-interaction process

$$\pi^+ + p \rightarrow K^+ + \Sigma^+.$$

Chapter 16
Particle Interactions and Cross Sections

16.1 Relativistic Approach to Interactions

In Particle Physics, we are studying the interactions between sub-microscopic particles that are usually moving with velocities close to the speed of light. We therefore need a synthesis of Quantum Physics and Special Relativity in order to describe these interactions.

There are two fundamental differences between non-relativistic and relativistic Quantum Physics:

1. The Schroedinger equation describes a particle moving under a certain potential. In Special Relativity there is no such thing as a potential since this would imply an instantaneous action at the distance.
2. Since energy can be converted into mass and vice versa, particle number in a relativistic scattering event is *not* conserved – particles can annihilate with each other and new particles can be created. A Schroedinger wavefunction, which is normalized such that the probability of finding a particle somewhere in space is always unity, is clearly an unsuitable tool for describing such processes.

We therefore seek to modify the non-relativistic concept of potential in such a way that relativistic invariance is satisfied.

The potential due to a force, whose force carrier (gauge boson) has mass M, is given by the Yukawa potential [105]:

$$V(r) = \frac{g^2}{4\pi r} e^{-Mcr/\hbar}, \tag{16.1}$$

where g^2 is a measure of the strength of the interaction that generates this potential. The quantum matrix element, $\mathcal{A}(q)$, of this potential between an initial state of a free particle with momentum p and a final state of a free particle with momentum $p + q$ (confined to a box of volume V) is

© The Author(s), under exclusive license to Springer Nature Switzerland AG 2021
A. Belyaev, D. Ross, *The Basics of Nuclear and Particle Physics*, Undergraduate Texts in Physics, https://doi.org/10.1007/978-3-030-80116-8_16

$$\mathcal{A}(q) = \frac{1}{V} \int d^3 r \frac{g^2}{4\pi r} \exp\left\{ \frac{(i q \cdot r - Mcr)}{\hbar} \right\} = \frac{1}{V} \frac{g^2 \hbar^2}{\left(q^2 + M^2 c^2\right)} . \tag{16.2}$$

This is *not* a relativistically invariant expression as the momentum transfer, q, changes under a Lorentz transformation. The expression of the RHS of (16.2) can be modified to cast it into a relativistically invariant quantity, by altering it to

$$\mathcal{A}(E_q, q) = \frac{1}{V} (g \hbar c)^2 \Delta(E_q, q, M) , \tag{16.3}$$

where

$$\Delta(E_q, q, M) = -\frac{1}{\left(E_q^2 - q^2 c^2 - M^2 c^4 \right)} . \tag{16.4}$$

$\Delta(E_q, q, M)$ is a function both of the momentum transfer, q, and the energy transfer, E_q, i.e. the energy difference between the initial-state and final-state particle. The quantity E_q also changes under a Lorentz transformation in such a way that $E_q^2 - q^2 c^2$ is relativistically invariant. In the non-relativistic limit $E_q \ll |q| c$, so that we recover the non-relativistic expression (16.2).

$\Delta(E_q, q, M)$ is the *"propagator"* of a particle with mass M. It is interpreted as the quantum amplitude for the propagation of a particle of mass M, energy E_q, and momentum q. Interactions between particles mediated by a force, whose force carrier (gauge boson) has mass M, are therefore described by the exchange of the force carriers between the interacting particles, transferring energy E_q and momentum q between them. We do not need to introduce a potential, which acts instantaneously at a distance. The interacting particles couple to the gauge bosons at a single point in space and a single instant in time. Everything is described in terms of relativistically invariant quantities. A change in momentum that occurs in non-relativistic dynamics when a force is applied is generalized to a change in both momentum and energy, in a manner that leads to a quantum amplitude that is invariant under Lorentz transformations.

16.1.1 Virtual Particles

A free particle of mass M with energy E_q and momentum q obeys the relativistic energy–momentum relation

$$E_q^2 = q^2 c^2 + M^2 c^4 .$$

However, for the propagator described above, the energy and momentum transfers do *not* obey this relation. In fact, if they did, the propagator would diverge!

We are rescued by Heisenberg's uncertainty principle and the fact that the propagation of the gauge boson takes place over a very short time period – the time taken for the gauge boson to pass between the two interacting particles. The uncertainty principle tells us that over a sufficiently short period of time there is an uncertainty in energy. This means that if a particle only exists for a very short time we no longer have the usual relation between energy, momentum and mass, because the energy is not precisely determined. Such particles, which are exchanged rapidly between the interacting particles, are not actually observed. For this reason they are called *"virtual particles"* and because their energy and momentum do not obey the relativistic energy–momentum relation they are said to be *"off mass-shell"*.

16.1.2 Relativistic Momentum Transfer

In Chap. 13 we introduced the relativistically invariant variable s, which in the CM frame is equal to the total energy of the incoming (or outgoing) particles. In the same way, the relativistically invariant variable t is used to generalize the momentum transfer. This quantity [101] is defined as

$$t = E_q^2 - q^2 c^2. \tag{16.5}$$

In terms of t, the propagator is

$$\Delta(t, M) = -\frac{1}{t - M^2 c^4}. \tag{16.6}$$

In the CM frame, t is very simply related to the scattering angle, θ_{CM}, between the directions of the initial and final particle momentum. p_1 and p_3.

$$t = -2p_{CM}^2 c^2 (1 - \cos\theta_{CM}) = -\left(2p_{CM}c \sin\left(\frac{\theta_{CM}}{2}\right)\right)^2, \tag{16.7}$$

where p_{CM} is the magnitude of the momentum of the particles in the CM frame. In this frame, $\sqrt{-t}/c$ is the momentum transfer (see (2.7)). The (negative) value of t ranges from $-4p_{CM}^2 c^2$ to zero.

The magnitude, p_{CM}, of the momentum of the incident particles in the CM frame can be related to s using the relation between energy and momentum

$$\sqrt{s} = E_1 + E_2 = \sqrt{p_{CM}^2 c^2 + m_a^2 c^4} + \sqrt{p_{CM}^2 c^2 + m_b^2 c^4},$$

where m_a and m_b are the masses of the two incoming particles. After some algebra we can write p_{CM} in a manifestly Lorentz invariant form, as a function of s, m_a and m_b.

$$p_{CM} = \frac{\lambda^{1/2}\left(s, m_a^2 c^4, m_b^2 c^4\right)}{2c\sqrt{s}}, \tag{16.8}$$

where the function $\lambda(x, y, z)$, given by

$$\lambda(x, y, z) = x^2 + y^2 + z^2 - 2xy - 2xz - 2yz,$$

is known as the *"Källén function"*. It occurs frequently in relativistic kinematics. In the ultra-relativistic limit, $\sqrt{s} \gg m_a c^2, m_b c^2$, the expression for p_{CM} simplifies greatly to

$$p_{CM} \approx \frac{\sqrt{s}}{2c} \tag{16.9}$$

and (16.7) simplifies to

$$t \approx -\frac{s}{2}\left(1 - \cos\theta_{CM}\right). \tag{16.10}$$

Note that the scattering angle is frame dependent and, for example, takes a different value in the CM frame from its value in the Lab frame, but the quantity t remains the same in all frames.

In inelastic processes, it is possible for the mass of the outgoing particle to be different from that of the incoming particle. This happens in weak-interaction processes in which a quark can change flavour and also in strong interactions in which the outgoing quark is bound to a different hadron from that of the incoming quark. In such cases the relation (16.7) between t and momentum transferred does not hold, because the CM momenta of the incoming and outgoing particles differ. Nevertheless, the relativistically invariant quantity, t, is a useful variable. Moreover, in the ultra-relativistic limit, where $\sqrt{s}$ is much larger than the rest energy of *all* the external particles, (16.10) remains valid.

16.2   Feynman Diagrams

We therefore have a relativistic expression for the scattering amplitude of two particles, a and b (masses m_a and m_b) with initial energies and momenta $(E_1, \boldsymbol{p}_1)$, $(E_2, \boldsymbol{p}_2)$ and final energies and momenta $(E_3, \boldsymbol{p}_3)$, $(E_4, \boldsymbol{p}_4)$. The amplitude can be represented diagrammatically as shown in Fig. 16.1, which represents the scattering of two (different) particles due to the exchange of a gauge boson (force carrier) with mass M. This is known as a *"Feynman diagram"* (or *"Feynman graph"*) [89]. The amplitude for the process is obtained by applying a set of *"Feynman rules"* for each vertex, where particles interact with each other, and each internal line (the propagation of the virtual off mass-shell particle). The

Fig. 16.1 Feynman diagram representing the scattering of two particles, a and b. The internal line is the gauge boson, G, which is exchanged between the particles a and b, and is represented in Feynman diagrams by a "wiggly" line as shown

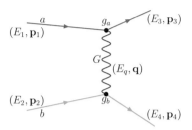

full set of Feynman rules takes into account the spins of the external and internal particles, which requires a detailed study of Quantum Field Theory.

Some of the Feynman rules for constructing the contribution to the amplitude from a Feynman diagram are:

- Include a (dimensionless) factor of $-g_a/\sqrt{\hbar c}$ at each vertex involving particle interacting with an exchanged gauge boson, with coupling constant g_a.
- Conserve energy and momentum at each vertex.
- Include a propagator factor:

$$\Delta(E, \boldsymbol{p}, M) = -\frac{1}{\left(E^2 - \boldsymbol{p}^2 c^2 - M^2 c^4\right)},$$

for each internal particle of mass M carrying energy E and momentum $\boldsymbol{p}$.
- Include a factor of $(\hbar c)^{3/2}/\sqrt{V}$ for each outgoing particle.

For the process described by the Feynman diagram of Fig. 16.1, conservation of energy and momentum at each vertex leads to

$$E_q = (E_3 - E_1) = (E_2 - E_4)$$

$$\boldsymbol{q} = \boldsymbol{p}_3 - \boldsymbol{p}_1 = \boldsymbol{p}_2 - \boldsymbol{p}_4.$$

This is equivalent to the conservation of energy and momentum in the scattering process:

$$E_1 + E_2 = E_3 + E_4$$

$$\boldsymbol{p}_1 + \boldsymbol{p}_2 = \boldsymbol{p}_3 + \boldsymbol{p}_4.$$

Taking the product of the vertex factors and the propagator, we reproduce the amplitude given by (16.3), with the g^2 replaced by the product of the coupling constants of particles a and b. For example, for an electromagnetic scattering of two particles with electric charges q_a and q_b, the factor g^2 is replaced by $q_a q_b/\varepsilon_0$.

This is actually not quite correct unless all the particles have zero spin (gauge bosons have spin one). A proper treatment yields

$$\mathcal{A}(s,t) = -\frac{\hbar^2 c^2 g_a g_b}{V} \eta(s,t,m_a,m_b) \frac{1}{(t-M^2c^4)} , \qquad (16.11)$$

where the function η is a relativistic correction factor, which depends on s,t and the masses and spins of the participating particles. We quote the result for η in the case of spin-$\frac{1}{2}$ particles exchanging a spin-one gauge boson. In this case the result given to us by Quantum Field Theory is[1]

$$\eta(s,t,m_a,m_b) = \frac{\sqrt{4(s-m_a^2c^4-m_b^2c^4)^2 + 2t^2 + 4st}}{4m_a m_b c^4}. \qquad (16.12)$$

In the non-relativistic limit, $s \approx (m_a c^2 + m_b c^2)^2$, and $|t| \ll s$ so that $\eta \to 1$, whereas in the ultra-relativistic limit, in which the masses are negligible, we have

$$\eta \approx \frac{\sqrt{4s(s+t) + 2t^2}}{4m_a m_b c^4}, \quad \sqrt{s} \gg m_a c^2, m_b c^2. \qquad (16.13)$$

We note that the amplitude is a function of the relativistically invariant kinematic variables, s and t, as well as the masses of the participating particles.

16.2.1 Multiple Feynman Graphs

Many processes have more than one Feynman diagram associated with them. An example is shown in Fig. 16.2, which describes the scattering of two identical particles.

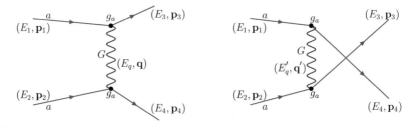

Fig. 16.2 Feynman diagram representing the scattering of two *identical* particles, a. The internal gauge boson can be exchanged between the particles in two different ways

---

[1]The amplitude actually depends on the spin states of the initial and final particles. What is quoted here is the RMS value over all possible spin configurations. This is the relevant quantity for experiments that do *not* measure the spins of the external particles.

The internal particles must be attached to the external particles in *all* possible ways. Figure 16.1 represents the scattering of two different particles in which the internal gauge boson couples the initial particle with energy and momentum (E_1, p_1) to the final particle with energy (E_3, p_3) at one end, and the other initial particle with energy and momentum (E_2, p_2) to the other final particle with energy (E_4, p_4) at the other end. This diagram also appears in Fig. 16.2a. In the case of identical particles, it is also possible for the gauge boson to couple the initial particle with energy and momentum (E_1, p_1) to the final particle with energy (E_4, p_4) at one end, and to couple the initial particle with energy and momentum (E_2, p_2) to the final particle with energy (E_3, p_3) at the other as shown in Fig. 16.2b. For diagram (b), the internal particle carries energy and momentum $(E_{q'}, q')$, given by

$$E_{q'} = E_1 - E_4 = E_3 - E_2$$

$$q' = p_1 - p_4 = p_3 - p_2.$$

The scattering amplitude is the sum of the contributions from the two graphs. When the square modulus of the amplitude[2] is taken, in order to calculate the cross section, there is a quantum interference term, namely the product of the contribution from the one Feynman diagram with the complex conjugate of the contribution from the other Feynman diagram. If the contributions from the two diagrams in Fig. 16.2 are $\mathcal{A}_{(a)}$ and $\mathcal{A}_{(b)}$, respectively, then the square modulus of the amplitude is given by

$$|\mathcal{A}|^2 = |\mathcal{A}_{(a)}|^2 + |\mathcal{A}_{(b)}|^2 + \mathcal{A}_{(a)}^* \mathcal{A}_{(b)} + \mathcal{A}_{(b)}^* \mathcal{A}_{(a)}. \tag{16.14}$$

16.3 Cross Section

Using Fermi's golden rule (7.51) the reaction rate, λ, for a given process is given by

$$\lambda = \frac{2\pi}{\hbar} |\mathcal{A}|^2 \rho_E,$$

ρ_E being the density of states. The cross section, σ, for the process is the reaction rate per unit incident flux, F:

$$\sigma = \frac{1}{F} \frac{2\pi}{\hbar} |\mathcal{A}|^2 \rho_E. \tag{16.15}$$

---

[2]In general, a quantum amplitude is a complex number.

16.3.1 Relativistic Density of States

The number of states for which the final particle a has a momentum p_3 is the phase space element (3.11)

$$V \frac{d^3 p_3}{(2\pi\hbar)^3}.$$

By conservation of momentum, the momentum of the other final-state particle is fixed once the momentum p_3 is determined.

The density of states, ρ_E, is the integral over phase space, i.e. the integral over the momentum p_3 with a δ-function that constrains the energy, E_3, to respect energy conservation:

$$\rho_E = V \int \frac{d^3 p_3}{(2\pi\hbar)^3} \delta(E_3 + E_4 - E_1 - E_2). \tag{16.16}$$

This is *not* a relativistically invariant expression. In order to render it invariant, it needs to be multiplied by a correction factor[3]

$$\frac{m_a m_b c^4}{E_3 E_4}.$$

This factor tends to unity in the non-relativistic limit where $E_3 \approx m_a c^2$, $E_4 \approx m_b c^2$.

We take the direction of the incident particles to be the z-axis. In spherical polar coordinates, $p_3 = (p_3, \theta, \phi)$, θ is the angle between p_3 and the z-axis – the direction of incidence. θ is therefore the scattering angle. The integration measure, $d^3 p_3$, is expressed as

$$d^3 p_3 = p_3^2 dp_3 d\Omega \equiv p_3^2 dp_3 \, |d\cos\theta| \, d\phi.$$

In this coordinate system, the (relativistic) density of states is

$$\rho_E = \frac{m_a m_b c^4 V}{(2\pi\hbar)^3} \int \frac{p_3^2}{E_3 E_4} dp_3 \, d\Omega \delta(E_3 + E_4 - E_1 - E_2). \tag{16.17}$$

In the CM frame p_3 is the magnitude of momentum of *either* of the final-state particles. In this frame, using the relativistic expressions for E_3, E_4 in terms of p_3, we get

---

[3]It can be shown that for a particle with momentum p and energy E, the measure $d^3 p/2E$ is invariant under Lorentz transformations (see problem 16.4).

$$\rho_E = \frac{m_a m_b c^2 V}{(2\pi\hbar)^3} \int \frac{p_3^2}{\sqrt{(p_3^2 + m_a^2 c^2)(p_3^2 + m_b^2 c^2)}} dp_3 \, d\Omega$$

$$\times \delta\left(c\sqrt{p_3^2 + m_a^2 c^2} + c\sqrt{p_3^2 + m_b^2 c^2} - \sqrt{s}\right). \quad (16.18)$$

The integral over p_3 can be performed using the energy-conserving δ-function and after some algebra we obtain

$$\rho_E = \frac{m_a m_b c^2 V}{(2\pi\hbar)^3} \frac{p_{\text{CM}}}{\sqrt{s}} \int d\Omega, \quad (16.19)$$

with p_{CM} given by (16.8). In the ultra-relativistic limit this becomes

$$\rho_E \approx \frac{1}{2} \frac{m_a m_b c V}{(2\pi\hbar)^3} \int d\Omega . \quad (16.20)$$

16.3.2   Flux Factor

The incident particles are confined to a volume, V. This means that the flux – the number of particles crossing unit area in unit time – is

$$F = \frac{|\boldsymbol{v}_1 - \boldsymbol{v}_2|}{V}, \quad (16.21)$$

where $\boldsymbol{v}_1$, $\boldsymbol{v}_2$ are the velocities of the two incident particles.

Once again, in order to render this quantity relativistically invariant, it needs to be multiplied by a factor that tends to unity in the non-relativistic limit. With this relativistic correction factor the flux is

$$F = \frac{E_1 E_2 |\boldsymbol{v}_1 - \boldsymbol{v}_2|}{V m_a m_b c^4} = \frac{|E_2 \boldsymbol{p}_1 - E_1 \boldsymbol{p}_2|}{V m_a m_b c^2} . \quad (16.22)$$

In the CM frame, where $\sqrt{s} = E_1 + E_2$ and $\boldsymbol{p}_1 = -\boldsymbol{p}_2 = \boldsymbol{p}_{\text{CM}}$, this becomes

$$F = \frac{\sqrt{s}}{V m_a m_b c^2} p_{\text{CM}}. \quad (16.23)$$

Using (16.8), we have the relativistic expression for the incident flux

$$F = \frac{\lambda^{1/2}\left(s, m_a^2 c^4, m_b^2 c^4\right)}{2 V m_a m_b c^3} . \quad (16.24)$$

Writing F as a function of s and the masses demonstrates that it is a relativistically invariant quantity. This expression for F simplifies considerably in the ultra-relativistic limit, to give

$$F \approx \frac{s}{2Vm_am_bc^3}, \quad (\sqrt{s} \gg m_ac^2, m_bc^2) .$$ (16.25)

Piecing together (16.15), (16.20) and (16.25), we arrive at the expression for the differential cross section with respect to solid angle, Ω, for the elastic process

$$a + b \rightarrow a + b$$

$$\frac{d\sigma(ab \rightarrow ab)}{d\Omega} = \frac{V^2m_a^2m_b^2c^4}{(2\pi)^2\hbar^4s}|\mathcal{A}(s,t)|^2, \quad (\sqrt{s} \gg m_ac^2, m_bc^2) .$$ (16.26)

In the CM frame, the infinitesimal interval of solid angle, $d\Omega$, is given by

$$d\Omega = d\phi \, |d \cos \theta_{\mathrm{CM}}| .$$

Integrating over the azimuthal angle, ϕ, simply gives a factor of 2π since the differential cross section is independent of ϕ. Very often the differential cross section is quoted with respect to the variable t. Using (16.10) we replace $|d \cos \theta_{\mathrm{CM}}|$ by $2dt/s$ to obtain[4]

$$\frac{d\sigma(ab \rightarrow ab)}{dt} = \frac{V^2m_a^2m_b^2c^4}{\pi\hbar^4s^2}|\mathcal{A}(s,t)|^2 .$$ (16.27)

The total cross section, σ, is obtained by integrating t over its range $-s < t < 0$,

$$\sigma(ab \rightarrow ab) = \frac{V^2m_a^2m_b^2c^4}{\pi\hbar^4s^2} \int_{-s}^{0} dt \, |\mathcal{A}(s,t)|^2 .$$ (16.28)

Since t is relativistically invariant, the differential cross section given by (16.28) is valid in *any* frame of reference. In the case where a and b are spin-$\frac{1}{2}$ particles and the internal particle has spin one, we can substitute (16.11) and (16.12) for the scattering amplitude, $\mathcal{A}$, to obtain (in the ultra-relativistic limit)

$$\frac{d\sigma(ab \rightarrow ab)}{dt} = \frac{g_a^2g_b^2}{8\pi s^2} \frac{\left(2s(s+t) + t^2\right)}{\left(t - M^2c^4\right)^2} .$$ (16.29)

---

[4]We write $|dt|$ simply as dt, since it is always assumed that a differential cross section is quoted with respect to a positive infinitesimal interval.

We notice that the factors of the volume, V, have cancelled as expected (this is an arbitrary volume chosen to normalize the particle wavefunctions). The external factors of the masses have also cancelled. This is also expected, as we sometimes have massless particles, such as photons, either in the initial or final states (or both).

16.4 Crossing

The Feynman rule for the representation of an antiparticle is to draw it as a particle with negative energy, moving in the opposite direction, i.e.

$$(E, \boldsymbol{p}) \;\rightarrow\; -(E, \boldsymbol{p})$$

This means that the arrows on the external lines indicate the direction of particle flow, so that for antiparticles they point in the *opposite* direction to the direction of motion of the antiparticle.

The Feynman diagram for the process in which a particle a annihilates against its antiparticle, $\bar{a}$, emitting a gauge boson, which then creates a different particle b and its antiparticle $\bar{b}$, is shown in Fig. 16.3. The arrow on the $\bar{a}$ antiparticle is outgoing, despite the fact that it represents an incoming antiparticle. This means that for the purpose of calculating the scattering amplitude, the energy E_2 of the antiparticle is replaced by $-E_2$, and the momentum by $-\boldsymbol{p}_2$. Likewise the arrow on the outgoing antiparticle $\bar{b}$ is incoming, indicating that the energy E_4 is to be replaced by $-E_4$ and the momentum by $-\boldsymbol{p}_4$.

The Feynman diagram shown in Fig. 16.3 is identical to the diagram shown in Fig. 16.1 (on its side). The scattering amplitude $\mathcal{A}_{a\bar{a}\rightarrow b\bar{b}}$ is therefore obtained to the amplitude $\mathcal{A}_{ab\rightarrow ab}$ by the substitutions

$$\left(E_3, \boldsymbol{p}_3\right) \;\leftrightarrow\; -\left(E_2, \boldsymbol{p}_2\right).$$

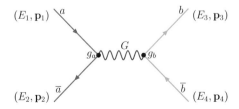

Fig. 16.3 Feynman diagram for the annihilation of a particle a and its antiparticle and the creation of a particle b and its antiparticle. The arrows on the external lines indicate the direction of particle flow. For antiparticles the arrows point in the *opposite* direction to the direction of motion of the antiparticle

In terms of the relativistically invariant quantities s and t, this is equivalent to the interchange of s and t so that

$$\mathcal{A}_{a\bar{a}\to b\bar{b}}(s,t) = \mathcal{A}_{ab\to ab}(t,s). \tag{16.30}$$

The relation (16.30) is known as *"crossing symmetry"*. By applying it to (16.11) and (16.12) we can immediately write down the scattering amplitude for the process

$$a + \bar{a} \to b + \bar{b}.$$

$$\mathcal{A}_{a\bar{a}\to b\bar{b}}(s,t) = -\frac{\hbar^2 g_a g_b}{4m_a m_b c^2 V} \frac{\sqrt{4\left(t - m_a^2 c^4 - m_b^2 c^4\right) + 2s^2 + 4st}}{\left(s - M^2 c^4\right)}. \tag{16.31}$$

The density of states and the flux factor are slightly modified by the fact that both the incident particles have mass m_a, whereas both the final particles have mass m_b. In the ultra-relativistic limit, this makes no difference and we obtain the differential cross section

$$\frac{d\sigma(a\bar{a} \to b\bar{b})}{dt} = \frac{g_a^2 g_b^2}{8\pi s^2} \frac{\left(2t(s+t) + s^2\right)}{\left(s - M^2 c^4\right)^2}. \tag{16.32}$$

16.4.1   Resonances

We note that the cross section predicted by (16.32) diverges when $s = M^2 c^4$. This does not make sense and occurs because at this value of s we would be at the peak of a resonance due to the production and decay of the gauge boson. The resonance has a width Γ and the propagator for the internal gauge boson needs to be modified by adding an imaginary part proportional to Γ to the denominator:

$$\Delta(s,M) \to -\frac{1}{\left(s - M^2 c^4\right) - iM\Gamma c^2}. \tag{16.33}$$

A strong hint for the reason behind this modification is discussed in Appendix 16. With this modification we have the behaviour of the cross section in the region $s \sim M^2 c^4$

$$\sigma \propto |\Delta(s,M)|^2 \propto \frac{1}{\left(s - M^2 c^4\right)^2 + M^2 \Gamma^2 c^4}$$

$$\approx \frac{1}{4M^2 c^4 \left(\left(E_{CM} - Mc^2\right)^2 + \Gamma^2/4\right)}. \tag{16.34}$$

We see that we get a cross section proportional to the Breit–Wigner distribution described in Chap. 12, as expected in the region of a resonance.

16.5   Multi-Particle Final States

The annihilation process described above is one example of a process in which the final-state particles differ from the incident particles. Provided the CM energy is sufficient, it is possible to produce more than two particles in the final state. With the incident energies of modern accelerators, there are usually several particles in the final state in each scattering event.

Figure 16.4 shows the Feynman diagrams for one such scattering process, namely

$$a + b \rightarrow a + b + c + \bar{c}. \tag{16.35}$$

We note that there are now five different Feynman diagrams and the contributions from each one must be added to obtain the scattering amplitude. Each diagram involves the exchange of three internal virtual particles – so the contribution from each diagram contains a product of three different propagators. Not all of these virtual particles are gauge bosons. They are, however, always off mass-shell since it is necessary that energy and momentum are conserved at each vertex. The energy and momentum of the internal particles do not satisfy the relation between energy,

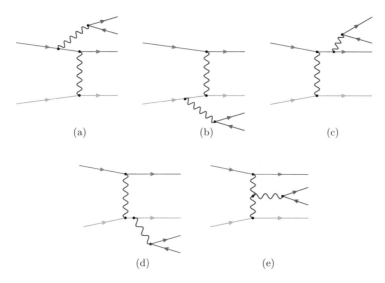

Fig. 16.4 The Feynman diagrams for a process in which there are three particles and an antiparticle in the final state

momentum and mass for a free particle.[5] The scattering amplitude is extremely complicated and needs to be calculated using a computer. A number of computer programs have been developed over the last 50 years, which compute scattering amplitudes for events with several final-state particles and/or antiparticles.

Figure 16.4e is a diagram that involves the interaction of three gauge boson. In most cases, gauge bosons *do* possess such self-couplings. For example, there is a coupling between W^+, W^- and Z-bosons. On the other hand, photons do not couple to themselves, so that if the process (16.35) is an electromagnetic process, Fig. 16.4e will not appear.

The expression for the density of states generalizes to the integral over the phase space of $(n-1)$ of the n final-state particles with an energy-conserving δ-function. In the CM frame this is

$$\rho_E = \int \prod_{i=1}^{(n-1)} \left\{ \frac{d^3 p_i}{(2\pi\hbar)^3} \frac{m_i c^2}{E_i} \right\} \frac{m_n c^2}{E_n} \delta \left(\sum_i^n E_i - \sqrt{s} \right).$$ (16.36)

Again, this is a very complicated integral, which usually has to be carried out numerically, using one of several available computer programs for performing multi-dimensional integrals.

For a given value of s, the momenta of the n final-state particles are described in terms of $(3n-4)$ variables, after accounting for the conservation of momentum (three components) and energy. The amplitude is independent of the overall azimuthal direction of the final-state particles, so that integration over that azimuthal angle simply introduces a factor of 2π – leaving $(3n-5)$ non-trivial variables. Experiments measure differential cross section with respect to some of these variables and the remaining variables are integrated out when the integral over phase space is performed. There are many different possible choices for the measured variables. Very often the transverse momentum, $p_T^{(X)}$, of a particular final-state particle, X, is specified. The transverse momentum is the magnitude of the momentum of particle X perpendicular to the direction of incidence of the scattering particles. Although this transverse momentum is not a relativistically invariant quantity, it is invariant under boosts in the direction of incidence of the scattering particles and, for example, is the same in the Lab frame and CM frame. The transverse momentum of long-lived final-state particles can be measured in a tracker detector.

In experiments that search for a short-lived intermediate particle that can only be observed as a resonance, it is useful to measure the relativistically invariant quantity

$$s_{ij} \equiv (E_i + E_j)^2 - c^2 (p_i + p_j)^2,$$

---

[5]The one exception is the case where the combined energy of the c and $\bar{c}$, in their CM frame, is exactly equal to the rest energy of the gauge boson, in which case there is a resonance peak.

where i and j are two of the final-state particles into which the intermediate particle can decay. $\sqrt{s_{ij}}$ is the CM energy in the i, j channel, i.e. the combined energy of particles i and j in their CM frame, where $\boldsymbol{p}_i + \boldsymbol{p}_j = \boldsymbol{0}$. The intermediate particle appears as a resonance in the differential cross section $d\sigma/ds_{ij}$, with a peak value of s_{ij} equal to the square of its rest energy. In an experiment in which the momenta of two final-state particles, i and j, are measured, it is possible to construct both the CM energy, s_{ij}, as well as the transverse momentum $p_{T(ij)} = (\boldsymbol{p}_i + \boldsymbol{p}_j)_T$, of the i-j system. This enables the determination of the double differential cross section

$$\frac{d^2\sigma}{ds_{ij}dp_{T(ij)}}.$$

This not only has a resonance (Breit–Wigner) structure due to the production and decay of a short-lived intermediate particle, X, but also gives the distribution in the transverse momentum, $p_T^{(X)}$, of the intermediate particle. This can then be compared with the theoretical prediction for the transverse momentum distribution obtained from the corresponding Feynman diagrams.

Summary

- In relativistic Quantum Physics, the potential between two interacting particles is replaced by the propagator that is the amplitude for the exchange of a force carrier particle (gauge boson) between the interacting particles.
- A scattering event can be represented by one or more Feynman diagrams consisting of vertices and propagators. The vertices contain factors that depend on the spins of the interacting particles. Each internal line has a propagator associated with it. Energy and momentum must be conserved at each vertex. The contribution of each diagram to the scattering amplitude can be determined from the vertex rules and propagators.
- The amplitudes are relativistically invariant. For 2-body to 2-body scattering, processes can be described in terms of the Mandelstam variables s and t. $\sqrt{s}$ is the total energy of the incident (or final-state) particles in the CM frame, and $\sqrt{-t}$ is the momentum transferred between the particles in the CM frame.
- The scattering amplitude for a related process in which an incoming particle is replaced by an outgoing antiparticle and vice versa is obtained by interchanging the energy and momentum of one particle with the negative of the energy and momentum of the other particle. For a 2-body final state this simply involves the interchange of the variables s and t.
- The cross section is the product of the reciprocal of the flux, the square modulus of the scattering amplitude and the density of states. Both the flux factor and the density of states acquire corrections in order to render them relativistically invariant.

- The density of states is obtained by integrating over the phase space of all but one of the final-state particles (in order to conserve momentum), with the insertion of a δ-function that conserves the total energy.
- Feynman diagrams can be drawn, which describe scattering in which there are several particles in the final state. The same Feynman rules for the vertices and internal lines apply for these diagrams, enabling the amplitude for such scatterings to be calculated. The density of states is also generalized to an integral over the phase space all but one of the final-state particles.

Problems

Problem 16.1 Draw the Feynman diagrams for the elastic electromagnetic process

$$a + \bar{a} \rightarrow a + \bar{a},$$

where a is a particle with electric charge q and $\bar{a}$ is its antiparticle.

Problem 16.2 Show that for the scattering of two particles with masses m_a and m_b and total energy, $\sqrt{s}$ in the CM frame, the energies, E_a and E_b, of the incident particles are

$$E_a = \frac{\left(s + m_a^2 c^4 - m_b^2 c^4\right)}{2\sqrt{s}}$$

$$E_b = \frac{\left(s + m_b^2 c^4 - m_a^2 c^4\right)}{2\sqrt{s}}.$$

[Hint: In the CM frame $E_a^2 - E_b^2 = (m_a^2 - m_b^2)c^4$.]

Problem 16.3 Show that in the ultra-relativistic limit the total cross section, σ, for the electromagnetic process

$$e^+ + e^- \rightarrow \mu^+ + \mu^-$$

is

$$\sigma = \frac{4\pi\alpha^2\hbar^2 c^2}{3s}.$$

Problem 16.4 For a particle with mass m, energy E and momentum $\boldsymbol{p}$, show that

$$\frac{\partial E}{\partial p_z} = c^2 \frac{p_z}{E}.$$

For a Lorentz transformation along the z-axis,

$$(E, p) \rightarrow (E', p'),$$

show that

$$\frac{d^3 p'}{E'} = \frac{d^3 p}{E}.$$

Appendix 16 – Propagator for an Unstable Particle

This is *not* a derivation of (16.33), which requires the use of Quantum Field Theory, but a heuristic argument aimed at explaining how the propagator for an unstable particle is consistent with general ideas of Quantum Physics.

We can represent the particle propagator (16.4) in integral form

$$\Delta(E_q, q, M) = -\frac{i}{\hbar} \int_0^\infty \frac{dt}{E_q} e^{-iE_q t/\hbar} \cos\left(\frac{\sqrt{(q^2 + M^2 c^2)} ct}{\hbar}\right). \qquad (16.37)$$

Taking t to refer to time, the factor $e^{-iE_q t/\hbar}$ is the time dependence of the wavefunction for the particle with energy E_q. This wavefunction is normalized so that the probability of the particle being "somewhere" in space is unity. However, if the particle is unstable and has a lifetime τ, then this probability is attenuated, i.e. the normalization of the wavefunction is attenuated by a time-dependent factor

$$e^{-t/2\tau'}$$

where $\tau' = E_q \tau/Mc^2$, is the dilated lifetime in the frame in which a particle of mass M has energy $E_q(= \gamma Mc^2)$.

This exponential decay factor can be included by making the substitution

$$E_q \rightarrow E_q - i\frac{Mc^2\hbar}{2E_q \tau} = E_q - i\frac{Mc^2}{2E_q}\Gamma$$

(using the relation $\tau = \hbar/\Gamma$ between lifetime and width).

The integral representation for the propagator (16.37) then becomes

$$\Delta(E_q, q, M) = -\frac{i}{\hbar} \int_0^\infty \frac{dt}{E_q - iMc^2\Gamma/(2E_q)} e^{-iE_q t/\hbar} e^{-Mc^2\Gamma t/(2\hbar E_q \tau)}$$

$$\times \cos\left(\frac{\sqrt{(q^2 + M^2 c^2)} ct}{\hbar}\right). \qquad (16.38)$$

Performing the integral over t, and keeping only terms linear on the width Γ in the denominator, we get a propagator that is proportional to the relativistic Breit–Wigner distribution

$$\Delta\left(E_q, q, M\right) = -\frac{1}{\left(E_q^2 - q^2 c^2 - M^2 c^4\right) - i M c^2 \Gamma}. \tag{16.39}$$

Chapter 17
Weak Interactions

17.1 W- and Z-Boson Interactions

The weak force carriers are the neutral Z and the charged $W^\pm$ bosons that have masses 91.2 GeV/c^2 and 80.4 GeV/c^2, respectively. These are spin-one gauge bosons. The $W^\pm$ and Z-bosons were discovered at CERN at the Super Proton–Antiproton Synchrotron (Sp$\bar{\text{p}}$S) collider of protons and antiprotons at $\sqrt{s} = 540$ GeV. The W-boson was discovered in January 1983 [6, 7] and the Z-boson was discovered a few months later in May 1983 [8, 9] by the UA1 experiment (led by Carlo Rubbia) and the UA2 experiment (led by Pierre Darriulat). This discovery was one of the major successes of CERN.

All Standard Model fermions are involved in weak interactions. The Feynman diagrams for Z-boson interactions with Standard Model fermions are shown in Fig. 17.1, where q_i and ℓ_i stand for quarks and leptons. The index $i = 1, 2, ...6$ corresponds to flavours u, d, c, s, t, b quarks for q_i and to $\nu_e, e, \nu_\mu, \mu, \nu_\tau, \tau$ for ℓ_i. These Feynman diagrams give rise to the so-called "*neutral weak currents*" that are conserved in the Standard Model so that the flavours of the fermions do not change in the interactions with the Z-boson.

Since the W-boson carries electromagnetic charge, it couples to different quark flavours, namely to quark pairs and lepton pairs with vertices shown in Fig. 17.2, where (q, q') correspond to the (u, d), (c, s), (t, b) quark pairs and (ν, ℓ) to (ν_e, e), (ν_μ, μ), (ν_τ, τ) lepton pairs, respectively. The arrow on the W-boson line indicates the direction of the W^+, but the same vertex applies for a W^- flowing in the opposite direction. Likewise the arrow on the fermion line indicates a direction of the fermion but could also refer to an antifermion flowing in the opposite direction. These diagrams correspond to the so-called "*charged currents*".

These pairs of quarks or leptons are $I^W = \frac{1}{2}$ multiplets of "*weak isospin*". The neutrinos and the quarks with electric charge $Q = +\frac{2}{3}e$ have third component of weak isospin, $I_3^W = +\frac{1}{2}$, whereas the charged leptons and the quarks with electric

© The Author(s), under exclusive license to Springer Nature Switzerland AG 2021
A. Belyaev, D. Ross, *The Basics of Nuclear and Particle Physics*, Undergraduate
Texts in Physics, https://doi.org/10.1007/978-3-030-80116-8_17

Fig. 17.1 Feynman diagrams representing Z-boson interactions with Standard Model fermions: quarks (left) and leptons (right)

Fig. 17.2 Feynman diagrams with W-boson interactions with Standard Model fermions: quarks (left) and leptons (right)

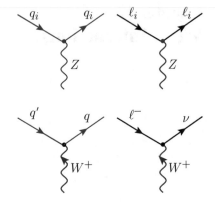

charge $Q = -\frac{1}{3}e$ have $I_3^W = -\frac{1}{2}$. This must be contrasted with *"strong isospin"* under which the strong interactions are invariant, but which involve only u- and d-quarks. *All* fundamental fermions are members of doublets of this weak isospin. The weak interactions are approximately invariant under this isospin in the sense that if we neglect the mass difference between the $W^\pm$ and Z-bosons, then these gauge bosons can be considered to be an $I^W = 1$ triplet of weak isospin. A vertex involving two fermions and a weak gauge boson has a total weak isospin $I^W = 0$ and is invariant under weak isospin transformations.

A weak isospin doublet of quarks (with three different colour states) and a corresponding weak isospin doublet of leptons are classed together in a *"generation"* or *"family"* of fundamental fermions. The three generations are therefore

$$1. \ \left\{ u^i, \ d^i \ (i = r, g, b), \ \nu_e, \ e^- \right\}$$

$$2. \ \left\{ c^i, \ s^i \ (i = r, g, b), \ \nu_\mu, \ \mu^- \right\}$$

$$3. \ \left\{ t^i, \ b^i \ (i = r, g, b), \ \nu_\tau, \ \tau^- \right\}.$$

One should note that W-bosons can couple to charged quark currents involving quarks of different generations meaning that the W-boson can effect transitions between quarks from different generations, as we shall discuss below.

The strength of the weak interactions and the properties of W- and Z-boson interactions with fermions deserve a closer look. When describing weak interactions, it is useful to define the ratio $M_W/M_Z = c_W \equiv \cos\theta_W$, where θ_W is the *"Weinberg angle"* (or weak mixing angle). As discussed in Chap. 7, only left-chiral fermions or right-chiral antifermions are involved in the interactions with W-boson. The coupling constant of left-chiral fermions with W-boson, g_L^W, is related to the electron charge, e, by

$$g_L^W = \frac{e}{\sqrt{2}s_W} = \frac{g_W}{\sqrt{2}}, \quad \text{(where } s_W \equiv \sin\theta_W \simeq 0.47\text{).} \tag{17.1}$$

This relation emerges from the Standard Model of weak and electromagnetic (electroweak) interactions, developed in the 1960s by Sheldon Glashow, Steven Weinberg and Abdus Salam [92–94]. Note that the existence of the W- and Z-bosons and their masses was predicted by the electroweak theory[1] 15–20 years before they were discovered experimentally.

From now on we use the coupling constant g_W, which is related by (17.1) to the electron charge. The coupling of right-chiral fermions or left-chirality antifermions to W-bosons, g_R^W, is zero. The case of interactions with Z-boson is slightly more complicated since in Standard Model the Z-boson mediates a mixture of left-chiral weak interactions and electromagnetic interactions and therefore interacts with fermions of *both* left- and right-chirality.

The coupling constant of the Z-boson to left-chirality fermions with electric charge Q is given by

$$g_L^Z = \frac{g_W}{c_W} \left(I_3^W - Q s_W^2 \right) \tag{17.2}$$

and to right-chirality fermions with electric charge Q

$$g_R^Z = -\frac{g_W s_W^2}{c_W} Q. \tag{17.3}$$

Weak interactions are short range. This can be seen from the weak-interaction potential that, for example, in case of W-boson exchange takes the form (see (16.1)):

$$V_W(r) = \frac{g_W^2}{4\pi\varepsilon_0 r} e^{-M_W c r/\hbar} \quad , \tag{17.4}$$

from which one can deduce that the range, r_W, of the weak force is

$$r_W = \frac{\hbar}{M_W c} = \frac{0.197\,[\text{GeV fm}]}{80.4\,[\text{GeV}]} \simeq 2.5 \times 10^{-3} \text{ fm} . \tag{17.5}$$

Note that in weak interactions both quark number (baryon number) and lepton number are strictly conserved.

The weak force is associated with the β-decay processes, which is driven by the weak interactions. A neutron decays into a proton because a d-quark in the neutron converts into a u-quark emitting virtual W^--boson that then decays into an electron and antineutrino as shown in Fig. 17.3. The amplitude for such a decay is proportional to

---

[1]This electroweak theory is also known as *"Quantum Flavourdynamics"* (QFD).

Fig. 17.3 Diagram for neutron decay via weak interactions

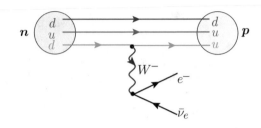

$$\frac{g_W^2}{2(E_q^2 - q^2c^2 - M_W^2c^4)},$$

where $g_W/\sqrt{2}$ is the strength of the coupling of the W^- to the quarks or leptons. q is the momentum transferred between the neutron and proton and E_q is the energy transferred. The values of $|q|c$ and E_q are of the order of a few MeVs (the Q-value of the β-decay) and so we can neglect it in comparison with M_Wc^2 that is 80.4 GeV. Thus the amplitude is proportional to

$$\frac{g_W^2}{2M_W^2c^4}.$$

The coupling g_W is not very small. In fact it is about twice as large as the electron charge, e (recalling the relation between g_W and e (17.1)). Weak interactions are called "weak" because of the large mass term in the denominator, which suppresses an amplitude of weak processes at low energies. At energies reached at modern high energy accelerators, $|q|$ can be of the order of, or even much greater than, M_Wc. In such cases weak interactions are stronger than electromagnetic interactions and almost comparable with strong interactions.

17.2   Cabibbo Theory

Particles containing strange quarks, e.g. $K^\pm$, K^0, Λ etc., cannot decay into non-strange hadrons via the strong interactions, which have to conserve strangeness. They can, however, decay via the weak interactions. This is possible because the W-boson not only couples a u-quark to a d-quark but can also couple (with a weaker strength) a u-quark to an s-quark, so we have a usW vertex with coupling $g_W \sin\theta_c/\sqrt{2}$, whilst the udW coupling is actually $g_W \cos\theta_C/\sqrt{2}$ as shown in Fig. 17.4. θ_C is the *"Cabibbo angle"* and takes the value $\theta_C = 13.04°$ ($\sin\theta_C \approx 0.23$).

The same usW coupling allows a strange hadron to decay into non-strange hadrons and (sometimes) leptons. For example, the decay

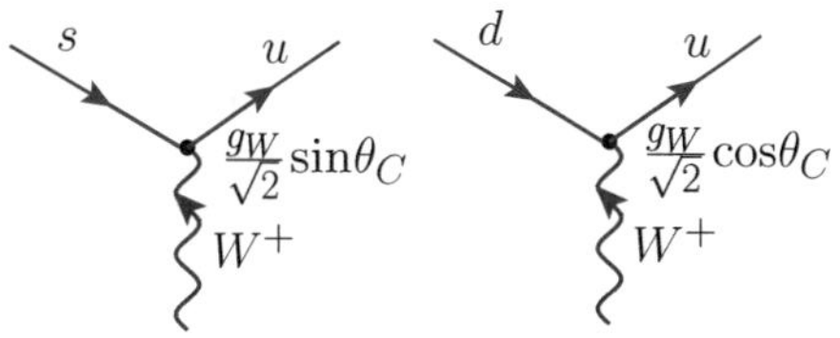

Fig. 17.4 Feynman diagram for usW (left) and udW (right) interactions

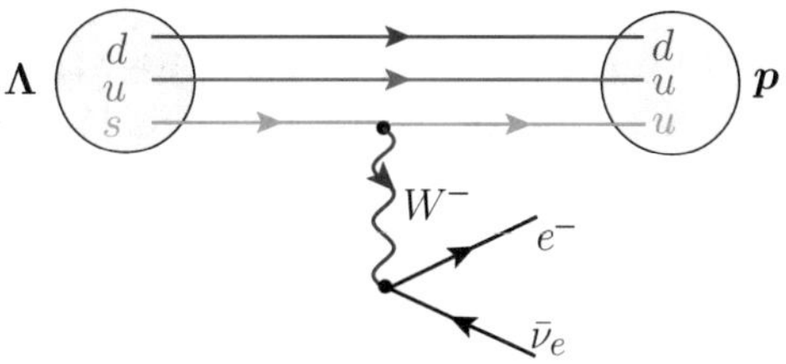

Fig. 17.5 Diagram for Λ-baryon decay into proton, $\Lambda \to p + e^- + \bar{\nu}_e$, via usW vertex

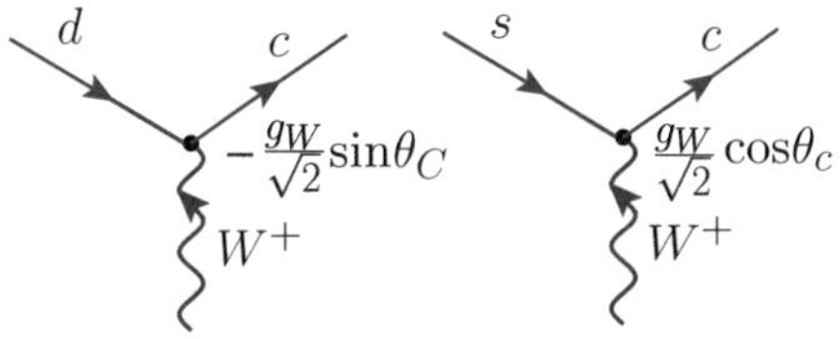

Fig. 17.6 Feynman diagram for cdW (left) and csW (right) interactions

$$\Lambda \to p + e^- + \bar{\nu}_e$$

occurs when an s-quark converts into a u-quark and emits a W^- that then decays into an electron and antineutrino. The Feynman diagram for this is presented in Fig. 17.5.

Likewise, the c-quark has a coupling to the s-quark with coupling $g_W \cos\theta_C/\sqrt{2}$ and a coupling to a d-quark with coupling $-g_W \sin\theta_C/\sqrt{2}$ with the corresponding Feynman diagrams presented in Fig. 17.6. This implies that charm hadrons are more likely to decay into hadrons with strangeness, because the coupling between a c-quark and an s-quark is larger than between a c-quark and a d-quark.

The strength of the interactions between a pair of up-type quarks (u, c), and down-type quarks (d, s) and a W-boson can be pieced together in a matrix form as follows:

$$\frac{g_W}{\sqrt{2}} \begin{pmatrix} \bar{u} & \bar{c} \end{pmatrix} \begin{pmatrix} \cos\theta_C & \sin\theta_C \\ -\sin\theta_C & \cos\theta_C \end{pmatrix} \begin{pmatrix} d \\ s \end{pmatrix} W^+ . \tag{17.6}$$

All particles are deemed to be incoming, but d, s could also refer to outgoing antiquarks and likewise $\bar{u}, \bar{c}$ could refer to outgoing quarks and the incoming W^+ could be an outgoing W^-. This 2×2 matrix is called the "*Cabibbo matrix*". It is described in terms of a single parameter, the Cabibbo angle [123]. This matrix mixes first and second generations of quarks. Without such mixing, the strength of the udW and csW interactions would be simply $g_W/\sqrt{2}$, but there would be *no* interactions between quarks of different generations and therefore no strangeness-changing or charm-changing weak decays.

Nowadays, we know that there are three generations of quarks, all of which have been discovered, and we know that all generations are mixed. The mixing of quarks of the first generation with the third one is of the order of 10^{-3}, whilst the mixing of the second generation of quarks with the third one is of the order of 10^{-2}. Therefore, the "Cabibbo matrix" is extended to a general 3×3 "*Cabibbo–Kobayashi–Maskawa*" (CKM) matrix V_{ij} (with $i = u, c, t$ and $j = d, s, b$) [124]:

$$\frac{g_W}{\sqrt{2}} \left(\bar{u} \; \bar{c} \; \bar{t} \right) \begin{pmatrix} V_{ud} & V_{us} & V_{ub} \\ V_{cd} & V_{cs} & V_{cb} \\ V_{td} & V_{ts} & V_{tb} \end{pmatrix} \begin{pmatrix} d \\ s \\ b \end{pmatrix} W^+. \tag{17.7}$$

Quantum-mechanical constraints lead to the conclusion that there are only four independent parameters by which this CKM matrix can be parametrized. These are three angles and one complex phase, responsible for the charge conjugation and parity ("*CP-symmetry*") violation, which we discuss in Chap. 21. Comparing the CKM matrix with the Cabibbo matrix we see that to a good approximation, $V_{ud} \approx V_{cs} \approx \cos\theta_C$ and $V_{us} \approx -V_{cd} \approx \sin\theta_C$.

17.3 Leptonic, Semi-leptonic and Hadronic Weak Decays

Weak interactions play a key role in light meson decays. The classic example is $\pi^+ \to \mu^+ + \nu_\mu$ decay. The Feynman diagram for this decay is shown in Fig. 17.7. π-mesons are the lightest hadrons and therefore cannot decay to any other hadron. The π^+ has mass 139.6 MeV/c^2.[2] It can decay into leptons (to e^+ or μ^+ that are lighter than the π^+) through W-boson exchange, as shown in Fig. 17.7. This diagram shows that the π^+, being a bound state of a u-quark and a $\bar{d}$-antiquark, decays to μ^+ and ν_μ through the production and decay of a virtual W^+. The decay width for this process is given by

$$\Gamma(\pi^+ \to \mu^+ + \nu_\mu) = \pi \frac{\alpha_W^2}{16} \cos^2\theta_C \frac{f_{\pi^+}^2 m_\mu^2 m_\pi}{M_W^4 c^2} \left(1 - \left(\frac{m_\mu}{m_\pi} \right)^2 \right)^2, \tag{17.8}$$

where we have introduced the dimensionless constant α_W in analogy with the electromagnetic fine structure constant, α:

---

[2]The lightest meson is π^0 from the same isospin triplet that has the slightly smaller mass of 135 MeV/c^2.

Fig. 17.7 Diagram for $\pi^+ \rightarrow \mu^+ + \nu_\mu$ decay. The incoming antiquark, represented by the outgoing (blue) line, carries the opposite colour to the quark (i.e. "anti-blue")

$$\alpha_W \equiv \frac{g_W^2}{(4\pi\varepsilon_0\hbar c)} = \frac{\alpha}{s_W^2} = \frac{\alpha}{\left(1 - M_W^2/M_Z^2\right)}. \tag{17.9}$$

There is a simple relation between α_W, the W-mass and the weak-interaction Fermi coupling constant, G_F, introduced in Sect. 7.3.2:

$$G_F = \frac{\pi\alpha_W}{\sqrt{2}M_W^2c^4}. \tag{17.10}$$

The constant f_{π^+} is the *"pion decay constant"*. This is a dimensionful constant that parametrizes the wavefunction of the u-quark and $\bar{d}$-antiquark inside the pion. It takes the value $f_{\pi^+} = 130.4$ MeV. In (17.8) the factor of M_W^4 in the denominator comes from the square of the W-boson propagator, whilst the term $\left(1 - m_\mu^2/m_\pi^2\right)^2$ comes from the integral over phase space. The decay width is proportional to the muon mass squared for the reason we will discuss in Sect. 17.5. For the decay channel $\pi^+ \rightarrow e^+ + \nu_e$, m_μ is replaced by m_e, so that the contribution to the total decay width from this channel is negligibly small. Because of the M_W^4 suppression, the decay width of the π^+ is extremely small, even for the dominant decay mode into a muon. Inserting numbers into (17.8) yields a value for the width, $\Gamma = 2.5 \times 10^{-8}$ eV, which makes the π^+ long-lived with a lifetime $\tau_{\pi^+} = 2.6 \times 10^{-8}$s.

Because the $W^\pm$ couples either to quarks or to leptons, decays of heavier hadrons (e.g. strange mesons) can either be:

- Leptonic – the final state consists only of leptons
- Semi-leptonic – the final state consists of both hadrons and leptons
- Hadronic – the final state consists of only hadrons

For strange baryons only semi-leptonic and hadronic decays are possible because baryon number is strictly conserved – so there must be a baryon in the final state. Lepton number is also strictly conserved, which means that a charged lepton is always accompanied by its antineutrino (or a charged antilepton by its neutrino) in the final state.

Let us consider examples of K^+-meson decay. The same Feynman diagram vertex (see Fig. 17.4) that describes conversion of an s-quark into a u-quark and emitting a W^- is responsible for the creation of a W^+ from the annihilation of an $\bar{s}$-antiquark with a u quark. This process defines the leptonic decay of K^+,

Fig. 17.8 Diagram for
$K^+ \rightarrow \mu^+ + \nu_\mu$ leptonic
decay

Fig. 17.9 Diagram for
$K^+ \rightarrow \mu^+ + \nu_\mu + \pi^0$
semi-leptonic decay

Fig. 17.10 A diagram for
$K^+ \rightarrow \pi^0 + \pi^+$ hadronic
decay. There is another
diagram in which the
final-state u-quarks are
interchanged

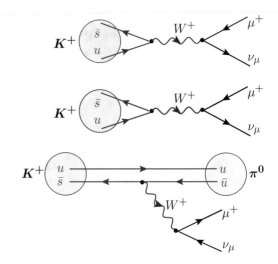

$$K^+ \rightarrow \mu^+ + \nu_\mu.$$

The Feynman diagram for this decay is shown in Fig. 17.8. This process is analogous to charged pion decay discussed above with f_{π^+} replaced by f_{K^+}, which takes the value 156.1 MeV, and $\cos^2 \theta_C$ replaced by $\sin^2 \theta_C$.

In case of semi-leptonic decay

$$K^+ \rightarrow \mu^+ + \nu_\mu + \pi^0$$

the $\bar{s}$-antiquark is converted into $\bar{u}$-antiquark, which binds with the u-quark from the K^+ to form a π^0 and emits a virtual W^+. This virtual W^+ splits into μ^+ and ν_μ as shown in Fig. 17.9.

The Feynman diagram for the hadronic K^+ decay

$$K^+ \rightarrow \pi^0 + \pi^-$$

is shown in Fig. 17.10. In this process an $\bar{s}$-antiquark from the K^+ is converted into a $\bar{u}$-antiquark that binds with the u-quark from the K^+ to form a π^0. A virtual W^+ is emitted, which, in turn, decays into a u-quark and a $\bar{d}$-antiquark that bind to form a π^+. There is another Feynman diagram (not shown in Fig. 17.10) in which the roles of the two final-state u-quarks are interchanged. One should note that $m_K > 2m_\pi$, which is why this decay mode is energetically allowed.

In the case of weak-interaction baryon decay, we have already seen an example of a semi-leptonic decay, $\Lambda \rightarrow p + e^- + \bar{\nu}_e$. An example of a hadronic Λ decay is

$$\Lambda \rightarrow p + \pi^-.$$

Fig. 17.11 A diagram for
$\Lambda \rightarrow p + \pi^-$ hadronic
decay. There is another
diagram in which the
final-state u-quarks are
interchanged

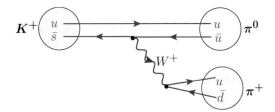

A Feynman diagram for this decay is shown in Fig. 17.11. This decay mechanism
is similar to the hadronic decay of K^+ except that in this case an s-quark from the
decaying Λ is converted into a u-quark, which binds with the other quarks in the Λ
to form a proton, emitting virtual W^-. The W^- then decays into a d-quark and $\bar{u}$-
antiquark that bind to form a π^-. There are other diagrams (not shown in Fig. 17.11)
such as one in which the final-state u-quarks are interchanged.

17.4 Flavour Selection Rules in Weak Interactions

In the exchange of a single $W^\pm$, an s-quark can be converted into a non-strange
quark, or vice versa (albeit with a coupling that is suppressed by $\sin\theta_C$), changing
the strangeness of the hadron by ± 1. It is highly unlikely that two strange quarks
would be converted into non-strange quarks in the same decay process – this would
require the exchange of two W-bosons. We therefore have a selection rule for weak
decay processes

$$\Delta S = \pm 1 , \tag{17.11}$$

so that hadrons with strangeness $S = -2$ that decay weakly must first decay into
a hadron with $S = -1$ (which in turn decays into non-strange hadrons). Thus, for
example, we have

$$\Xi^0 \rightarrow \Lambda + \pi^0 \begin{array}{l} \nearrow p + \pi^- + \pi^0 \\ \searrow n + \pi^0 + \pi^0 \end{array}$$

The same selection rules apply for changes in other flavours (charm, bottom).

17.5 Parity Violation

The parity violation observed in β-decay arises because the $W^\pm$ couples only to
quarks or leptons that have left-chirality, i.e. primarily to negative helicity states in
which the component of spin in their direction of motion is $-\frac{1}{2}\hbar$.

Fig. 17.12 Illustration of $K^- \rightarrow \mu^- + \bar{\nu}_\mu$ decay, $\boldsymbol{p}_\nu(\boldsymbol{p}_\mu)$ and $\boldsymbol{s}_\nu(\boldsymbol{s}_\mu)$ show the direction of the momentum and the spin of $\bar{\nu}_\mu(\mu^-)$, respectively. "RH" indicates that both muon and antineutrino have right-helicity

$W^\pm$ always couple to left-helicity neutrinos. For quarks and massive leptons the $W^\pm$ *can* couple to right-helicity states, but the coupling is suppressed by a factor mc^2/E where m is the particle mass and E is its energy. This factor follows from the ratio of W-boson coupling to left- and right-helicity leptons given by (7.40), which was discussed in Sect. 7.6.1. The suppression is much larger for relativistically moving particles. In the case of nuclear β-decay, the nucleus is moving non-relativistically, but the electron typically has an energy of a few MeVs (and a mass of 0.511 MeV/c^2), so there is a significant suppression of the coupling of W-boson to right-helicity electrons. This is what was observed in the experiment by C.S. Wu on $^{60}_{27}$Co [60]. For the coupling of $W^\pm$ to antiquarks or antileptons, the helicity preference is reversed – the $W^\pm$ always couples to positive helicity antineutrinos and predominantly to positive helicity e^+, μ^+, τ^+ or to antiquarks. The coupling of $W^\pm$ to right-helicity quarks or leptons or to left-helicity antiquarks or antileptons is suppressed by a factor of mc^2/E.

A striking example of the consequence of this preferred helicity coupling can be seen in the leptonic decay of a charged kaon discussed in the previous section. Let us consider $K^- \rightarrow \mu^- + \bar{\nu}_\mu$ process. The kinematics and spin configurations of the μ^- and $\bar{\nu}_\mu$ for this decay are shown schematically in Fig. 17.12. In the rest frame of the K^-, the μ^- and the $\bar{\nu}_\mu$ must move in opposite directions since the momentum of K^- is zero. The K^- has zero spin, so by conservation of angular momentum, the two decay particles must have opposite spin component in any one chosen direction. The configuration in which the spin of the antineutrino is opposite to its momentum cannot take place since in this case the antineutrino would then have left-helicity. The only possible configuration is for the spin of the antineutrino to be in the same direction to its momentum since it must have right helicity. By conservation of angular momentum, this implies that the μ^- has the same configuration, i.e. both the antineutrino and μ^- have the *same* helicity and therefore the $W^\pm$ couples to the right-helicity (lepton) μ^-. Such a coupling is suppressed by $m_\mu c^2/E_\mu$.

If we look at the decay mode $K^- \rightarrow e^- + \bar{\nu}_e$, the same argument would lead to a suppression (of the decay amplitude) of $m_e c^2/E_e$. Since $m_e \ll m_\mu$ we expect the decay into a positron to be heavily suppressed. For the partial widths we expect

$$\frac{\Gamma(K^- \rightarrow \mu^- + \bar{\nu}_\mu)}{\Gamma(K^- \rightarrow e^- + \bar{\nu}_e)} = \left(\frac{m_\mu}{m_e}\right)^2 = \left(\frac{105.6 \text{ MeV}/c^2}{0.511 \text{ MeV}/c^2}\right)^2 \approx 4.3 \times 10^4.$$

(17.12)

This coincides very closely with the experimentally observed ratio. Exactly the same physical arguments are applied to the decays $K^+ \rightarrow \mu^+ + \nu_\mu$ and $K^+ \rightarrow e^+ + \nu_e$ for which the left-helicity antilepton μ^+ or e^+ coupling to W-boson is suppressed as $m_\mu c^2/E_\mu$ or $m_e c^2/E_e$, respectively.

17.6   Processes with Z-Boson Exchange

As discussed in the beginning of this chapter, weak interactions are also mediated by an electrically neutral gauge boson, Z, which, like the W-boson, couples to both quarks and leptons but contrary to the W-boson does not change fermion flavour.

In that sense, the interactions of the Z are similar to those of the photon, but there are some important differences:

- The Z couples to neutrinos, whereas the photon does not (neutrinos have zero electric charge).
- The Z has a mass of 91.2 GeV/c^2, so the interactions are short range – like the interactions of the $W^\pm$.
- The Z also has a coupling of different strengths to left-helicity and right-helicity quarks and leptons (as discussed in detail at the beginning of this chapter) and so these interactions also violate parity.

Nevertheless, in any process where there can be photon exchange, there can also be Z exchange. In terms of Feynman diagrams for $e^+ e^-$ scattering into any pair of final-state particles (e.g. a fermion–antifermion pair), we have the diagram with the photon exchange shown in the upper diagram of Fig. 17.13 that comes together with the diagram with the Z-boson exchange shown in the lower diagram of Fig. 17.13. The diagram with the photon exchange has a propagator $1/s$, where $\sqrt{s}$ is the CM energy, whereas the second diagram (Z exchange) has a propagator $1/(s - M_Z^2 c^4)$. At relatively low CM energies for which $\sqrt{s} \ll M_Z c^2$, the second diagram is suppressed by the dominant $M_Z^2 c^4$ factor in the denominator of the propagator and may be neglected. But as $\sqrt{s}$ grows to become comparable to (or greater than) $M_Z c^2$ both of these diagrams become equally important.

Fig. 17.13 Feynman diagrams for $e^+ e^- \rightarrow f\bar{f}$ process via photon exchange (upper diagram) and Z-boson exchange (lower diagram)

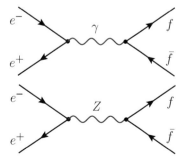

Fig. 17.14 Feynman
diagrams for ZW^+W^- and
γW^+W^- vertices

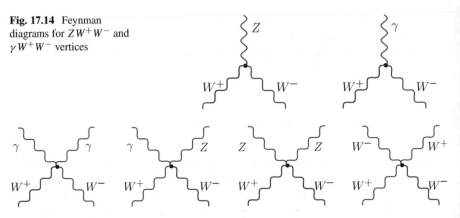

Fig. 17.15 Feynman diagram for quartic $\gamma\gamma W^+W^-$, γZW^+W^-, ZZW^+W^- and $W^+W^-W^+W^-$ interactions

Fig. 17.16 Feynman diagram
for $e^+ + e^- \rightarrow W^+ + W^-$
process via neutrino exchange

The Z and photon can both couple to $W^\pm$, with interaction vertices shown in Fig. 17.14. The interaction between the photon and $W^\pm$ is not surprising since the $W^\pm$ are charged and we would expect them to interact with photons, with coupling constant e. The interaction of $W^\pm$ with the Z is similar but has a different coupling constant $-g_W c_W$.

Besides triple gauge boson couplings, there are quartic couplings between photons, Z-bosons and $W^\pm$ bosons: $\gamma\gamma W^+W^-$, γZW^+W^-, ZZW^+W^-, and $W^+W^-W^+W^-$ with coupling constants e^2, $eg_W c_W$, $g_W^2 c_W^2$ and g_W^2, respectively. Feynman diagrams for these vertices are shown in Fig. 17.15. An important feature of weak interactions is that the gauge bosons interact with each other.

The coupling of the Z and photon to the $W^\pm$ was confirmed at the LEPII collider at CERN, where it was possible to accelerate electrons and positrons to energies sufficient to produce a W^+ and a W^- in the final state.

From the coupling of the W to electron and neutrino one can construct the Feynman diagram for the process

$$e^+ + e^- \rightarrow W^+ + W^- \ .$$

The contribution to this process involving neutrino exchange is shown in Fig. 17.16.

However, owing to the coupling of the Z and photon to $W^\pm$, we also have contributions from Feynman diagrams involving the exchange of a photon or Z-boson, which then couple to W^+W^-, as shown in Fig. 17.17. The data from LEPII

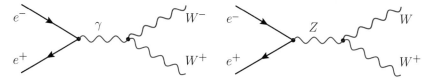

Fig. 17.17 Feynman diagram for $e^+ + e^- \rightarrow W^+ + W^-$ process via photon and Z-boson exchange

Fig. 17.18 Measurements of the cross section for $e^+ + e^- \rightarrow W^+ + W^-$ at LEPII, compared to the theory predictions with uncertainties indicated by the width of the cyan band [Reprint taken from [125] by permission from Elsevier]

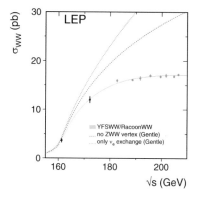

clearly show that all these diagrams have to be taken into account, in order to have agreement with the theoretical prediction. The square amplitude of the complete set of diagrams also includes important quantum interferences between contributions to the amplitude from different Feynman diagrams. In Fig. 17.18, the cross section, $\sigma(e^+e^- \rightarrow W^+W^-)$, is presented as a function of collider energy. The narrow cyan band gives the theoretical prediction with all diagrams (and interferences) taken into account (the width of the band indicates the theoretical uncertainty). The red dashed curve presents the theoretical prediction if there were no interacting Z-bosons, whilst the blue dashed curve is the contribution to the cross section from the neutrino exchange alone (Fig. 17.16). The fact that the cyan curve lies significantly below the red dashed curve indicates substantial negative interference between the contribution for the Z-boson and the contributions from neutrino and photon exchange. The black and green filled circles are the data points that are in a very good agreement with theoretical prediction when all diagrams are taken into account.

17.7   The Magnitude of Low-Energy Weak Processes

The relation (17.9) between the weak and electromagnetic coupling constants and the ratio of the W-mass to the Z-mass enables us to make an order of magnitude estimate of the rates for weak processes at low energies. At energies much less than $M_W c^2$, the amplitude for a $W^{\pm}$ exchange process is proportional to

$$\frac{\alpha_W}{2M_W^2 c^4},$$

and the rate is proportional to the square of this quantity.

For a weak decay rate we want to construct quantities with dimension of inverse time, so we need to multiply the square of this quantity by something with dimension of the fifth power of energy divided by $\hbar$. The only quantity proportional to the energy is the Q-value of the decay, Q_β, so we get an estimate

$$\text{Decay Rate} \sim \left(\frac{\alpha_W}{2M_W^2 c^4}\right)^2 \frac{Q_\beta^5}{\hbar}. \tag{17.13}$$

The pre-factor is actually quite small. For example, for muon decay $Q_\beta \approx m_\mu c^2$, and the muon decay rate is actually

$$\frac{1}{\tau_\mu} = \frac{1}{96\pi} \left(\frac{\alpha_W}{2M_W^2 c^4}\right)^2 \frac{\left(m_\mu c^2\right)^5}{\hbar}. \tag{17.14}$$

From the measured masses of the W and Z and the relation between the weak and electromagnetic coupling (17.9) we determine the muon lifetime to be

$$\tau_\mu = 2.3 \times 10^{-6}\, \text{s},$$

which is very close to the experimentally measured quantity[3] $\tau_\mu = 2.2 \times 10^{-6}\, \text{s}$. This demonstrates the success of the electroweak theory.

Summary

- The weak force carriers are the massive neutral Z and charged $W^{\pm}$ bosons. All Standard Model fermions, f, are involved in weak interactions. The $Zf\bar{f}$ vertices define neutral currents that are conserved, meaning that the fermion flavour does not change. The vertices with $Wf\bar{f}$ bosons lead to charged currents, involving quarks or leptons of different charges. Such currents change quark or lepton flavour.

---

[3]The small discrepancy between the theoretical prediction and the experimental value is due to small quantum correction to the value of α_W given by (17.9).

- Charged currents can also change quark generation. This is described by the CKM matrix. The mixing of quarks of the first two generations is one order of magnitude larger than the mixing between the second and third generations of quarks. The mixing between the first generation and third generation of quarks is suppressed by a further order of magnitude.
- Weak bosons interact with each other. The photon also interacts with charged $W^\pm$ bosons. The following interaction vertices exist:

  - Triple gauge boson interactions: ZW^+W^- and γW^+W^-
  - Quartic gauge boson interactions: $\gamma\gamma W^+W^-$, $\gamma Z W^+W^-$, ZZW^+W^-, and $W^+W^-W^+W^-$

- The strength of the weak interactions, g_W, is related to the electron charge, e, by $g_W = e/s_W$, where $s_W \equiv \sin\theta_W = \sqrt{1 - M_W^2/M_Z^2} \simeq 0.47$ so the strength of the weak charge is about twice as large as the strength of the electromagnetic one.
- Weak interactions are short range. The range is defined by $r_{W,Z} = \hbar/M_{W,Z}c$ and is of the order of 10^{-3} fm.
- Weak interactions are responsible for leptonic, semi-leptonic and hadronic decays of mesons and baryons. In particular, they are responsible for the leptonic decay of the π^+, which is the lightest charged meson.
- There is a selection rule for meson and baryon decay processes, related to the change of the strangeness: $\Delta S = \pm 1$. The same selection rule applies for charm and bottom flavours.
- Interactions with $W^\pm$ and Z-bosons violate parity: $W^\pm$ always couple to left-helicity neutrinos. For quarks and massive leptons the $W^\pm$ *can* couple to right-helicity states, but the coupling is suppressed by a factor mc^2/E. The Z also has a coupling of different strengths to left-helicity and right-helicity quarks and leptons and so these interactions also violate parity.
- One of the very striking consequences of parity violation can be seen in leptonic decays of spin-zero mesons. Since by conservation of (spin) angular momentum, decays of negatively charged mesons require the final-state charged lepton (with mass m) to have right-helicity and the W-boson couples to left-chirality fermions, the decay width is suppressed by a factor of $\left(mc^2/E\right)^2$. Similarly positively charged mesons must decay into left-helicity charged antileptons, so that their decay width is also suppressed by the same factor. This favours the meson decay into the heaviest energetically allowed lepton. This dependence has been experimentally confirmed.
- Weak interactions conserve both quark number (baryon number) and lepton number.

Problems

Problem 17.1 Draw Feynman diagrams for the process

$$e^+ + e^- \rightarrow \nu_e + \bar{\nu}_e.$$

Problem 17.2 In various theories beyond the Standard Model, force carriers are heavy bosons with the mass of the order of 1 TeV/c^2. Estimate the range of these forces.

Problem 17.3 Which of the following processes are forbidden and which are allowed by weak interactions? State your reason.

(a) $e^+ + e^- \rightarrow Z + Z$
(b) $p + p \rightarrow e^+ + e^+$
(c) $u + \bar{d} \rightarrow W^+ + Z$
(d) $u + \bar{d} \rightarrow \gamma + e^+$
(e) $e^+ + e^- \rightarrow n + n$
(f) $\gamma + e^+ \rightarrow W^+ + \bar{\nu}_e$
(g) $e^+ + e^- \rightarrow \gamma + \nu_e$
(h) $e^+ + e^- \rightarrow \tau^+ + \tau^-$
(i) $W^+ + W^- \rightarrow Z + Z$
(j) $e^+ + e^- \rightarrow \mu^+ + e^-$

[Assume that the energy of the initial particles is enough to produce final-state particles.]

Problem 17.4 The total decay width of the W-boson is 2.1 GeV. The ratio of its partial decay width to leptons to its partial decay width to hadrons is 0.5. Assuming that W-decay branching fraction to any quark flavour is the same, calculate the partial W-boson decay width to $u\bar{d}$-quark pair.

[For simplicity you can assume CKM matrix to be diagonal.]

Problem 17.5 The B^+ meson is a bound state of a u-quark and a $\bar{b}$-antiquark, with zero spin. Calculate the ratio of the partial widths

$$\frac{\Gamma(B^+ \rightarrow \tau^+ \nu_\tau)}{\Gamma(B^+ \rightarrow e^+ \nu_e)},$$

using spin and helicity arguments, and explain your result.
$\left[m_{B^+} = 5.3 \text{ GeV}/c^2, \ m_\tau = 1.78 \text{ GeV}/c^2, \ m_{m_e} = 0.51 \text{ MeV}/c^2 \right]$

Chapter 18
The Higgs Mechanism and the Higgs Boson

There is one more particle predicted by the electroweak theory which was discovered most recently, completing the entire particle set of the Standard Model, as presented in Fig. 1 of the the Historical Introduction. The name of the particle is the *"Higgs boson"*, which was discovered by the ATLAS and CMS collaborations at the LHC, announced in July 2012 [10, 11]. This particle is named after Peter Higgs, who in 1964 [126] along with five other theoretical physicists – Robert Brout and François Englert [127] and Gerri Guralnik, Richard Hagen and Tom Kibble [128] – propose a mechanism (*"Higgs mechanism"*) by which particles acquire their mass.

18.1 The Higgs Field and the Mechanism of Particle Mass Generation

The basic idea is that there exists a scalar field corresponding to a spin-zero particle, ϕ, called the *"Higgs field"* which has a constant non-zero value, $\langle \phi \rangle$, everywhere in space. This constant value, which has dimension of energy, is called the *"vacuum expectation value"*. For all other particles the value of their associated field in a state in which there are no particles (the *"vacuum state"*) is zero, but the Higgs field is unique in that even for a vacuum the field has a non-zero constant value. This value has not always been non-zero. According to the modern Quantum Field Theory there was a transition between $\langle \phi \rangle = 0$ and $\langle \phi \rangle \neq 0$ which took place at very high temperatures at the early stage of the evolution of the Universe. There is a function of the Higgs field, $V(\phi)$, called the *"Higgs potential"*. This is *not* to be thought of as a potential for a force between two separated particles, which, as explained in Chap. 16, is not consistent with Special Relativity, but rather it is proportional to the energy density due to the Higgs field. The vacuum state is the state for which this energy density is a minimum. The value of ϕ for which $V(\phi)$ is minimum is then its vacuum expectation value, $\langle \phi \rangle$.

© The Author(s), under exclusive license to Springer Nature Switzerland AG 2021
A. Belyaev, D. Ross, *The Basics of Nuclear and Particle Physics*, Undergraduate
Texts in Physics, https://doi.org/10.1007/978-3-030-80116-8_18

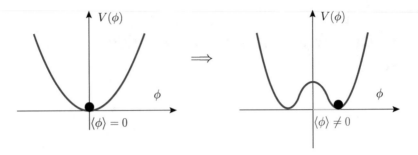

Fig. 18.1 Schematic illustration of the evolution of the vacuum from $\langle\phi\rangle = 0$ state (left) to the $\langle\phi\rangle \neq 0$ state (right)

Depending on the shape of the potential, the minimum of $V(\phi)$ can take place at $\langle\phi\rangle = 0$ as illustrated in Fig. 18.1(left) which corresponds to the early Universe epoch, or the minimum can take place at $\langle\phi\rangle \neq 0$ as happened at a later stage of the evolution of the Universe at lower temperatures, when the shape of $V(\phi)$ evolved to that of a "Mexican hat", schematically presented in Fig. 18.1 (right).

The constant $\langle\phi\rangle$ and the Higgs field itself play crucial role. When $\langle\phi\rangle = 0$, all particles are massless and travel with velocity c. When $\langle\phi\rangle \neq 0$, this background Higgs field interacts with nearly all other particles which are then slowed down – thereby moving with less than the speed of light and consequently acquiring a mass M:[1]

$$M = \frac{1}{c^2}\frac{g_H}{\sqrt{\varepsilon_0\hbar c}}\langle\phi\rangle, \qquad (18.1)$$

where g_H is the coupling of the particle to the Higgs field.[2] One should also note that in the same way that there are quanta of the electromagnetic field which are particles (photons), there must be quanta of the Higgs field. These are called *"Higgs particles"* or *"Higgs bosons"*. They must necessarily exist if the Higgs mechanism for generating masses for particles takes place (to be consistent with Quantum Field Theory). The Higgs field ϕ can be written as the sum of $\langle\phi\rangle$ and a field H, whose quantum excitations correspond to Higgs particles. Therefore any interaction of a particle with a Higgs particle is automatically accompanied by the interaction of the particle with the Higgs content in the vacuum, thereby generating a mass as given by (18.1).

---

[1]Massless particles *always* move with the speed of light (in a vacuum). Therefore any particle which moves with a lower velocity *must* be massive.

[2]g_H is taken to have the same dimension as electric charge, which is equal to the denominator factor $\sqrt{\varepsilon_0\hbar c}$, so that since $\langle\phi\rangle$ has dimension of energy, the RHS of (18.1) has the dimension of mass.

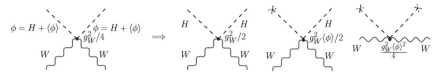

Fig. 18.2 The vertices for the interaction of two W-bosons with one or two Higgs particle(s) or with the background Higgs field. The cross at the end of a Higgs line indicates the vacuum expectation value of the Higgs field, rather than a Higgs particle

The consequence of the Higgs mechanism can be illustrated, for example, by the Higgs field interacting with W-boson, shown in Fig. 18.2.

The Higgs field ϕ, is the sum of the Higgs boson field, H, and the vacuum expectation value, $\langle\phi\rangle$. The quartic interaction of the Higgs field with the W-boson field, $WW\phi\phi$, with strength $g_W^2/4$, as shown in Fig. 18.2, (the cross at the end of the dashed lines indicates the vacuum expectation value of the Higgs field) generates three terms:

- $HHWW$ quartic interaction with strength $g_W^2/2^3$
- HWW triple interaction with the strength $g_W^2\langle\phi\rangle/2$
- A term involving only two W-bosons. This is the term that generates the W-mass. A demonstration of how these interactions generate a mass is presented in Appendix 18. Inserting the appropriate coupling constant into (18.1), the W-mass is

$$M_W = \frac{1}{2c^2}\frac{g_W}{\sqrt{\varepsilon_0\hbar c}}\langle\phi\rangle . \tag{18.2}$$

Since we know the values of M_W and $g_W = e/\sin\theta_W$ we can find the value of the vacuum expectation value $\langle\phi\rangle$:

$$\langle\phi\rangle = \frac{2M_W\sin\theta_W\sqrt{\varepsilon_0\hbar c}}{e} = \frac{2M_W\sin\theta_W}{\sqrt{4\pi\alpha}} = \frac{2\times 80.4\ \text{GeV}\times 0.47}{0.303} \simeq 250\ \text{GeV}. \tag{18.3}$$

It is important to stress that the mass of W-boson is proportional to $\langle\phi\rangle$ and g_W coupling.

Fermions also acquire their mass through their interactions with the Higgs boson. The strength of this interaction is characterized by the (dimensionless) *"Yukawa coupling"* constant, $Y_{\psi\psi\phi}$. In Fig. 18.3 we present an example of the Higgs mechanism for the $\bar{\psi}\psi\phi$ Higgs field interaction with a fermion denoted by ψ. When the Higgs field is replaced by its vacuum expectation value we get a term involving the fermion only, and this gives rise to a mass for the fermion in an analogous

---

[3]There is a factor of 2 in the Feynman rule for this vertex arising from the two ways of attaching the two external Higgs particles.

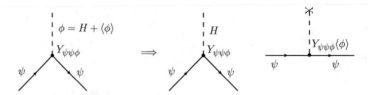

Fig. 18.3 Illustration of the effect of the Higgs mechanism for the $\bar{\psi}\psi\phi$ Higgs field interaction with fermions, which generates $\bar{\psi}\psi H$ vertex as well as $\bar{\psi}\psi$ bilinear term for the fermion mass

manner to the mass generation for the W-boson.[4] All particles couple to the Higgs field with couplings that are also proportional to their masses or the square of their masses. This is the core of the Higgs mechanism.

The fermion mass, m_ψ, is given by

$$m_\psi = \frac{1}{c^2} Y_{\psi\psi\phi} \langle\phi\rangle. \tag{18.4}$$

Each fermion has a separate Yukawa coupling constant. It is important to note that (18.4) implies that the Higgs boson couples with greater strength to fermions with larger mass.

The mass of the Higgs particle can be obtained from the curvature of the Higgs potential at its minimum. This gives us the expression for the Higgs mass (see Problem 18.4):

$$M_H = \frac{1}{c^2} \left(\frac{\partial^2}{\partial\phi^2} V(\phi)|_{\phi=\langle\phi\rangle} \right)^{1/2} = \frac{1}{c^2} \sqrt{2\lambda_\phi} \langle\phi\rangle, \tag{18.5}$$

where λ_ϕ is the (dimensionless) coupling constant for the interaction of a Higgs particle with other Higgs particles. We note that (part of) this mass comes from the coupling of a Higgs particle with the Higgs field (set to its vacuum expectation value), in analogy with the mass generation of other particles. This Higgs self-interaction also generates couplings between three or four Higgs bosons, with coupling constant λ_ϕ. The direct measurement of λ_ϕ is an extremely challenging experimental problem, and is one of the primary goals of modern Particle Physics. This measurement would be one of the tests of whether the Higgs boson's properties are consistent with the predictions of the Standard Model.

In summary, the Higgs mechanism generates the masses for all particles interacting with the Higgs field (including the Higgs boson), predicts the existence of Higgs boson and gives rise to interactions of all massive particles with the Higgs boson. In general, a Quantum Field Theory involving massive spin-one particles has serious

---

[4]Note that the coefficient in front of the term involving only two fermions is linear in mass, as opposed to the case of bosons, for which the term is quadratic in mass.

problems and generates unwanted infinities in calculated quantities. However, in 1971, Gerard 't Hooft [129], under the influence of his Ph.D. supervisor Tini Veltman, demonstrated that if the masses of these spin-one particles are generated by the Higgs mechanism, these inconsistencies do not occur.

18.2   The Properties of the Higgs Boson

The Higgs boson was discovered in 2012 and the following properties have been confidently established:

1. It has spin zero. This is consistent with the theoretical predictions since the vacuum expectation value has to be invariant under Lorentz transformations – so that it is the same in all frames of reference. The field of a spin-zero particle is Lorentz invariant, whereas the field of a particle with non-zero spin transforms under a Lorentz transformation.
2. The Higgs boson couples to $W^{\pm}$ and Z (which are consequently massive).
3. It does *not* couple directly to photons (which are massless) so it is uncharged.
4. It does not couple directly to gluons (which are massless) and so it does not take part directly in the strong interactions.
5. Its coupling to massive particles is proportional to the particle mass.
6. Its mass is measured to be $125 \, \text{GeV}/c^2$ to better than 1% accuracy.

Higgs boson decay is dominated by the most massive particles, for which the decay is kinematically allowed, because its coupling to a given particle is proportional to that particle's mass. The t-quark mass is $175 \, \text{GeV}/c^2$, exceeding the Higgs mass, so it cannot decay into a $t\bar{t}$ pair. The next most massive quark is the b-quark so the Higgs boson predominantly decays into a $b\bar{b}$ pair (whose combined mass is less than that of the Higgs), shown in diagram of Fig. 18.4a. The Higgs boson also decays to $\tau^+\tau^-$ pair, the dominant leptonic decay channel since τ-lepton is the most massive amongst the leptons. The corresponding diagram is shown in Fig. 18.4b. Although the Higgs boson is not sufficiently massive to decay into on-mass-shell W^+ and W^- bosons or two on-mass-shell Z-bosons, it can decay to one real and one virtual boson, e.g. WW^* or ZZ^* pair (where W^* and Z^* denote virtual W and virtual Z bosons, respectively) followed by the almost immediate decay of the virtual gauge boson into a fermion-antifermion pair, as shown by Feynman diagrams in Fig. 18.4c and d, respectively.

Although the Higgs boson does not couple to massless particles such as photons and gluons directly, it can decay into them via intermediate loop of virtual massive particles such as top-quark, bottom-quark or W-boson as shown in Fig. 18.5. The main contribution to such a decay mechanism comes from loops containing the most massive particles, since they have the strongest coupling to the Higgs boson. In the case of $H \rightarrow \gamma\gamma$ decay the main contribution therefore comes from the most massive quarks: t, b and from the most massive lepton – τ as well as from the W-boson as shown in Fig. 18.5a. All these particles carry an electric charge since they

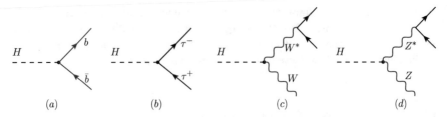

Fig. 18.4 Feynman diagrams for Higgs boson decay into $b\bar{b}$ pair (**a**), $\tau^+\tau^-$ pair (**b**) as well as WW^* (**c**) and ZZ^* (**d**) pairs followed by off-mass-shell W^* and Z^* decay into a fermion pair

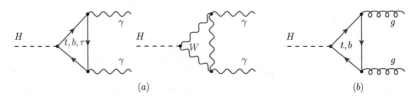

Fig. 18.5 Feynman diagrams for loop-induces Higgs boson decay processes to pair of photons (γ) (**a**) and to pair of gluons (g) (**b**)

must interact with the photon. In the case of $H \to gg$ decay, only quarks appear in the loop since these particles must be strongly interacting in order to interact with the gluon (see Fig. 18.5b).

The total Higgs boson width, predicted in the Standard Model is 4.2 MeV corresponding to a lifetime of 1.6×10^{-22} s. Despite the fact that the Higgs boson is very short-lived, its width is very small in comparison with its mass – it is even two orders of magnitude smaller than the resolution of ECAL of CMS or ATLAS detectors, so it cannot be measured directly by current experiments.

The theoretically predicted branching ratios for various Higgs boson decay channels are shown in Fig. 18.6 as a function of the Higgs boson mass. One can see that indeed $H \to b\bar{b}$ channel is dominant, followed by WW, gg, $\tau\tau$, $c\bar{c}$, ZZ, $\gamma\gamma$, $Z\gamma$ and $\mu\mu$ channels given in the descending order of their partial decay widths (at a Higgs mass of 125 GeV/c^2). It is important to stress that although the decay channel with the highest branching ratio gives the highest event rate, it is not necessarily the best channel in which to discover the Higgs boson, since as well as a high signal rate, the rate for the background processes should be low enough to observe the signal. We discuss this in the next section.

18.3 Higgs Boson Production: Discovery and Future Prospects

The highest Higgs boson production rate originates from the process involving the most massive fermions and gauge bosons which eventually have the strongest

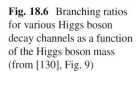

Fig. 18.6 Branching ratios for various Higgs boson decay channels as a function of the Higgs boson mass (from [130], Fig. 9)

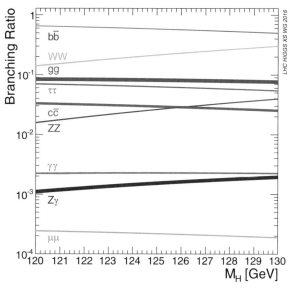

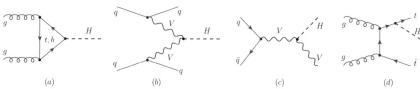

Fig. 18.7 The main Higgs boson production processes (**a**)-(**d**)

coupling to the Higgs particle. The main Higgs boson production processes at the LHC with their corresponding Feynman diagrams are presented in Fig. 18.7. These processes, in descending order of cross section, include:

- Gluon fusion process, $gg \rightarrow H$, in which the Higgs boson is produced from the loop-induced process, mainly from the interaction of the top quark in the loop, as shown in Fig. 18.7a, together with a subdominant contribution from the bottom quark.
- Weak-boson fusion process, $qq \rightarrow qqVV \rightarrow qqH$, in which the Higgs is produced from weak-boson fusion $VV \rightarrow H$, where V stands for W- or Z-boson, whilst q stands for quark or antiquark, as shown in Fig. 18.7b.
- Associated production with a gauge boson (also called "Higgs-strahlung"), $q\bar{q} \rightarrow VH$, the Feynman diagram for which is presented in Fig. 18.7c. In this process the Higgs boson is produced in association with a W- or Z-boson from $q\bar{q}$ annihilation to a virtual W- or Z-boson, respectively.
- Associated production with top-quarks. One of the Feynman diagrams contributing to this process is shown in Fig. 18.7d. In this process the Higgs boson is radiated from the top-quark line.

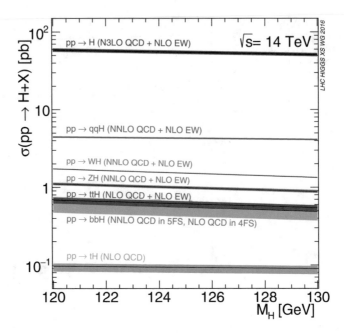

Fig. 18.8 The cross sections for the Higgs boson main production processes versus the Higgs boson mass (from [130], Fig. 177). Theoretical uncertainties are represented by the widths of the cross section bands

The respective predicted cross sections for these processes are presented in Fig. 18.8 as a function of the Higgs boson mass. Note, that even though the process $gg \rightarrow H$ is loop induced, it has the highest cross section, owing to the very large Yukawa coupling of the t-quark to the Higgs. The event rate for a given signal signature (i.e. specific signal final state) is given by the Higgs boson production cross section multiplied by the corresponding branching ratio (or product of branching ratios in case of several particle decays in the final state). For example, if we are interested in the study of the process

$$q\bar{q} \rightarrow Z(\rightarrow \mu^+\mu^-) + H(\rightarrow b\bar{b}) \rightarrow \mu^+\mu^- b\bar{b}, \tag{18.6}$$

then

$$\sigma(q\bar{q} \rightarrow Z + H \rightarrow \mu^+\mu^- b\bar{b}) = \sigma(q\bar{q} \rightarrow ZH) \times \mathrm{BR}(Z \rightarrow \mu^+\mu^-) \times \mathrm{BR}(H \rightarrow b\bar{b}). \tag{18.7}$$

Let us note that the *observation* of the signal from the particle we are looking for does not necessarily comes from the highest signal rate. The point is that the decision about the observability of the signal is taken using *both* – number of signal (S) and background (B) events. It is based on the probability of the background fluctuation to the level of the observed $S + B$ events under the hypothesis that there is no signal ("*null hypothesis assumption*"), i.e. that the excess of S events is entirely due to statistical fluctuation of the background. Such a probability is called the "*p-value*", which can be expressed in terms of number of standard deviations, σ, for a normal (Gaussian) distribution with mean value μ. The *p*-value for the observation of x events is given by

$$p(x) = \frac{1}{\sigma \sqrt{2\pi}} \int_x^\infty \exp\left\{-\frac{1}{2}\left(\frac{x'-\mu}{\sigma}\right)^2\right\} dx'. \qquad (18.8)$$

The number of standard deviations indicates the level of the "*signal significance*", α, which, for a sufficiently large number of signal and background events, takes the form

$$\alpha = \frac{S}{\sqrt{S+B}} = \sqrt{L}\frac{\sigma_S}{\sqrt{\sigma_S+\sigma_B}}, \qquad (18.9)$$

where L is the integrated luminosity, whilst σ_S and σ_B are signal and background cross sections, respectively. In the last part of the (18.9) we have used (13.8), relating the number of events, luminosity and the cross section. In Particle Physics a signal is confidently considered to have been established if $\alpha \geq 5$ (or 5σ), when one can claim discovery. This value of significance corresponds to a $p = 2.9 \times 10^{-7}$ which is the extremely small probability that this signal is mimicked by a statistical fluctuation of the background. One can see from (18.9) that the background plays an important role in the signal discovery.

Let us stress that Higgs boson discovery was not based on the process with the highest signal event rate, which would be the $gg \rightarrow H \rightarrow b\bar{b}$ process. The leading process for the Higgs boson discovery was, in fact, $H \rightarrow \gamma\gamma$, mainly coming from the process $gg \rightarrow H \rightarrow \gamma\gamma$. The corresponding Standard Model background rate for the production a $\gamma\gamma$ pair with a CM energy within 1–2% of the Higgs rest energy, 125 GeV, is relatively low, so that the S/B ratio for this process is high and this decay channel has the highest signal significance. The distribution of events in terms of the invariant CM energy of two final-state photons, $m_{\gamma\gamma}$, which was observed by the CMS collaboration in 2012 is presented in Fig. 18.9. A similar distribution was observed by the ATLAS collaboration. This two-photon signature has a significance of around 4σ. One can estimate this from the distributions from the distribution: in the 5 bins around the peak at 125 GeV, including the central one, there are about 270 signal events and 3600 background events, which, using (18.9), gives a significance of 4.3. From the combination of this signature with another very important signature, namely the 4-lepton signature from $H \rightarrow ZZ^* \rightarrow 4$ leptons, both ATLAS and CMS collaborations were able to claim Higgs boson discovery with significance just above 5σ.

From (18.9) one can also see that the signal significance is proportional to $\sqrt{L}$ (the square root of the integrated luminosity) but this does not mean that the signal can always be observed by increasing the integrated luminosity to sufficiently large values, particularly in cases where the signal is very small compared with the

Fig. 18.9 Invariant mass
distribution of two photons,
$m_{\gamma\gamma}$ reported by CMS
collaboration in 2012
(from [11], Fig. 3). "S/(S + B)
Weighted" means the
weighted average over several
sets of data distinguished by
various kinematic cuts – these
sets have different signals and
expected backgrounds

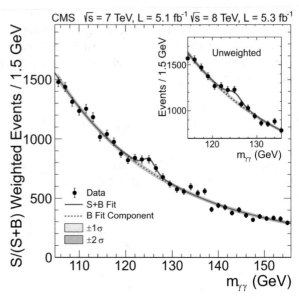

background. There is a certain lower limit for the ratio S/B, below which the signal
cannot be observed no matter how high the integrated luminosity is. This limit is
related to the theoretical and systematic background uncertainty which is typically
above 1%. If S/B is below this level, the signal cannot be observed. On the other
hand, an increase of the luminosity helps the further exploration of the properties
of the Higgs boson, from the observation of more and more production and decay
channels, which allow a more precise measurement of the Higgs boson couplings
with all Standard Model particles.

The current total integrated luminosity collected by LHC is about 140 fb^{-1}. Such
high luminosity has allowed the measurement of the Higgs boson couplings to top-
quark, W-, and Z-bosons to within about 10%, whereas the measurement of the
couplings to lighter particles, such as μ, τ and b-quark are accurate about 20–
30%, as one can see from Fig. 18.10. These measured couplings agree well with
the Standard Model predictions within the experimental uncertainties.

In the future, the LHC will be running at $\sqrt{s} = 14$ TeV CM energy and will reach
an integrated luminosity of about 3000 fb^{-1}. This will allow the LHC to measure
the above-mentioned Higgs boson couplings to within an accuracy of a few percent,
and also possibly have some sensitivity to HHH (triple Higgs) and even $HHHH$
(quartic Higgs) self-couplings.

Fig. 18.10 Upper panel: Higgs boson couplings to the Standard Model particles as reported by CMS collaboration in 2020 [131] (Fig. 14b). Lower panel: the ratio of the measured couplings of the Higgs boson to the Standard Model prediction, including the experimental uncertainties

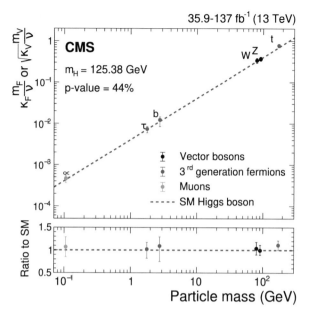

Summary

- The Higgs mechanism plays a crucial role in Particle Physics – due to this mechanism particles acquire their mass. This mechanism requires the Higgs field to have non-zero vacuum expectation value, $\langle \phi \rangle$. This is the unique property of the Higgs field, in contrast with the field for all other particles. The quanta of the Higgs field are Higgs bosons, which are excitations of the Higgs field. The interaction of the $\langle \phi \rangle$ part of the Higgs field with other particles (including Higgs boson) makes those particles massive.
- The Higgs mechanism predicts the existence of a massive Higgs boson within the framework of the Standard Model. Therefore the discovery of the Higgs boson in 2012 by the ATLAS and CMS collaborations was not only the discovery of the last particle of the Standard Model but also verification of the Higgs mechanism.
- The Higgs boson has zero spin and its coupling to any massive particle is proportional to the mass of that particle. Its own mass is measured to be $125 \, \text{GeV}/c^2$.
- Although the Higgs boson does not couple to massless particles such as photons and gluons directly, it can decay to them via an intermediate loop of virtual massive particles such as top-quark, bottom-quark or W-boson.
- The main Higgs boson production processes at the LHC are gluon fusion, weak-boson fusion, Higgs-strahlung and associated production with top-quarks processes.
- The most important processes which probe Higgs boson physics are *not* those with the highest signal rate, but those with the highest significance, defined

as $\alpha = S/\sqrt{S+B}$ (valid for high enough statistics for the number of signal and background events). The measured Higgs boson couplings to Standard Model particles agree well with the Standard Model predictions to within the experimental uncertainties.

Problems

Problem 18.1 Using Feynman rules for weak interactions and for Higgs boson interactions with fermions and gauge bosons, draw the Feynman diagrams for

$$b + u \rightarrow H + t + d$$

process. Neglect Higgs boson interaction with the first generation of fermions.

Problem 18.2 The branching ratio for $H \rightarrow \tau^+\tau^-$ is 6.3%, whilst for $H \rightarrow \mu^+\mu^-$ it is 0.022%. Using this information and the value of the muon mass, $m_\mu = 105.7$ MeV/c^2 derive the τ mass, m_τ.

Problem 18.3 The cross section of the Higgs boson production in $gg \rightarrow H$ process is 55 pb at the LHC with $\sqrt{s} = 14$ TeV.

(a) assuming the branching ratio of $H \rightarrow \gamma\gamma$ is 0.23% and that the number of background events is 10 times bigger than the number of signal events, estimate the integrated luminosity required for the observation of the $gg \rightarrow H \rightarrow \gamma\gamma$ process with the 5σ significance.

(b) using the branching ratio for Higgs boson decay to ZZ, BR($H \rightarrow ZZ$) = 2.6%, and the branching ratio of Z-boson decay to e^+e^- and $\mu^+\mu^-$ BR($Z \rightarrow e^+e^-$) = BR($Z \rightarrow \mu^+\mu^-$) = 3%, estimate the minimal integrated luminosity, required to measure the cross section of the process $gg \rightarrow H \rightarrow ZZ \rightarrow e^+e^-\mu^+\mu^-$ to within 10% accuracy. Neglect background for this estimation.

Problem 18.4 The Higgs potential, $V(\phi)$, is given by

$$V(\phi) = \frac{1}{4}\lambda_\phi \left(\phi^2 - \langle\phi\rangle^2\right)^2 .$$

Show that this is a minimum at $\phi = \langle\phi\rangle$ and that the mass of the Higgs is given by (18.5).

Appendix 18: Masses from the Higgs Mechanism

In the Standard Model there is a direct coupling between two W-bosons and two Higgs bosons with coupling constant $g_W^2/4\varepsilon_0\hbar c$. As shown in Fig. 18.2, when both

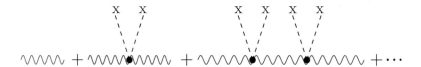

Fig. 18.11 Sum of interactions of W-boson with the background Higgs field (indicated by a dashed line with an x at the end)

the Higgs particles are replaced by the expectation value of the Higgs field, we get a term involving two W's only, which we can think of as the interaction of the W-boson with the background Higgs field, $\phi = \langle \phi \rangle$.

A massless W-boson has a propagator

$$-\frac{1}{E^2 - p^2c^2}.$$

Consider such a massless W-boson propagating and also interacting once, twice, *etc.* with the background Higgs field, as shown in Fig. 18.11.

Inserting the massless propagators and the vertex rule for the interaction of the W with the Higgs vacuum expectation value we have

$$\frac{-1}{E^2 - p^2c^2} + \frac{-1}{E^2 - p^2c^2}\left(\frac{-g_W^2\langle\phi\rangle^2}{4\varepsilon_0\hbar c}\right)\frac{-1}{E^2 - p^2c^2}$$

$$+\frac{-1}{E^2 - p^2c^2}\left(\frac{-g_W^2\langle\phi\rangle^2}{4\varepsilon_0\hbar c}\right)\frac{-1}{E^2 - p^2c^2}\left(\frac{-g_W^2\langle\phi\rangle^2}{4\varepsilon_0\hbar c}\right)\frac{-1}{E^2 - p^2c^2} + \cdots$$

Summing the geometric series we get

$$\frac{-1}{E^2 - p^2c^2 - g_W^2\langle\phi\rangle^2/4\varepsilon_0\hbar c}.$$

This is the propagator for a particle with mass M_W given by (18.2).

Other particles acquire their mass in a similar way.

Chapter 19
Electromagnetic Interactions

19.1 Electromagnetic Decays

There are some cases when particles could potentially decay via the strong interactions without violating flavour conservation. But there are cases when such decays are not energetically allowed for certain configurations of the masses of the initial and final-state particles. For example, the Σ^{*0} (mass 1385 MeV/c^2) can decay into a Λ (mass 1115 MeV/c^2) and a π^0 (mass 135 MeV/c^2). The quark content of the Σ^{*0} and Λ are the same and the π^0 consists of a superposition of quark-antiquark pairs of the same flavour. Σ^{*0} is a component of a (strong) isospin $I = 1$ multiplet, as is the pion, whereas Λ has isospin $I = 0$. Isospin is therefore conserved and the decay can proceed through the strong interactions. On the other hand, the Σ^0 whose mass is 1189 MeV/c^2, which is less than the combined mass of Λ and a pion (the lightest hadron), so the strong-interaction decay into a Λ plus a pion is energetically forbidden.

In such cases the decay can proceed via the electromagnetic interactions producing one or more photons in the final state. The dominant decay mode of the Σ^0 is

$$\Sigma^0 \rightarrow \Lambda + \gamma.$$

The quark content of the Σ^0 (u, d, s), and the Λ are the same, but one of the quarks emits a photon in the process, as shown in Fig. 19.1. In this decay the isospin is *not* conserved – the initial state has isospin $I = 1$, whereas the final state has isospin $I = 0$. But this decay is allowed, since electromagnetic interactions do not conserve strong isospin.

Because the electromagnetic coupling constant, e, is much smaller than the strong coupling constant, the rates for such decays are usually much smaller than the rates for decays which can proceed via the strong interactions. The lifetime of

© The Author(s), under exclusive license to Springer Nature Switzerland AG 2021
A. Belyaev, D. Ross, *The Basics of Nuclear and Particle Physics*, Undergraduate Texts in Physics, https://doi.org/10.1007/978-3-030-80116-8_19

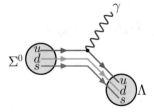

Fig. 19.1 One of the diagrams in which a quark emits a photon and the isospin of the bound state changes from $I = 1$ to $I = 0$. There are other contributions to the decay from photons emitted from the other quarks in the baryon

Fig. 19.2 Feynman diagram representing the decay of π^0 through a loop of u- or d-quarks

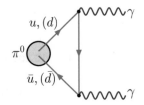

the Σ^0 is 7.4×10^{-20} s, whereas the Σ^{*0} has lifetime 1.8×10^{-23} s, corresponding to a width of 36 MeV.

19.1.1 Neutral Pion Decay

Another important example of electromagnetic decay is the decay of the π^0 into two photons.

$$\pi^0 \rightarrow \gamma + \gamma.$$

The decay of a pion into only one photon would not be possible by conservation of energy and momentum. For a π^0 decaying at rest, momentum is conserved so that the two photons have identical frequency and move in opposite directions. The π^0 is actually a superposition of a u-$\bar{u}$ quark-antiquark pair and a d-$\bar{d}$ quark-antiquark pair

$$|\pi^0\rangle = \frac{1}{\sqrt{2}} \left(|u\bar{u}\rangle - |d\bar{d}\rangle \right).$$

In either of these states, the quark can annihilate against the antiquark of identical flavour to produce two photons. Thus, although π^0 is electrically neutral, it can couple to two photons via a triangular loop of quarks. The Feynman diagram describing this mechanism is shown in Fig. 19.2.

The width for this decay process can be calculated using Quantum Field Theory and summing over the u and d contributions we get an expression for the π^0 decay width, $\Gamma(\pi^0 \to 2\gamma)$:

$$\Gamma(\pi^0 \to 2\gamma) = \frac{\alpha^2 m_\pi^3 c^6}{288\pi^3 f_\pi^2} N_c^2. \tag{19.1}$$

$N_c = 3$ is the number of quark colours. We expose this term explicitly in order to emphasize that there are three diagrams of the type shown in Fig. 19.2 – one for each quark colour. This introduces a factor of three in the decay amplitude, $\mathcal{A}(\pi^0 \to 2\gamma)$. The decay width is proportional to $|\mathcal{A}(\pi^o \to 2\gamma)|^2$ and so we get a factor of nine in the decay width from the three possible colour states of quarks.

We can also understand the factor of α^2. At the level of the matrix-element there are two couplings of quarks to a photon, each with a coupling proportional to e, so that there is a factor of e^4 in the decay width. The dimensionless factor proportional to e^4 is α^2. The factor f_π, which appears in the denominator of (19.1) is the pion decay constant, which enters into the expression for the weak-interaction decay width of a charged pion (see (17.8)). It is a (dimensionful) measure of the probability of finding a quark-antiquark pair inside the pion.

If we insert numbers $f_\pi = 130.4\,\text{MeV}$, we get a width of 7.74 eV, which is in total agreement the measured value of 7.82 ± 0.14 eV [132]. Had we not accounted for the three quark colours, the prediction would have underestimated the decay width by a factor of nine.

There are other possible decay channels for the π^0. The decay

$$\pi^0 \to e^+ + e^- + \gamma$$

can take place through a mechanism in which one or other of the emitted photons produces an electron–positron pair, as shown in Fig. 19.3. The photon which produces the electron–positron pair is an internal off mass-shell photon. The contributions to the decay amplitude from the Feynman diagrams in Fig. 19.3 have an extra power of e compared with the Feynman diagram in Fig. 19.2, and so we expect the decay width to be smaller by about a factor of α. The measured branching ratio

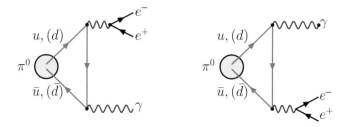

Fig. 19.3 Feynman diagrams representing the decay of π^0 to $\gamma e^+ e^-$

$$BR \equiv \frac{\Gamma(\pi^0 \to \gamma e^+ e^-)}{\Gamma(\pi^0 \to \gamma\gamma)}$$

is 0.017, which is of the order of magnitude of α.

The decay mode

$$\pi^0 \to 2e^+ + 2e^-$$

in which both of the photons are internal and produce an electron–positron pair is also possible. The branching ratio for this decay channel is 3×10^{-5}.

19.2 Electron–Positron Annihilation

Another striking piece of evidence that quarks come in three colours comes from the study of the process

$$e^+ + e^- \to \text{ hadrons.}$$

The cross section is the total cross section of electron–positron annihilation into all possible hadrons in the final state. This reaction proceeds though the production and decay of an intermediate virtual photon.[1] In terms of quarks this is equal to the cross section for the production of all possible quark-antiquark pairs. The Feynman diagram for electron–positron annihilation into a quark-antiquark pair is shown in Fig. 19.4. It is useful to compare this with the process

$$e^+ + e^- \to \mu^+ + \mu^- \, ,$$

whose Feynman diagram is shown in Fig. 19.5. At energies much larger than the rest energies of the final-state particles, the only difference between these two Feynman diagrams is the coupling of the final-state quarks or final-state muons to the photon, i.e. the electric charges of the quarks and the muons.

Fig. 19.4 Feynman diagrams representing the scattering of e^+e^- to a quark-antiquark pair

---

[1] As explained in Chap. 17, there is also a contribution from a virtual Z-boson exchange, but this is negligible for $\sqrt{s} \ll M_z c^2$.

Fig. 19.5 Feynman diagrams representing the scattering of e^+e^- to $\mu^+\mu^-$ pair

This means that for a quark of flavour i with electric charge e_{q_i} (in atomic units), the ratio of the amplitudes is

$$\frac{\mathcal{A}(e^+e^- \to q_i\bar{q}_i)}{\mathcal{A}(e^+e^- \to \mu^+\mu^-)} = e_{q_i}.$$

In order to calculate the ratio of total cross sections we square the amplitude and sum over all possible final-state quarks that can be produced, so that the ratio, R, of the total hadronic cross section to the cross section to a $\mu^+\mu^-$ pair is

$$R \equiv \frac{\sigma(e^+e^- \to \text{hadrons})}{\sigma(e^+e^- \to \mu^+\mu^-)} = \sum_i e_{q_i}^2. \tag{19.2}$$

How many quarks we sum over depends on the CM energy $\sqrt{s}$. If $\sqrt{s} < 2m_cc^2$, then there is insufficient energy to produce a c-quark and its antiquark, or any of the heavier quarks. In this energy region, only u, d and s quarks can be produced in the final state and we have

$$R = 3\left(e_u^2 + e_d^2 + e_s^2\right) = 3\left(\left(\frac{2}{3}\right)^2 + \left(\frac{-1}{3}\right)^2 + \left(\frac{-1}{3}\right)^2\right) = 2. \tag{19.3}$$

The factor of 3 is needed because we can produce final-state quark-antiquark pair in any of the three colour-anticolour combinations. The factor of 3 appears in the cross section, rather than the amplitude because, in principle, the three different colour-anticolour combinations are different distinct physical final states.

In the region $2m_cc^2 < \sqrt{s} < 2m_bc^2$ we can also produce a c-quark and $\bar{c}$-antiquark pair. This contribution has to be added to the cross section. Likewise, for $\sqrt{s} > 2m_bc^2$ we also have to include the production of b-quarks. Thus we expect this ratio, R, to have jumps as we cross thresholds in incoming energy which allow the production of more massive quarks. Such jumps can be seen in the experimental data [134] shown in Fig. 19.6. Apart from the resonances, discussed below, we see that the flat parts of the graph take the value $R = 2$ below the J/Ψ resonance which is the threshold for the production of charm quarks, $R = 3.33$ between the J/Ψ resonance and the Υ resonance, which is the threshold for the production of b-quarks, and $R = 3.67$ above the Υ resonance.

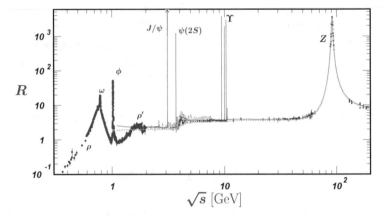

Fig. 19.6 Total cross section for electron–positron annihilation. (Reproduced from [133], Fig. 52.2)

19.2.1 Fragmentation

We are actually interested in the total cross section for $e^+ e^-$ to annihilate to produce hadrons, whereas what we have calculated is the total cross section to all quarks and their antiquarks.

What happens is that the quarks and antiquarks, which cannot be observed directly, interact with gluons in a complicated way and are converted into sets of ordinary hadrons. This process is called *"fragmentation"*. Its mechanism is not fully understood but several computer simulations have been developed which mimic this process fairly well. Although we have to rely on these computer simulations to be able to predict the cross section for the production of specific particles, we *do* know that the probability that a given final-state quark-antiquark pair will end up in one of all the possible final-state hadrons, after fragmentation, is unity. This implies that the total cross section over all possible final-state quark-antiquark pairs is equal to the total cross section over all possible[2] final-state hadrons.

In the CM frame, the final-state quark and antiquark are moving in opposite directions. What usually happens is that the process of fragmentation acting on the quark and antiquark separately leads to two narrow jets of particles moving in opposite directions, as we discuss in Chap. 20.

---

[2]"All possible hadrons" means all sets of hadrons whose total CM energy is less than $\sqrt{s}$, and whose total quantum numbers, e.g. electric charge, flavour, *etc.*, are zero so that such a final-state can couple to a virtual photon.

Fig. 19.7 Feynman diagrams showing ρ^0-meson vector dominance. The ρ^0 is represented as a quark-antiquark pair

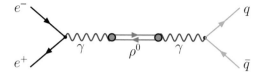

19.2.2 Resonances

As well as the almost constant value for the ratio R between energy thresholds (with jumps near each threshold), the quantity R is populated with resonances wherever $\sqrt{s}/c^2$ is equal to the mass of a neutral, spin-one particle that can couple directly to a photon.

The spin has to be the same as the spin of the photon (spin one), as there is a direct coupling between the photon and the resonant particle which must conserve angular momentum. Hadrons with spin one are often called *"vector mesons"*. The direct coupling of a vector meson to a virtual photon was first proposed by Jun Sakurai [135] and is known as *"vector meson dominance"*. Close to the resonance the total cross section for electron–positron scattering is dominated by the resonance of the vector meson, as depicted by the Feynman diagram of Fig. 19.7.

At low energies these vector mesons are particles, such as ρ^0, which consist of superpositions of a quark and antiquark of the same flavour (like in the case of π^0 but in a spin-one state). When $\sqrt{s} \approx m_\rho c^2$, the s dependence of the scattering amplitude is dominated by the ρ^0 propagator

$$\frac{1}{(s - m_\rho^2 c^4 - im_\rho \Gamma_\rho c^2)}$$

which gives rise to a resonance. The thresholds for the production of more massive quarks are also indicated by resonances for the production of bound states of the quark-antiquark pair corresponding to that threshold. When $\sqrt{s} \approx 2m_c c^2$ it is possible to create a resonance of the spin-one particle J/Ψ, which is a bound state of a c-quarks and a $\bar{c}$ antiquark, with mass $3.1\,\text{GeV}/c^2$. There are more resonances corresponding to excited states with the same quark content.

Likewise at the threshold $\sqrt{s} \approx 2m_b c^2$, there is a resonance for the Υ, which is a bound state of a b-quark and a $\bar{b}$-antiquark, with mass $9.5\,\text{GeV}/c^2$ – and also some further resonances corresponding to excited bound states.

19.3 Deep Inelastic Scattering

There is a further process involving the exchange of a virtual photon, which can probe the structure of the proton through electromagnetic interactions. This is *"deep*

Fig. 19.8 Electron–proton deep inelastic scattering. X stands for the final hadronic state

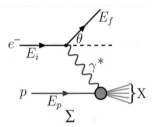

inelastic scattering" (DIS) of an electron and proton, as shown in Fig. 19.8, in which a virtual photon is exchanged between the incident electron and the proton[3].

The scattering is "deep", meaning that the energy of the exchanged photon is large enough to smash the proton and to produce several particles in the final state. Hence the scattering is also "inelastic". The hadronic subprocess is

$$\gamma^* + p \ \rightarrow \ X,$$

where γ^* indicates a virtual photon and X is *any* energetically allowed final hadronic state.

The hadronic final state is any possible set of final-state particles, whose quantum numbers sum up to give the quantum number of the proton, and whose production is energetically allowed. These final states are characterized by the total CM energy, $\sqrt{W}$, of the final-state hadrons. The measured cross section is the sum over *all* possible final states with total CM energy $\sqrt{W}$. For example, if the energy of the exchanged photon is sufficient to produce a hadron and a pion in the final state, then this hadron could be a proton with energy and momentum $(E_p, \boldsymbol{p}_p)$ and a π^0 with energy and momentum $(E_\pi, \boldsymbol{p}_\pi)$, and the value of the relativistically invariant quantity W is given by

$$W = \left(E_p + E_\pi\right)^2 - \left(\boldsymbol{p}_p + \boldsymbol{p}_\pi\right)^2 c^2.$$

Alternatively the final state could be a neutron and a π^+ with the same value of W. At the high energies at which these electron–proton scattering have been conducted there are many different sets of particles in a final state characterized by a given value of W.

These inelastic electron–proton scattering experiments were first carried out at SLAC from 1968, using electrons, which were accelerated to energies of up to 20 GeV, scattering against a fixed hydrogen target. Far greater CM energies were achieved at the HERA collider at DESY, which operated between 1992 and 2007 and collided 27.5 GeV beams of electrons or positrons against a beam of protons with energy 920 GeV.

---

[3]The target could also be a neutron. In what follows we consider only electron–proton scattering.

For elastic scattering, the kinematics of a scattering event is described by two variables: the square of the CM energy, s, of the incident particles and the square of the v momentum transfer, $-t$. For inelastic scattering, we also need to specify the square of the CM energy, W, of the hadronic final state. What is measured in the experiment is the initial electron energy E_i, the final electron energy, E_f and the scattering angle, θ. In discussing DIS one uses the positive quantity $Q^2 \equiv -t$. This is related to the to the measured quantities, E_i, E_f and θ by[4]

$$Q^2 = 4\frac{E_i E_f}{c^2} \sin^2 \left(\frac{\theta}{2} \right). \tag{19.4}$$

The quantity W is given by

$$W = 2\left(E_i - E_f\right)\left(E_p + P_p c\right) - 4E_f \left(E_i - P_p c\right) \sin^2 \left(\frac{\theta}{2} \right) + m_p^2 c^4, \tag{19.5}$$

where $P_p \equiv \sqrt{E_p^2/c^2 + m_p^2 c^2}$ is the (magnitude of) momentum of the incident proton. For a fixed target experiment we set P_p to zero and E_p to the proton rest energy, $m_p c^2$.

For square of the CM energy, s is given by

$$s = 2E_i(E_p + P_p c) + m_p^2 c^4. \tag{19.6}$$

If the CM energy is much greater than the rest energy of the proton, (19.6) may be simplified to $s = 4E_i E_p$ $(= 2E_i m_p c^2$ for a fixed target experiment).

It is conventional to express the dimensional quantities, s and W in terms of Q^2 and two dimensionless variables, x and y, defined as

$$x = \frac{Q^2}{Q^2 + W/c^2 - m_p^2 c^2}, \tag{19.7}$$

$$y = \frac{Q^2 c^2 + W - m_p^2 c^4}{s}. \tag{19.8}$$

The kinematically allowed range of x and y are

$$0 < x, y < 1.$$

The end-point $x = 1$ corresponds to elastic scattering.

---

[4]Henceforth we are assuming that the rest energy of the electron may be neglected in comparison with E_i, E_f or $\sqrt{W}$. At the collider experiment (HERA) one can also neglect the proton rest energy.

QED can be used to express the double differential cross section with respect to the variables x, y as

$$\frac{d^2\sigma}{dxdy} = \frac{4\pi\alpha^2\hbar^2}{Q^2xy}\left[xy^2F_1\left(x, Q^2\right) + (1 - y)F_2\left(x, Q^2\right)\right].\qquad(19.9)$$

The functions $F_1(x, Q^2)$ and $F_2(x, Q^2)$ contain information about the internal structure of the proton in terms of their constituent quarks. They are called "*structure functions*".

It turns out that these two functions are not totally independent, but approximately obey the relation

$$F_2\left(x, Q^2\right) \approx 2xF_1\left(x, Q^2\right),\qquad(19.10)$$

which is known as the "*Callan-Gross relation*" [136].

Furthermore, the structure functions depend strongly on the dimensionless variable, x, but only have a very mild dependence on the dimensionful variable Q^2. Complete independence of the structure functions on Q^2 was proposed by James ("Bj") Bjorken [137] and is known as "*Bjorken scaling*". In practice, this scaling is violated since the structure functions *do* vary with Q^2, but only as powers of $\log\left(Q^2\right)$. The variable x has a physical interpretation in terms of the distribution of quarks inside the proton. We will return to this in Chap. 20.

Summary

- Particles which are energetically forbidden from decaying via the strong interactions can decay electromagnetically, provided such decays are kinematically allowed, emitting one or more final-state photons or virtual photons which then decay into lepton–antilepton pairs. In electromagnetic decays isospin is *not* conserved.
- The decay width of π^0 into two photons can be calculated in terms of the charged pion decay constant. The process proceeds though a loop of u- or d-quarks which couple to two photons. The decay amplitude has a factor of three due to the three colours of quarks, leading to a factor of nine in the decay width.
- The cross section for the process

$$e^+ + e^- \rightarrow \text{hadrons}$$

is obtained by summing the cross section for the process

$$e^+ + e^- \rightarrow q_i + \bar{q}_i$$

over all possible (kinematically allowed) final-state quark-antiquark pairs. There is a factor of three in the cross section for the sum over all possible quark colours.

- Mesons with spin one (vector mesons) can couple directly to off mass-shell photons produced in electron–positron annihilation giving rise to resonances in the total cross section.
- The differential cross section for the deep inelastic scattering of an electron and proton can be described in terms of two structure functions $F_1(x, Q^2)$ and $F_2(x, Q^2)$, which are functions of the square of the momentum transfer, Q^2 and $x \equiv Q^2/(Q^2 + W/c^2)$, where W is the square of the CM energy of the final-state hadrons. The structure functions contain information about the distributions of quarks inside the proton.

Problems

Problem 19.1 Which of the following decays are electromagnetic?

(a) $\eta \rightarrow \pi^0 + \gamma + \gamma$,
(b) $\omega \rightarrow \pi^0 + \mu^+ + \mu^-$,
(c) $\rho^0 \rightarrow \pi^+ + \pi^-$,
(d) $\Sigma^+ \rightarrow p + \pi^0$?

Problem 19.2 Draw the Feynman graphs for the decay

$$\pi^0 \rightarrow e^+ + e^+ + e^- + e^-.$$

Explain why there are two diagrams.

Problem 19.3 Calculate the values of Q^2, x and y for the following DIS events:

(a) A fixed proton target experiment with initial electron energy 20 GeV, final electron energy 9 GeV and scattering angle 6.2°.
(b) A collider experiment with initial electron energy 27.5 GeV, initial proton energy 920 GeV, final electron energy 19.5 GeV and scattering angle 56°.

Chapter 20
Quantum Chromodynamics (QCD)

20.1 Gluons and Colour

In the same way that in weak interactions the weak gauge bosons $W^\pm$ can effect changes of flavour when they interact with quarks, in the case of strong interactions the strong gauge bosons (gluons) can effect changes of the colour of the quarks (but conserve flavour).

Thus we get interaction vertices of the type shown in Fig. 20.1 in which a red quark is converted into a blue quark of the same flavour, emitting a gluon. The fact that the flavour is unchanged is the reason why flavour is conserved in strong interactions. Note that unlike W- and Z-bosons or photons, gluons are represented in Feynman diagrams as "coiled" lines.

This theory of strong interactions, first proposed in 1973 by Harald Fritsch, Murray Gell-Mann and Heinrich Leutwyler [138] is called *"Quantum Chromodynamics"* (QCD), and together with the electroweak theory described in Chap. 17 forms the Standard Model, which describes, in terms of a Quantum Field Theory, all the fundamental forces of Nature with the exception of gravity.

The mathematical representation of the three quark colours is the symmetry group SU(3) introduced in Chap. 15 in the context of the eightfold way. The three quark colours are the three components of the fundamental building block of this SU(3) symmetry. There are eight SU(3) transformations – six from a change in quark colour and two from changes in the relative sign of the quark wavefunctions in the different colour states. Each such transformation has a different gluon associated with it, so there are eight different gluons. Note that unlike the case of the eightfold way for which the SU(3) symmetry is only approximate, colour SU(3) symmetry is exact! When a quark changes colour, the colour difference is carried off by the gluon. In Fig. 20.1, a quark in a red state is transformed into a quark in a blue state emitting a red–antiblue gluon. Colour is strictly conserved at every strong interaction vertex. Quarks in different colour states have identical mass, spin and all other properties.

© The Author(s), under exclusive license to Springer Nature Switzerland AG 2021   309
A. Belyaev, D. Ross, *The Basics of Nuclear and Particle Physics*, Undergraduate
Texts in Physics, https://doi.org/10.1007/978-3-030-80116-8_20

Fig. 20.1 A quark changes
its colour and emits a
colour-changing gluon (the
coiled line), which carries off
the colour difference

Fig. 20.2 Triple (**a**) and
quartic (**b**) self-couplings of
gluons

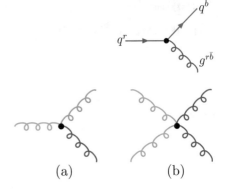

(a)   (b)

In the same way that the Z-bosons and $W^{\pm}$-bosons can couple to each other,
gluons can couple to each other with triple and quartic vertices as shown in Fig. 20.2.

In the case of weak and electromagnetic interactions, the strengths of the
couplings of the gauge bosons to quarks and leptons, which are controlled by the
electron charge, e, or the weak coupling, g_W, are much smaller than one:

$$\alpha = \frac{e^2}{4\pi\varepsilon_0\hbar c} = \frac{1}{137}, \quad \alpha_W = \frac{g_W^2}{4\pi\varepsilon_0\hbar c} \approx \frac{1}{30}.$$

This means that we can calculate the rates for weak and electromagnetic processes
using perturbation theory, i.e. the amplitudes can be expanded as a power series
in α or α_W, with the *"leading order (LO)"* term arising from Feynman diagrams
containing the minimum number of interactions required for the process under
consideration to take place. The next term in the expansion, the *"next to leading
order (NLO)"* term, gives rise to small corrections of order α, for electromagnetic
processes, and α_W for weak processes. However, for the strong interactions inside
a nucleus, the coupling constant is too large to make such an expansion possible.
It is for this reason that we cannot, even in principle, calculate the energy levels of
nuclei, or the masses of hadrons using a perturbative expansion in powers of the
strong coupling constant.

20.2   Running Coupling

However, under certain conditions a perturbative expansion of QCD can be used. It
is convenient to work in terms of α_s where

$$\alpha_s = \frac{g_s^2}{4\pi\varepsilon_0\hbar c},$$

Fig. 20.3 Feynman diagram with a loop of charged particles, which screens the charge of the particle to which the photon is attached

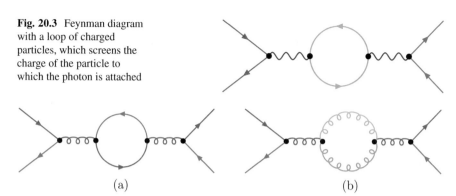

Fig. 20.4 Feynman diagrams showing a correction to the exchange of a gluon between two quarks from (**a**) the production of a quark and antiquark, and (**b**) the production of an extra pair of gluons

g_s being the coupling of the gluons to quarks or the coupling of gluons to each other.

It turns out that the QCD coupling depends on the momentum transfer in the process under consideration and at sufficiently high energies, with correspondingly large momentum transfers, Q, the strong coupling becomes small.

In contrast, in the case of Electromagnetism, the effective charge of a charged particle becomes larger at high energy scales (large momentum transfer), as a result of screening. When an electric charge is probed by another charge, the virtual photon exchanged between them can sometimes create a pair of charged particles (a particle and its antiparticle), which exist for a short while before annihilating each other again. Diagrammatically we would represent this as shown in Fig. 20.3. The effect is to surround the probed charge by a cloud of charged particles that act as a screen – reducing the effective measured charge. As the energy or momentum scale increases, the probe penetrates further into this screen so that the measured charge increases. When we write $\alpha = 1/137$, this value refers to the strength of the electromagnetic coupling in processes that take place at relatively low energy (much smaller than the electron rest energy) with low momentum transfer. At energies of the order of the rest energy of the Z-boson for which we can have momentum transfers at the scale $Q \sim M_Z c$ the value is closer to $1/129$. The high energy photon probes "deeper" into the charged particle, thereby experiencing a larger effective electric charge.

In the case of QCD, as well as an exchanged gluon producing a quark–antiquark pair, which screens the strong coupling, as shown in Fig. 20.4a, we can also have processes in which a cloud of gluons can be produced by the exchanged virtual gluon, because gluons interact with each other (unlike photons). Thus we have Feynman diagrams like that shown in Fig. 20.4b. The effect of self-interaction of the gluon, i.e. the production of an extra pair of gluons, surrounds a quark with a cloud of gluons. Unlike the production of an intermediate quark–antiquark pair, an intermediate gluon pair leads to anti-screening so that at large momenta the effective coupling decreases.

Therefore, the coupling "constant" is replaced by the *"running coupling"* $\alpha_s(Q)$, which is a function of the momentum transfer, Q. The dependence of the running coupling on Q is determined by the *"β-function"*, which is a function of the running coupling, defined as follows:

$$\beta\left(\alpha_s(Q)\right) = \frac{Q}{2}\frac{d\alpha_s(Q)}{dQ}. \tag{20.1}$$

For Electromagnetism, β is positive so that α increases with increasing Q, whereas in QCD it is negative, leading to a decrease in α_s as Q increases.

For QCD, provided $\alpha_s(Q)$ is small, we have an expansion of β as a power series in α_s,

$$\beta(\alpha_s) = \frac{\alpha_s^2}{4\pi}\beta_0 + \mathcal{O}(\alpha_s^3). \tag{20.2}$$

The coefficient β_0 was calculated in 1973 by David Politzer [139], and independently by David Gross and Frank Wilczek [140], and found to be given by

$$\beta_0 = -11 + \frac{2}{3}n_f. \tag{20.3}$$

Here n_f means the number of active flavours and is used in the same way as in the calculation of the cross section for the process $e^+e^- \to$ hadrons, discussed in Chap. 19. It is the number of quark flavours that can be pair-produced by a virtual gluon with momentum Q, i.e. the number of flavours for which the combined mass of a quark–antiquark pair, $2m_q$, is less than Q/c. Thus for values of Q below $2m_c c$, where m_c is the mass of the c-quark, n_f is equal to three. If Q lies between $2m_c c$ and $2m_b c$ where m_b is the mass of the b-quark, then $n_f = 4$ etc. The term proportional to n_f comes from Fig. 20.4a in which an intermediate quark–antiquark pair is produced. The contribution from fermion loops is clearly proportional to the number of such pairs that can be created.

The first term on the RHS of (20.3) comes from the interaction of the gluons with each other (Fig. 20.4b) producing a gluon cloud that decreases the effective coupling with increasing Q, as indicated by the minus sign. This sign means that the effective coupling decreases with increasing Q so that it *can* be used as an expansion parameter in a perturbation series. This vitally important property of QCD is known as *"asymptotic freedom"*.

The solution to the differential equation (20.1) is

$$\alpha_s(Q) = \frac{\alpha_s(\mu)}{\left(1 - (\beta_0/4\pi)\alpha_s(\mu)\ln(Q^2/\mu^2)\right)}, \tag{20.4}$$

where $\alpha_s(\mu)$ is the value of α_s at some reference momentum scale (it serves as the integration constant for the differential equation). Usually, this is taken to be

$\mu = M_Z c$, since the value of α_s was measured very accurately at LEP at this scale and its value was found to be

$$\alpha_s(M_Z c) = 0.118 \pm 0.001.$$

This is a fairly small number. We also see that since β_0 is negative, the effective coupling becomes even smaller for momentum scales above $M_Z c$. The β-function goes through a jump as Q crosses a threshold for the production of a new quark–antiquark pair, and this discontinuity needs to be accounted for when calculating the running of the coupling across such a threshold. In other words, (20.4) with β_0 set to $\beta_0^{(n_f)}$, corresponding to the number of active flavours, can only be used up to the next threshold at $Q = Q_T$, for the production of the next flavour quark–antiquark pair. For values of Q above that threshold momentum (20.4) is replaced by

$$\alpha_s(Q) = \frac{\alpha_s(Q_T)}{\left(1 - (\beta_0^{(n_f+1)}/4\pi)\alpha_s(Q_T)\ln(Q^2/Q_T^2)\right)}, \qquad (20.5)$$

where $\beta_0^{(n_f+1)}$ is the value of β_0 above the threshold for the next active flavour. A similar procedure is invoked in order to calculate the value of the running coupling below the b-quark or c-quark threshold.

We note from (20.4) that the dependence of the running coupling on Q is logarithmic – the dependence is fairly slow. Nevertheless, over a large range of Q from a few GeV/c to a few TeV/c, there is a substantial change in $\alpha_s(Q)$. Experimental measurements of $\alpha_s(Q)$ over this large range of momentum scales agree well with the theoretical prediction, as can be seen from Fig. 20.5. We see that for Q greater than a few GeV/c,[1] $\alpha_s(Q)$ is small enough that we would expect a calculation using perturbation theory to be fairly reliable. In many cases, in addition to the LO perturbative results, the NLO and in some cases even the NNLO results have been computed.

Below these momentum or energy scales, such as the rest energies of light hadrons or the binding energies of nuclei, we cannot use perturbation theory for QCD.

---

[1] At relatively low values of Q, the uncertainty in the theoretical prediction due to the small experimental error on $\alpha_s(M_Z c)$ is amplified as can be seen from the gap between the solid lines in Fig. 20.5.

Fig. 20.5 Graph of the effective strong coupling, $\alpha_s(Q)$, against energy Q [133]. The solid line is the theoretical prediction from the β-function. The measured values from various different experiments agree well with the prediction. The single fit parameter is the value of $\alpha_s(Q = M_Z c)$

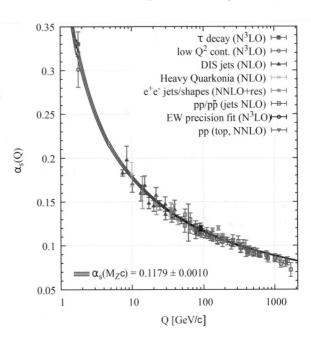

20.3   Quark Confinement

We have seen that the weak interactions are short range because the gauge bosons $W^{\pm}$ and Z are massive and so the weak potential is of the Yukawa type whose exponential fall-off with distance has an exponent proportional to M_W.

In the case of QCD, the gluons are massless, so we might expect the strong interactions to be long range (as in the case of electromagnetic interactions mediated by massless photons), whereas we know that the strong interactions have a range of a few fm.

The answer to this puzzle is the converse of asymptotic freedom. At large momentum, where we are probing short distances, the effective coupling decreases. Conversely at large quark separations the effective coupling increases and the binding between them gets stronger.

It is not possible to isolate a single quark or gluon. The only states that can exist in isolation are bound states of quarks, antiquarks and gluons that are in a colour singlet state, i.e. they are invariant under colour SU(3) transformations. As discussed in Chap. 15, this means that a baryon wavefunction is an antisymmetric superposition of states in which the three quarks have different colours and a meson is a superposition of states in which the quark and antiquark have opposite colours.

Consider a meson, which is a quark–antiquark state of the opposite colour, (e.g. green and anti-green), as indicated in Fig. 20.6. Classically, we can consider the strong potential generated by the exchange of many gluons between the quark and antiquark as an elastic string, in which the attractive force becomes stronger as the

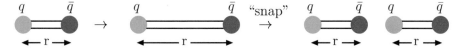

Fig. 20.6 A meson consisting of a quark and antiquark of opposite colours (anti-green is represented by purple)

quark and antiquark are pulled apart so that the string is stretched. Eventually, the string will "snap", but as it does so there is enough elastic energy in the string to produce an antiquark at the end of the part of the string containing the quark, with opposite colour and a quark of opposite colour at the end of the part of the string containing the antiquark. So we end up with two mesons, both of which are colour singlets (colourless), but we do not succeed in isolating a single quark or antiquark.

The only hadron states that we can observe are colourless (colour singlet) states, which is a consequence of quark confinement. Its exact mechanism is not understood, but numerical studies in QCD have confirmed that this confinement does indeed take place (see Sect. 20.9).

20.4   Quark–Antiquark Potential and Heavy Quark Bound States

Bound states of a c-quark and $\bar{c}$-antiquark (known as *"charmonium"*) have masses of about twice the mass, m_c, of the c-quark. At momentum scales $Q \sim 2 m_c c$, the running coupling is $\alpha_s \sim 0.25$, which is small enough to use perturbation theory to obtain energy levels for the charmonium mesons.

For bound states of heavy flavour quarks (c-quarks or b-quarks), the relatively small value of the strong coupling means that the binding energies of these hadrons are small compared with the rest energy of the constituent quarks. The quarks move non-relativistically inside the mesons and one can attempt to solve the Schroedinger equation using a suitable potential. This potential should contain a confining term, i.e. a term that increases with the distance, r, between the quark and antiquark increases, in addition to a Coulomb-like term that decreases with r.

An example of such a potential that yields good results for the spectrum of charmonium mesons is

$$V(r) = -\frac{4}{3}\frac{\alpha_s}{r} + kr. \tag{20.6}$$

The first term is the usual Coulomb-like potential that dominates at short distances where the potential can be viewed as the exchange of massless gluons – the factor of 4/3 is associated with the number of quark colours and the number of gluons. The

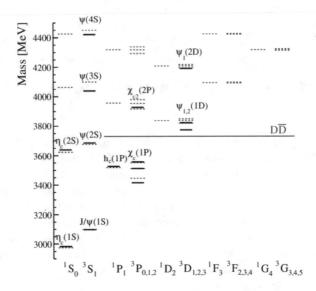

Fig. 20.7 The charmonium spectrum [141]. The black lines are the experimental values and the red dashed lines are the theoretical predictions, using a confining potential of the form (20.6) The blue line is the threshold for the production of the lightest charm meson (D) and its antiparticle

term kr increases with increasing separation and represents confinement. By fitting the data, k is found to be of the order of $1\,\mathrm{GeV\,fm^{-1}}$ ($\sim 10^5\,\mathrm{N}!$).

Using this potential and making corrections for relativistic effects and spin–orbit coupling, the mass spectrum of charmonium (J/Ψ and its excited states) can be obtained with a high degree of accuracy, as shown in Fig. 20.7. A similarly successful spectrum can be obtained for the spectrum, of bound states of a b-quark and $\bar{b}$-antiquark (Υ and its excited states).

20.5   Three Jets in Electron–Positron Annihilation

When we were considering the process

$$e^+ + e^- \rightarrow \text{hadrons}$$

the Feynman diagram considered was the annihilation of an electron–positron pair producing a quark–antiquark final state (see Fig. 19.4). In the CM frame, in order to conserve momentum the final-state quark and antiquark must have equal and opposite momentum. These quarks fragment to produce two jets of hadrons moving in opposite directions. Such two-jet events were observed in the electron–positron annihilation experiment carried out at DESY in the late 1970s. A typical event is shown in Fig. 20.8a.

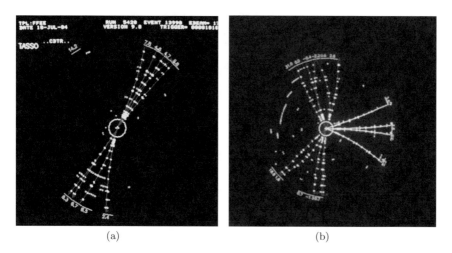

Fig. 20.8 (**a**) A two-jet event detected by the TASSO detector at DESY. (**b**) A three-jet event detected by the TASSO detector at DESY. (Credit: Oxford PPU)

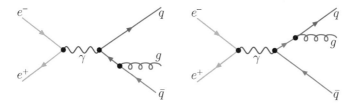

Fig. 20.9 The Feynman diagrams for the process $e + e^- \to q\bar{q}g$

Because quarks interact with gluons, one can also have Feynman diagrams in which the quark, or antiquark radiates a gluon. In such cases there are three particles in the final state. The gluon also fragments into a hadron jet and so we get three jets of particles. The occurrence of three-jet events resulting from the interaction of quarks with gluons was predicted in 1977 by John Ellis, Mary K. Gaillard and Graham Ross [142]. The corresponding Feynman diagrams are shown in Fig. 20.9. They calculated the cross section for three-jet production as a function of the jet energies, which, using momentum conservation, determines the angular distribution of the three jets.

The first such events were observed by the TASSO detector at DESY in 1979 [143]. A typical event is shown in Fig. 20.8b. Since gluons cannot be isolated and observed directly, the observation of three-jet events was taken to be the discovery of the gluon. It was the first piece of evidence that gluons existed and coupled to quarks. Furthermore, the angular distribution of the jets matched the calculation of [142].

Four- and five-jet events have now also been observed. At LEP energies (100–200 GeV) the running coupling is small ($\alpha_s \sim 0.1$). For this reason three-jet events

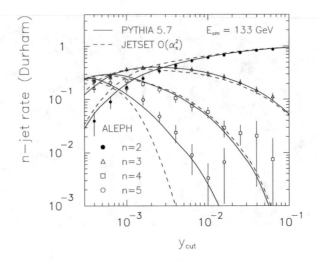

Fig. 20.10 Data on multi-jet production in e^+e^- annihilation from the א detector at LEP. The different lines represent the prediction of perturbative QCD together with different models (as shown) to account for the fragmentation of the final-state quark, antiquark and gluon into physical hadrons. [Reprinted from [144] by permission from Springer Nature]

are rarer than two-jet events because the amplitude for the process contains a factor of the gluon coupling, g_s, and so the rate is expected to be suppressed relative to the two-jet rate by a factor of order $\alpha_s \propto g_s^2$. Likewise four- and five-jet events are even rarer.

The exact definition of a jet depends on how big an angle between particles constitutes a single jet. This is parametrized by the kinematic variable, y_{cut}, which is a measure of the maximum angle between any two particles in a single jet. Because of the small running coupling, perturbative QCD can be used to calculate the number of jets as a function of this variable y_{cut}.

Some correction to the result from pure perturbative QCD has to be made in order to account for the process of fragmentation. The different curves shown in Fig. 20.10 give the results for different models used to simulate this fragmentation process. Nevertheless, the agreement between the data and the QCD theory, shown in Fig. 20.10, is impressively good.

20.6   Sea Quarks and Gluon Content of Hadrons

The quarks (and antiquarks) inside hadrons are bound together by exchanging gluons. Thus, as well as having the quarks inside hadrons, there will be gluons, radiated from the quarks or antiquarks that make up the hadron. These gluons can in turn create quark–antiquark pairs (which exist for a very short time and then annihilate). Thus, for example, a "snapshot" of a π^+, consisting of a u-quark and a

Fig. 20.11 An example of the content of a π^+ meson at any one time. In this case there are three gluons and a quark–antiquark pair. In reality there would be many more sea quarks and gluons inside a hadron

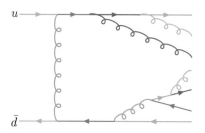

$\bar{d}$-antiquark, could be as shown in Fig. 20.11 where we can see that as well as the quark and antiquark that form the meson (known as *"valence quarks"*) there are also three gluons and an extra quark–antiquark pair. These extra quark–antiquark pairs are called *"sea quarks"*.

The QCD picture of a hadron is therefore one of the valence quarks that determine the properties (quantum numbers) of the hadron, which are "swimming" in a "sea" of quarks, antiquarks and gluons. Pairs of sea quarks carry no net flavour (i.e. they are of opposite electric charge, strangeness, *etc.*) so that they have no effect on the properties of the hadron in which they occur.

20.7   Parton Distribution Functions

Quarks, antiquarks and gluons are collectively known as *"partons"*. If we consider a relativistically moving hadron (with energy $E \approx pc$), some fraction, x, will be carried by a parton of each possible type. The probability that a fraction, x, of the momentum of the hadron, h (e.g. a proton), is carried by a parton of type i is called the parton distribution function and is written as

$$f_i^h(x)$$

where i can mean a gluon, or a quark or antiquark of a given flavour.

In terms of classical physics, we would expect the probability of finding a particular type of parton with fraction x of the hadron's momentum to be a function of x only. Unfortunately, Quantum Physics spoils this classical picture. In order to determine the momentum of a parton inside a hadron, it has to be observed by scattering a probe off it. The probability to measure a particular momentum fraction, x, depends on the momentum, Q, transferred to the hadron by the probe. The parton distribution functions, $f_i^h(x, Q^2)$ (by convention, parton distributions are written as a function of the square of the momentum transfer, Q^2, rather than Q), are therefore functions of Q^2 as well as x.

More specifically, a quark or antiquark emitted from a hadron with fraction x of the hadron momentum can radiate a gluon, which carries off some of the momentum, so that the momentum measured by the probe is $x' < x$, as shown

Fig. 20.12 A Feynman
diagram representing the
process in which a quark
emitted from a hadron
radiates a gluon before being
observed by a probe carrying
a momentum transfer Q

Fig. 20.13 The parton
distribution functions for
different partons inside a
proton at $Q^2 = 100\,\mathrm{GeV}^2$

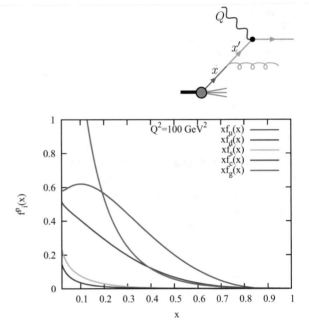

in Fig. 20.12. The amplitude for such a gluon emission depends on the momentum
transferred, Q. Likewise a gluon emitted from a hadron can produce a quark–
antiquark pair, each of which will carry off a smaller fraction of the hadron's
momentum, than the initial gluon. For a sufficiently large Q, for which the running
coupling, $\alpha_s(Q)$, is sufficiently small, the Q-dependence of the parton distribution
functions can be calculated perturbatively, using a technique derived independently
by Yuri Dokshitser [145], Vladimir Gribov and Lev Lipatov [146], and by Guido
Altarelli and Giorgio Parisi [147].

Although the Q-dependence of the structure functions can be calculated using
perturbative QCD, it is not possible to calculate their x-dependence. They have to
be inferred by examining experimental data. Several groups have been involved in
projects to extract parton distributions from all the available data. These different
groups use different parametrizations with which to fit data, but they all give
approximately the same results. An example of these parton distributions inside a
proton (as a function of x) at $Q^2 = 100\,\mathrm{GeV}^2/\mathrm{c}^2$ [148] is shown in Fig. 20.13. We
note first that the parton distribution function for a u-quark is about double that of a
d-quark. This is expected because the proton consists of two valence u-quarks and
one valence d-quark. The s- or c-quarks are only present as sea quarks and we can
see that the probability to find an s-quark or c-quark inside the proton is smaller than
that of the valence quarks and only has support at small values of x – only a small
fraction of the proton's momentum are carried off by sea quarks. At low x the gluon
distribution dominates the quark distributions, but at larger x the gluon distribution
becomes smaller than the valence quark distributions.

Once these parton distribution functions are known, QCD can be used to predict scattering cross sections for other processes.[2]

20.8   Factorization

Perturbative QCD can be used to calculate cross sections at the parton level, provided that the energy is large enough so that the scale of momentum transfer scale of the process, Q, is large enough for $\alpha_S(Q)$ to be sufficiently small that it can be used as an expansion parameter in a perturbative calculation.

For, example we can calculate the cross section for the processes

- $g + g \rightarrow q + \bar{q}$
- $g + g \rightarrow g + g$
- $q + q \rightarrow q + q$
- $q + \bar{q} \rightarrow q + \bar{q}$

  etc.

Let us denote the calculated differential cross section for two partons of type i and j to go into two other partons with transverse momentum p_T transverse to the incoming partons by

$$\frac{d\hat{\sigma}_{ij}(\hat{s})}{dp_T},$$

where $\sqrt{\hat{s}}$ is the CM energy of the incoming partons.

What we are really interested in is a process in which the initial states are not partons (which cannot be isolated in a laboratory owing to confinement) but initial-state hadrons such as a proton and an antiproton.

In order to obtain the differential cross section for proton–antiproton scattering into two jets of final-state hadrons with transverse momentum p_T we can invoke the factorization theorem.

If we pull a parton of type i from one of the incoming protons, with a fraction x_1 of the momentum of the parent proton, and a parton of type j from the antiproton, with a fraction x_2 of its momentum, then the square of the CM energy of the two partons, assumed to be moving ultra-relativistically, is given by (see problem 20.4)

$$\hat{s} = x_1 x_2 s, \tag{20.7}$$

where s is the square of the CM energy of the incoming proton and antiproton.

---

[2]Conversely, QCD calculations for cross sections that have been measured so far can be used in conjunction with QCD experimental data in order to extract the parton distribution functions.

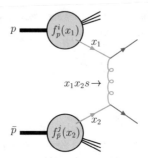

Fig. 20.14 Contribution to the differential cross section for proton–antiproton scattering due to the scattering of a quark pulled out of the proton with fraction x_1 of the proton momentum and another quark pulled out of the antiproton with fraction x_2 of the antiproton momentum

QCD factorisation tells us that if $f_i^p(x_1, p_T^2)$ and $f_j^{\bar{p}}(x_2, p_T^2)$ are the parton distribution functions for partons i and j, at momentum scale p_T, inside the proton and antiproton, respectively, then the contribution to the proton–antiproton differential cross section due to this particular parton scattering is

$$\frac{d\sigma(p\bar{p})}{dp_T}\bigg|_{ij} = \int_0^1 dx_1 \int_0^1 dx_2 f_i^p(x_1, p_T^2) f_j^{\bar{p}}(x_2, p_T^2) \frac{d\hat{\sigma}_{ij}(\hat{s})}{dp_T}, \qquad (20.8)$$

where $d\hat{\sigma}_{ij}(\hat{s})/dp_T$ is the differential cross section with respect to transverse momentum for incoming partons i and j with CM energy given by (20.7). In other words it is the differential cross section for the scattering of two partons of type i and j carrying fractions x_1 and x_2 of the momentum of the parent scattering hadrons summed (integrated) over all possible values of momentum fractions.

For example if the parton level scattering is quark–quark scattering, this contribution is represented diagrammatically in Fig. 20.14. Provided the transverse momentum, p_T, is sufficiently large so that the running coupling at that momentum scale is small, then the partonic cross section, $\hat{\sigma}$, can be calculated using perturbative QCD [149], with scattering amplitudes obtained from Feynman diagrams such as that shown in Fig. 20.14.

The total differential cross section for proton–antiproton scattering is obtained by summing over all possible parton types that can be pulled out of the incoming particles (quarks, antiquarks, of all possible flavours, and gluons).

Thus we finally obtain an expression for the proton–antiproton differential cross section

$$\frac{d\sigma_{p\bar{p}}(s)}{dp_T} = \sum_{i,j} \int_0^1 dx_1 \int_0^1 dx_2 f_i^p(x_1, p_T^2) f_j^{\bar{p}}(x_2, p_T^2) \frac{d\hat{\sigma}_{ij}(x_1 x_2 s)}{dp_T}, \qquad (20.9)$$

where the sum over i, j means sum over all possible parton types and flavours.

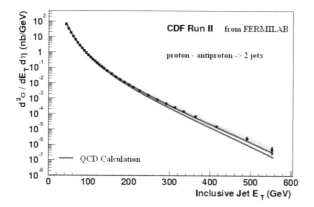

Fig. 20.15 Differential cross section for the process $p\bar{p} \rightarrow 2$ jets, as a function of transverse energy E_T, measured by the CDF collaboration at the Tevatron. (Figure taken from [150])

QCD calculations based on this factorisation theorem agree well with experiment, as can be seen from Fig. 20.15, which shows the differential cross section for the scattering of a proton and antiproton into two jets with transverse energy, $E_T \equiv cp_T$. The data were obtained by the CDF collaboration at the Tevatron accelerator [150].

The cross section for different hadronic scattering processes can be calculated from expressions analogous to (20.9) with appropriate parton distribution functions and appropriate partonic differential cross sections.

Only a single parton from each hadron takes part in the parton scattering process. The other partons in the incoming hadrons finally fragment into hadrons, which are moving almost in the same direction as the incoming hadrons (and are usually not observed because they get lost in the beam pipe of the accelerator). The partons that participate in the scattering fragment into several hadrons, which usually form two widely separated jets, as shown in Fig. 20.15.

20.8.1 Deep Inelastic Scattering Revisited

One immediate application of QCD factorization is the relationship between parton distribution functions and the structure functions for deep inelastic electron–proton scattering that were defined in Chap. 19 (see (19.9)). In this case there is only one initial-state hadron, so we only need one parton distribution function, which must be either a quark or antiquark, since electrons do not scatter off the electrically neutral gluons. The total electron–proton cross section is then given by

$$\frac{d\sigma(ep \rightarrow X)}{dxdy} = \sum_i \int_0^1 dx' f_i^p(x', Q^2) \frac{d\hat{\sigma}_{e\,q_i \rightarrow e\,q_i}(x's)}{dx'dy}, \tag{20.10}$$

where (see (19.7) and (19.8))

$$Q^2 = sxy. \tag{20.11}$$

Using (20.11) we have the relation

$$\frac{d\hat{\sigma}_{e\,q_i \to e\,q_i}(x's)}{dy} = \frac{1}{sx} \frac{d\hat{\sigma}_{e\,q_i \to e\,q_i}(x's)}{dQ^2}. \tag{20.12}$$

$d\hat{\sigma}_{e\,q_i \to e\,q_i}(x's)/dQ^2$ is the differential cross section with respect to the invariant square of the momentum transfer, $-t$, for the elastic scattering process

$$e + q_i \;\rightarrow\; e + q_i,$$

with t set equal to $-Q^2 c^2$. This differential cross section is given by (16.29). Conservation of energy and momentum leads to the constraint on x'

$$x' = x \equiv \frac{Q^2 c^2}{Q^2 c^2 + W}.$$

Using (20.11) and (20.12) we may express (16.29) in the form

$$\frac{d\hat{\sigma}_{e\,q_i \to e\,q_i}(x's)}{dy} = \begin{cases} \frac{2\pi\alpha^2\hbar^2}{Q^2 y} e_{q_i}^2 \left(1 + (1-y)^2\right) & \text{if } x' = x \\ 0 & \text{if } x' \neq x, \end{cases} \tag{20.13}$$

where e_{q_i} is the electric charge of the quark or antiquark with label i, in nuclear units. It is convenient to write (20.13) as a double differential cross section with respect to y and x', with the condition $x = x'$ expressed as a δ-function:

$$\frac{d\hat{\sigma}_{e\,q_i \to e\,q_i}(x's)}{dx'dy} = \frac{2\pi\alpha^2\hbar^2}{Q^2 y} e_{q_i}^2 \left(1 + (1-y)^2\right) \delta\left(x' - x\right). \tag{20.14}$$

Inserting (20.14) into (20.10) and integrating over x', we get

$$\frac{d\sigma(ep \to X)}{dxdy} = \frac{2\pi\alpha^2\hbar^2}{Q^2 y} e_{q_i}^2 \, f(x, Q^2) \left(1 + (1-y)^2\right). \tag{20.15}$$

Comparing (20.15) with (19.9), we get the expressions for the structure functions in terms of the parton distribution functions for u-, d- and s-quarks and their antiquarks.

$$F_2(x, Q^2) = 2x F_1(x, Q^2) = \sum_i e_{q_i}^2 x f_i^p(x, Q^2)$$

$$= x \left(\frac{4}{9} \left(f_u^p(x, Q^2) + f_{\bar{u}}^p(x, Q^2) \right) + \frac{1}{9} \left(f_d^p(x, Q^2) + f_{\bar{d}}^p(x, Q^2) \right) \right.$$

$$\left. + \frac{1}{9} \left(f_s^p(x, Q^2) + f_{\bar{s}}^p(x, Q^2) \right) + \frac{4}{9} \left(f_c^p(x, Q^2) + f_{\bar{c}}^p(x, Q^2) \right) \right). \quad (20.16)$$

Since quark and antiquark distributions in the sea are identical, and there are no valence s-quarks or c-quarks inside the proton, we may set $f_s^p = f_{\bar{s}}^p$ and $f_c^p = f_{\bar{c}}^p$ in (20.16). The observation of structure functions in the early deep inelastic scattering experiments at SLAC in 1968 was the first indication that protons were not point particles, but had a substructure in which the virtual photon was scattering off individual quarks.

If the parton distribution functions were independent of the momentum transfer, the Bjorken scaling, discussed in Chap. 19, would be exact, i.e. the structure functions would be functions of x only. That this is not the case can be seen from the experimental data for structure functions, taken from various different collaborations. These structure functions are presented in Fig. 20.16 as functions of Q^2 for several different values of x. If Bjorken scaling were exact, all these lines would be horizontal. As can be seen, they display a mild dependence on Q^2, changing rather little over a range of Q^2 spanning two orders of magnitude. This dependence fits very well to the theoretical determination of the Q^2 dependence of the parton distribution functions [145–147].

20.9 Lattice QCD

Perturbative QCD can be used to calculate quantities such as scattering amplitudes at high energies, provided all momentum scales are large. For quantities that involve relatively low momenta, or static quantities such as particle masses, perturbative QCD is unsuitable. Instead, a great deal of progress has been made using *"lattice QCD"*, in which the continuum of space and time is replaced by a lattice of discrete sites. The fields associated with quarks and antiquarks are assumed only to exist on the lattice sites. In this way, QCD field theory becomes analogous to multi-particle Quantum Mechanics used in crystallography. The time evolution of the fields at the lattice sites is sampled at discrete time slices, so that the lattice is four-dimensional – spanning time as well as the three directions of space. There are still a large number of variables to deal with and high performance computers are employed to carry out the necessary numerical calculations.

In this lattice approach, gluons are incorporated by inserting a gluon field on the links between adjacent lattice sites, as shown in Fig. 20.17. The field equations are solved numerically and physical quantities are calculated from a numerical analysis

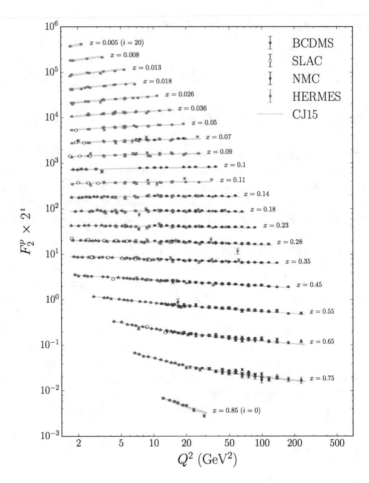

Fig. 20.16 The structure function F_2^p for several different values of x, plotted against Q^2, showing data from various different collaborations [151]. (In order to aid visibility, a factor of two in the normalization has been introduced between adjacent values of x.)

Fig. 20.17 A two-dimensional section of a lattice, with lattice spacing a. The volume of the lattice is L^3. The blue circles are the lattice sites on which the quark fields exist and the red arrows are the links on which the gluon fields exist

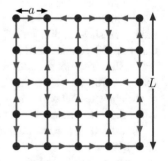

of the fluctuations of quark fields on various lattice sites, in a background of the gluon field.

This description of QCD on a discrete lattice was proposed by Ken Wilson [152] who showed that this provided a mechanism for the computation of QCD at large values of the coupling, where quark confinement was expected to manifest itself. Later, numerical studies [153] showed that this mechanism works and that the gluon fields do indeed generate a confining potential. This potential has been used to determine the spectrum of charmonium, discussed in Sect. 20.4, with remarkable success [154].

The rapid development of high performance computers, together with sophisticated and efficient computational algorithms, has enabled several groups to perform accurate calculations of a very large number of physical quantities, such as meson and baryon masses. Typical recent lattice simulations employ lattices with 64 sites in each direction and 128 time slices. These dimensionful quantities are determined by comparing their Compton wavelengths with the lattice spacing a. For example for the mass of the pion, m_π, lattice calculations determine the ratio $m_\pi ca/\hbar$. Therefore, what one really calculates are ratios of meson or baryon masses. The lattice spacing can be deduced by comparing the lattice calculation results with experiment for the mass of one (standard) particle and the masses of all the others can then be deduced. For a lattice with 64 lattice sites in each direction, the lattice spacing is found to be of order 0.1 fm. Errors due to the non-zero lattice spacing or the finiteness of the lattice volume can be reduced by performing the calculation using lattices of different sizes and then extrapolating to zero lattice spacing ($a \to 0$) or infinite volume ($L \to \infty$).

Apart from hadron masses, there have been successful calculations of the leptonic decay constants of mesons, discussed in Chap. 17, as well as semi-leptonic decay amplitudes. Some progress has also been made on the determination of parton distribution functions.

The lattice QCD approximation is valid at momentum scales that probe distances that are much larger than the lattice spacing. For a lattice spacing of order 0.1 fm, this means momenta that are much smaller than 2 GeV/c. On the other hand, we can see from Fig. 20.5 that at $Q = 2\,\text{GeV}/c$, $\alpha_S(Q) \approx 0.3$, which is small enough to get fairly good results using perturbative QCD. Thus we see that between perturbative QCD and lattice QCD, reliable results for physical quantities spanning (almost) the entire range of momenta (or energy) can be obtained.

Summary

- The Quantum Field Theory that describes strong interactions is QCD, in which quarks undergo transformations of colour, emitting or absorbing a gluon. These transformations between the three quark colours are described mathematically by SU(3) symmetry. There are eight gluons corresponding to the eight SU(3) colour transformations. Colour is conserved at each interaction between quarks and gluons.

- Gluons interact with other gluons as well as with quarks.
- The effective strong coupling decreases with increasing momentum transfer, so that at high energies perturbative QCD can be used. The rate of change of the strong coupling with momentum transfer is given by the β-function.
- Since quarks and gluons are confined, the effective potential between quarks, generated by the gluon field, increases with increasing quark separation. Such a confining potential can be used to calculate the spectrum of mesons consisting of heavy quarks (and antiquarks).
- The emission of a gluon off a quark or antiquark produced in electron–positron annihilation leads to three-jet events, which were observed at DESY in 1979. This was taken as the discovery of the gluon.
- As well as the main constituent quarks and antiquarks, known as valence quarks, a hadron will also contain gluons and quark–antiquark pairs known as sea quarks. These arise as a result of the strong interactions between quarks and gluons inside the hadron.
- A parton distribution function is the probability of finding a parton (quark, antiquark or gluon) inside a hadron with a fraction x of its momentum – in a frame in which the hadron is ultra-relativistic. These parton distribution functions depend not only on x but also on Q, the momentum transferred to the hadron in a given experiment. For sufficiently large Q, this dependence can be calculated using perturbative QCD, whereas the x-dependence has to be obtained from experimental fits.
- Measurable cross sections can be obtained from the factorization theorem. They are obtained by multiplying the partonic cross section, calculated in perturbative QCD, by the parton distribution functions for the incoming hadrons with fractions x_1 and x_2 of the momenta of the incoming hadrons, integrating over x_1 and x_2 and summing over all parton types and flavours.
- Low momentum or static quantities can be calculated quite accurately using numerical simulation of quark and gluon fields on a discrete lattice.

Problems

Problem 20.1 Draw the Feynman diagrams for the partonic process

$$q + \bar{q} \rightarrow g + g.$$

Problem 20.2 The value of $\alpha_s(Q)$ at $Q = M_Z c = 91.2\,\mathrm{GeV/c}$ is 0.118. Calculate the value of $\alpha_s(Q)$ at:

(a) At the threshold for the production of a t-quark $\bar{t}$–antiquark pair, $Q = 350\,\mathrm{GeV/c}$.
(b) At the momentum scale of the LHC, $Q = 14\,\mathrm{TeV/c}$.

Problem 20.3 Isospin symmetry implies

- $f_u^p = f_d^n$ and $f_d^p = f_u^n$.
- The sea quark distributions are the same for the proton and the neutron.

Use this to explain how the structure functions, F_2^p from deep inelastic electron–proton scattering and F_2^n from deep inelastic electron–neutron scattering, can be used to determine the difference between f_u^p, the u-quark distribution, and f_d^p, the d-quark distribution in a proton.

Problem 20.4 In a colliding beam experiment, two hadrons collide with CM energy $\sqrt{s}$. Show that the square of the CM energy $\hat{s}$ of a parton with fraction x_1 of the momentum of one hadron and a parton with fraction x_2 of the momentum of the other hadron is given by

$$\hat{s} = x_1 x_2 s.$$

[Assume that all particles are moving ultra-relativistically so that the masses of the hadrons and quarks may be neglected.]

Chapter 21
Parity, Charge Conjugation and CP

21.1 Intrinsic Parity

In analogy with the parity of nuclear bound states, hadrons, which are bound states of quarks (and antiquarks), also have parity. This is called *"intrinsic parity"*, η, and under a parity inversion, the wavefunction for a particle $\{P\}$ acquires a factor $\eta_{\{P\}}$:

$$\mathcal{P}\Psi_{\{P\}}(\mathbf{r}) \; = \; \Psi_{\{P\}}(-\mathbf{r}) \; = \; \eta_{\{P\}}\Psi_{\{P\}}(\mathbf{r}), \tag{21.1}$$

where $\mathcal{P}$ denotes the parity inversion operator. Applying the parity operator twice must bring us back to the original state, so that $\eta_{\{P\}}$ can take the values $+1$ (even parity) or -1 (odd parity).

The lighter baryons (for which there is zero orbital angular momentum) have positive intrinsic parity. On the other hand, antiquarks have the opposite parity from quarks.[1] This means that the light antibaryons have negative parity. It also means that the light mesons, such as pions and kaons, which are bound states of a quark and an antiquark with zero orbital angular momentum have negative intrinsic parity. The lightest spin-one mesons, such as the ρ-meson, also have zero orbital angular momentum, and thus they too have negative intrinsic parity – they have spin one because of the alignment of the spins of the (valence) quark and antiquark.

A two-particle state, whose wavefunction has orbital angular momentum quantum number ℓ, has a parity η_{AB} equal to the product of the intrinsic parities η_A and η_B of the two particles, multiplied by a factor of $(-1)^\ell$ from the two-particle wavefunction:

$$\eta_{AB} = \eta_A \times \eta_B \times (-1)^\ell \; . \tag{21.2}$$

---

[1] The fact that fermions and antifermions have opposite parity follows from the Dirac equation.

© The Author(s), under exclusive license to Springer Nature Switzerland AG 2021
A. Belyaev, D. Ross, *The Basics of Nuclear and Particle Physics*, Undergraduate
Texts in Physics, https://doi.org/10.1007/978-3-030-80116-8_21

For more massive (higher energy) particles, the quarks can be in non-zero orbital angular momentum states so that both baryons and mesons with higher masses can have either parity.

The parity is always conserved in strong-interaction processes. A consequence of this is seen in the decay

$$\rho^0 \rightarrow \pi^+ + \pi^- .$$

Since ρ mesons have spin one and pions have spin zero, the final pion state must have $\ell = 1$. The ρ has negative intrinsic parity and so do the two pions. The orbital angular momentum $\ell = 1$ means that by (21.2), the parity of the final state is

$$\eta_\pi^2 (-1)^1 = -1,$$

so that the parity is conserved. On the other hand, two π^0's cannot be in an $\ell = 1$ state. The reason for this is that pions are bosons and the wavefunction for two identical pions must be symmetric under interchange, whereas the wavefunction for an $\ell = 1$ state of two pions is antisymmetric under the interchange of the two pions. This means that the decay mode (governed by the strong interactions)

$$\rho^0 \rightarrow \pi^0 + \pi^0$$

is forbidden.

The hadronic decays of a K^+ into pions do not conserve strangeness and are therefore mediated by weak interactions. Decays such as

$$K^+ \rightarrow \pi^+ + \pi^0$$

and

$$K^+ \rightarrow \pi^+ + \pi^0 + \pi^0, \text{ or } \pi^+ + \pi^+ + \pi^-$$

all occur. Since the K^+ has spin zero, the final state must have zero orbital angular momentum, so that overall angular momentum is conserved. The final two-pion state has even parity, whereas the final three-pion state has odd parity. The fact that the negative parity K^+ can decay into *either* a parity even final state with two pions *or* a parity odd final state with three pions is a demonstration that the weak interactions do not conserve parity. It was originally assumed that there were two distinct particles, the τ-particle, which decays into three pions, and the θ-particle, which decays into two. It was later observed that these "two" particles have identical mass and identical lifetime and were in fact the same particle (K^+). Its decay into either two pions or three pions was the first observed manifestation of parity violation in weak interactions and was discovered before C.S. Wu's experiment [60] on the β-decay of $^{60}_{27}\text{Co}$.

21.2   Charge Conjugation

"Charge conjugation" is the operation of replacing a particle $\{P\}$ by its antiparticle, $\{\overline{P}\}$:

$$\mathcal{C}\Psi_{\{P\}} = \Psi_{\{\overline{P}\}}.$$

For example,

$$\mathcal{C}\Psi_{\pi^+} = \Psi_{\pi^-}$$

$$\mathcal{C}\Psi_p = \Psi_{\overline{p}}.$$

Some mesons are their own antiparticles such as π^0 or J/Ψ (a quark–antiquark pair of the same flavour). In this case, the particle is in an eigenstate of charge conjugation with charge conjugation quantum number η_C

$$\mathcal{C}\Psi_{\pi^0} = \eta_C \Psi_{\pi^0},$$

where η_C can take the values ± 1 (again using the fact that the application of the charge conjugation operator twice must bring us back to the original state).

A photon has charge conjugation $\eta_C = -1$. This is because under charge conjugation electric charges change sign. Electric and magnetic fields therefore also change sign under charge conjugation, and the field whose quantum excitations are photons is the electromagnetic field. We know that the π^0 can decay into two photons via electromagnetic interaction, which are invariant under charge conjugation

$$\pi^0 \rightarrow \gamma + \gamma.$$

This forces the charge conjugation of π^0 to be $\eta_C = +1$.

The spectra of charmonium (c–$\bar{c}$ bound states) or bottomonium (b–$\bar{b}$ bound states) contain both positive and negative charge conjugation states.

21.3   CP

Like parity, charge conjugation is conserved by the strong and electromagnetic interactions but *not* by the weak interactions. On the other hand, the weak interactions are (almost) invariant under the combined operations of charge conjugation *and* parity inversion, known as *"CP"*.

The weak interactions allow a left-handed chirality electron to convert into a neutrino emitting a W^-, but *not* a right-handed chirality electron. For highly relativistic electrons, this means that only electrons with left-handed helicity partake in weak interactions. For the antiparticle, the positron, the situation is the opposite: left-handed positrons do *not* partake in weak interactions, so that charge conjugation symmetry is violated. However, right-handed positrons *do* partake in weak interactions, with exactly the same strength (coupling constant) as the left-handed electrons. Thus, we see that there is a symmetry of the weak interactions under the *combined* operations of charge conjugation and a parity transformation. Charge conjugation converts an electron into a positron, whereas a parity transformation converts a positron (moving ultra-relativistically) with left-handed helicity to one with right-handed helicity.

If it were possible to repeat the experiment of C.S. Wu [60] using the antiparticle of $^{60}_{27}\text{Co}$ ($^{60}_{27}\overline{\text{Co}}$ – which decays into $^{60}_{28}\overline{\text{Ni}}$ emitting a positron and a neutrino), one would find that the positrons tended to be emitted in the same direction as the spin of the antinucleus (whereas in the original experiment, the electrons tended to be emitted in the opposite direction from the spin of the nucleus). The fact that these two experiments have different charged lepton distributions is a manifestation of the violation of charge conjugation symmetry by the weak interactions, since the experiment performed on the antinucleus produces a different result from the experiment performed on the nucleus. As explained in Sect. 7.6, the configuration in which the particles are mainly emitted in the same direction as the spin of the nucleus is the parity transform of the configuration in which they are mainly emitted in the opposite direction. So again we would see that the weak interactions are symmetric under the combined operations of charge conjugation and parity reversal.

21.4 $K^0 - \overline{K^0}$ Oscillations

The invariance of the weak interactions under CP has consequences for the K^0 and $\overline{K^0}$ mesons.

Applying the parity and charge conjugation operators to the wavefunction Ψ_{K^0} for the K^0,

$$\mathcal{P}\Psi_{K^0} = -\Psi_{K^0} \tag{21.3}$$

and

$$\mathcal{C}\Psi_{K^0} = \Psi_{\overline{K^0}}, \tag{21.4}$$

so that under the combined transformation

$$\mathcal{CP}\Psi_{K^0} = -\Psi_{\overline{K^0}}. \tag{21.5}$$

This means that the "particles" K^0 and $\overline{K^0}$ are not eigenstates of CP. But if CP is conserved, then the energy eigenstates (i.e. particles with well-defined mass) must also be eigenstates of CP (the operator $\mathcal{CP}$ commutes with the weak Hamiltonian, H_{WK}). These eigenstates of CP are

$$\Psi_{K_L} = \frac{1}{\sqrt{2}} \left(\Psi_{K^0} + \Psi_{\overline{K^0}} \right), \quad CP = -1 \tag{21.6}$$

and

$$\Psi_{K_S} = \frac{1}{\sqrt{2}} \left(\Psi_{K^0} - \Psi_{\overline{K^0}} \right), \quad CP = +1, \tag{21.7}$$

where L and S stand for "long" and "short", indicating that they have, respectively, long and short lifetimes, for reasons we discuss below. These particles ("*mass eigenstates*") are therefore not pure K^0 or $\overline{K^0}$ states, but quantum superpositions of the two. These superposition states exist because the Hamiltonian, H_{WK}, has a non-zero matrix element between a K^0 and a $\overline{K^0}$ state:

$$\langle K^0 | H_{WK} | \overline{K_0} \rangle \neq 0 \,,$$

and this matrix element effects transitions between a K^0 and a $\overline{K^0}$. If the time scale for such transitions is τ (the transition rate $\sim 1/\tau$), then from Heisenberg's uncertainty principle (applied to the rest energy of the kaons), there is an uncertainty, $\Delta m_K \sim \hbar/\tau c^2$ in the mass of K^0 or $\overline{K^0}$. On the other hand, the masses of K_L and K_S can be known exactly.

The allowed hadronic decays of K_L are

$$K_L \rightarrow \pi^0 + \pi^0 + \pi^0, \quad \text{or} \quad \pi^0 + \pi^+ + \pi^-.$$

By conservation of angular momentum, the three pions in the final state must have zero orbital angular momentum (the kaon has zero spin), and so the parity of the final state is the product of the parity of the three pions, $(-1)^3 = (-1)$. Because of this and since these three pion final states are invariant under charge conjugation, these states are eigenstates of CP with eigenvalue -1.

Likewise, for K_S, we have

$$K_S \rightarrow \pi^0 + \pi^0, \quad \text{or} \quad \pi^+ + \pi^-.$$

Again, there is no orbital angular momentum, so the parity of the final state is $(-1)^2 = +1$, and it is invariant under charge conjugation, so that the overall CP is $+1$.

The lifetime of the K_S is shorter than that of the K_L, because the Q-value for the decay into only two pions is larger than that for a decay into three pions ($m_K -$

$2m_\pi > m_K - 3m_\pi$). K_L has a lifetime of 5×10^{-8} s, whereas the lifetime of K_S is only 9×10^{-11} s.

On the other hand, we can distinguish a K^0 ($\bar{s}$-d bound state) from a $\overline{K^0}$ (s-$\bar{d}$ bound state) from their semi-leptonic decay modes

$$K^0 \rightarrow \pi^- + \mu^+ + \nu_\mu$$

$$\overline{K^0} \rightarrow \pi^+ + \mu^- + \overline{\nu_\mu}.$$

If, at time $t = 0$, we have a pure K^0, this is a superposition of the K_L and K_S states

$$\Psi_{K^0}(t = 0) = \frac{1}{\sqrt{2}} \left(\Psi_{K_L} + \Psi_{K_S} \right).$$

The K_L and K_S have slightly different masses[2]

$$\Delta m_K \equiv m_{K_L} - m_{K_S} = 3.5 \times 10^{-12} \text{ MeV}/c^2.$$

K_L and K_S therefore have different energies, which means that their wavefunctions have different frequencies of oscillation.

Applying the Schroedinger equation to obtain the time dependence of the wavefunction, which at time $t = 0$ represents a pure K^0 state, we obtain a wavefunction which contains oscillations between the wavefunction for a K^0 and the wavefunction for a $\overline{K^0}$. If, at some later time, t, the particle decays semi-leptonically, the probabilities $P(K^0, t)$ or $P(\overline{K^0}, t)$ of observing a K^0 decay (decay products (μ^+, π^-, ν_μ)) or $\overline{K^0}$ decay (decay products (μ^-, $\pi^+ \overline{\nu_\mu}$)) are given by[3]

$$P(K^0, t) = \frac{1}{4} \left(e^{-\Gamma_L t/\hbar} + e^{-\Gamma_S t/\hbar} \right) + \frac{1}{2} e^{-(\Gamma_L + \Gamma_S)t/2\hbar} \cos\left(\frac{\Delta m_K c^2 t}{\hbar} \right)$$

$$P(\overline{K^0}, t) = \frac{1}{4} \left(e^{-\Gamma_L t/\hbar} + e^{-\Gamma_S t/\hbar} \right) - \frac{1}{2} e^{-(\Gamma_L + \Gamma_S)t/2\hbar} \cos\left(\frac{\Delta m_K c^2 t}{\hbar} \right),$$

$$(21.8)$$

where Γ_L and Γ_S are the decay widths of K_L and K_S, respectively.

These probabilities as functions of time are plotted in Fig. 21.1. At time $t = 0$, we start with a K^0 so that the probability of finding K_0 is one and the probability of finding $\overline{K^0}$ is zero. At later time, the probability of finding a $\overline{K^0}$ increases and overtakes the probability of finding a K^0 and then decreases. In other words, as

---

[2] A recent lattice calculation [155] yields the value $(5.5 \pm 1.7) \times 10^{-12}$ MeV/c^2.

[3] An outline derivation is given in Appendix 21.A.

Fig. 21.1 The probability that at time t, a particle which is initially a pure K^0 state is observed to be in a state K^0 (red) or $\overline{K^0}$ (blue). At large values of t, much greater than the lifetime of K_S (but still much less than the lifetime of K_L), the probability to observe either particle is 1/4, with a probability of 1/2 that the particle has decayed

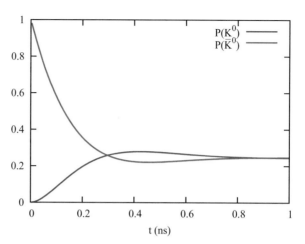

time progresses, there are oscillations between the K^0 and $\overline{K^0}$ states. The oscillation period frequency is $\tau = 2\pi\hbar/\Delta m_K c^2$, so that the uncertainty in the masses of the observed K^0 or $\overline{K^0}$ is of the order of the mass difference between K_L and K_S.

Whereas it may be difficult to imagine particles that are neither K^0 nor $\overline{K^0}$ but a superposition of the two, these oscillations constitute a striking example of the effects of quantum interference. The existence of two particles with different lifetimes and slightly different masses – K_L and K_S (rather than K^0 and its antiparticle $\overline{K^0}$) – was proposed in 1955 by Murray Gell-Mann and Brahm Pais [156], and the existence of these two mass states was observed at BNL in 1958 [157]. Quantum superpositions of other neutral mesons and their antiparticles also occur for mesons containing a c-quark (D^0-mesons) or a b-quark, (B^0-mesons). B^0-$\overline{B^0}$ oscillations were observed directly at DESY in 1987 [158] and D^0-$\overline{D^0}$ mixing was observed both at SLAC and at KEK[4] in 2007 [159, 160].

21.5 CP Violation

In 1964, Jim Cronin and Val Fitsch [161] observed a few decays of K_L into two pions (two-pion final states created at times much larger than the lifetime of K_S). Such a decay, in which the surviving K_L, whose CP $= -1$, decays into a CP $= +1$ final state, indicated that CP invariance was violated to a very small extent by the weak interactions. In other words, there is a very small CP violating component of the weak-interaction Hamiltonian, H_{WK}, which does *not* commute with the operator $\mathcal{CP}$, leading to the inequality

---

[4]Kō Enerugī Kasokuki Kenkyū Kikō laboratory situated at Tsukuba, Japan.

$$\langle K^0 | H_{\mathrm{WK}} | \overline{K_0} \rangle \equiv \langle \overline{K^0} | (\mathcal{CP})^{-1} H_{\mathrm{WK}} \mathcal{CP} | K^0 \rangle \neq \langle \overline{K^0} | H_{\mathrm{WK}} | K_0 \rangle .$$

The mass eigenstates are then

$$|K_L\rangle = \frac{1}{\sqrt{2\left(1+|\epsilon|^2\right)}} \left((1+\epsilon)|K^0\rangle + (1-\epsilon)|\overline{K^0}\rangle \right)$$

$$|K_S\rangle = \frac{1}{\sqrt{2\left(1+|\epsilon|^2\right)}} \left((1-\epsilon)|K^0\rangle - (1+\epsilon)|\overline{K^0}\rangle \right), \qquad (21.9)$$

where the CP violating parameter ϵ is given by (see Appendix 21.A)

$$\frac{1+\epsilon}{1-\epsilon} = \frac{\langle \overline{K^0} | H_{\mathrm{WK}} | K^0 \rangle}{\langle K^0 | H_{\mathrm{WK}} | \overline{K^0} \rangle}. \qquad (21.10)$$

We can see by applying (21.5) that these mass eigenstates are not eigenstates of $\mathcal{CP}$. The CP violating parameter, ϵ, takes the value $\epsilon = 2.2 \times 10^{-3}$ [161]. This leads to indirect CP violation, i.e. CP is violated because the particles are superpositions of CP eigenstates and therefore oscillate between their CP $= +1$ and CP $= -1$ components.

In addition to this, there is direct CP violation, in which there are direct CP violating transitions, effected by the weak-interaction Hamiltonian, between states with opposite CP. This direct CP violation is parametrized by another variable, ϵ', which is three orders of magnitude smaller than ϵ. It was first observed at CERN in 1988 [162]. The presence of this direct CP violation plays an essential role in the explanation provided by Andrei Sakharov [163] for the dominance of matter over antimatter in the Universe.

Particles and antiparticles have identical mass and identical total decay widths. However, their partial widths into a specific decay channel are only equal if CP is conserved. In 2001, a discrepancy between the decay rates of B^0 and its antiparticle $\overline{B^0}$ into the same final state $J/\Psi + K_S$ was observed at SLAC [164], signalling CP violation in the B-meson system (mesons containing a b-quark). In 2019, the LHCb collaboration observed CP violation in the D-meson system (mesons containing a c-quark) [165].

21.5.1 CP Violation and the CKM Matrix

Let us take a closer look at the mechanism for $K^0 - \overline{K^0}$ mixing. The relevant Feynman diagrams are shown in Fig. 21.2. q_i and q_j run over the positively charged quark flavours, u, c, t. The process involves the exchange of two W-bosons (with opposite charge). At each vertex with an incoming quark, an outgoing quark and a

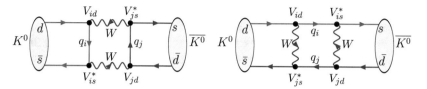

Fig. 21.2 Feynman diagrams contributing to weak interaction induced transitions between K^0 and $\overline{K^0}$, q_i, q_j run over u-, c- and t-quarks

W-boson, there is an element of the CKM matrix if the outgoing quark is positively charged or an element of the complex conjugate of the CKM matrix if the incoming quark is positively charged. The contribution to the amplitude from these Feynman diagrams for given values of i and j depends on the masses of the quarks q_i and q_j. We may therefore write

$$\langle \overline{K^0} | H_{\mathrm{WK}} | K^0 \rangle = \sum_{ij} \mathcal{A}_{ij} V_{id} V_{is}^* V_{jd} V_{js}^*, \qquad (21.11)$$

where $\mathcal{A}_{ij}$ is the contribution from the Feynman diagrams of Fig. 21.2, without the CKM matrix elements. These amplitudes are extremely small because they involve the exchange of two W-bosons and are therefore proportional to G_F^2. Together with the suppression due to the small off-diagonal elements of the CKM matrix, this leads to the very small difference, Δm_K, in the masses of K_L and K_S.

The Feynman diagrams, which determine the matrix element $\langle K^0 | H_{\mathrm{WK}} | \overline{K^0} \rangle$, are the diagrams in Fig. 21.2 but with the quarks and antiquarks interchanged, i.e. with the arrows on the quark lines reversed. For such diagrams, the CKM matrix elements are replaced by their complex conjugates (and vice versa) giving

$$\langle K^0 | H_{\mathrm{WK}} | \overline{K^0} \rangle = \sum_{ij} \mathcal{A}_{ij} V_{id}^* V_{is} V_{jd}^* V_{js}. \qquad (21.12)$$

From (21.10), we see that the CP violating parameter, ϵ, is proportional to the difference between these two matrix elements. If the CKM matrix elements were all real, then these two matrix elements would be identical and there would be no CP violation. The CP violating parameter, ϵ, is proportional to linear sums of products of the imaginary parts of the CKM matrix elements:

$$\epsilon = \sum_{ij} a_{ij} \Im m \left\{ V_{id} V_{is}^* V_{jd} V_{js}^* \right\}. \qquad (21.13)$$

If there were only two generations of quark flavours, the 3×3 CKM matrix would be replaced by the 2×2 Cabibbo matrix. As we show in Appendix 21.B, the Cabibbo matrix is uniquely determined by one parameter (the Cabibbo angle) and there is no complex element in the matrix. On the other hand, for the case of three generations of quark flavours, the 3×3 CKM matrix is uniquely defined by four parameters – three angles of rotation in three dimensions – and a complex phase. From the observation of CP violation, Makato Kobayashi and Toshide Maskawa [124] predicted the existence of a third generation of quarks, 4 years before the discovery of the Υ meson – a bound state of the third generation b-quark and a $\bar{b}$-antiquark [166].

Summary

- Particles possess intrinsic parity which determines how their wavefunction behaves under a parity transformation.
- Charge conjugation is the replacement of a particle by its antiparticle. Neutral particles, which are their own antiparticles (such as the π^0 or the η), are eigenstates of charge conjugation with eigenvalue ± 1.
- Weak interactions are neither invariant under parity transformations nor under charge conjugation. However, to a very good approximation, they are invariant under the combined transformation, CP, of parity inversion and charge conjugation.
- The CP invariant weak interactions can effect transitions between neutral particles and their antiparticles (such as K^0 and $\overline{K^0}$), leading to superpositions of these two states which are eigenstates of CP. These CP eigenstates have different decay widths and slightly different mass. As they propagate, the probabilities of observing the particle or the antiparticle (through their different decay channels) oscillate in time.
- There is a very small amount of CP violation in the weak interactions. As a result of this, the mass eigenstates are not quite pure CP eigenstates.
- The mixing between particle and antiparticle can be expressed in terms of the elements of the CKM matrix. CP violation is possible provided some of these matrix elements are complex. This cannot happen for the 2×2 Cabibbo matrix, which mixes only two quark generations, but it can occur for the CKM matrix which mixes three quark generations.

Problems

Problem 21.1 Draw the Feynman diagrams for $D^0 - \overline{D^0}$ mixing and indicate the relevant CKM matrix elements.

Problem 21.2 The meson η is in the same representation of the eightfold way as the pions. How does this particle transform under parity inversion, charge conjugation and CP? Explain why η decays very often into three pions but a decay into two pions has not been seen.

Problem 21.3 In one of the original experiments to measure the mass difference between K_L and K_S, a state of pure K^0 was prepared at time $t = 0$ and the asymmetry between the number, N_{K^0}, of K^0 decays and the number, $N_{\overline{K^0}}$, of $\overline{K^0}$ decays,

$$A(t) = \frac{N_{K^0}(t) - N_{\overline{K^0}}(t)}{N_{K^0}(t) + N_{\overline{K^0}}(t)},$$

was determined at several later times.

It was found that $A = 0$ at time $t = 2.7 \times 10^{-10}$ s and again at $t = 8.1 \times 10^{-10}$ s. The value of A was found to be $A = -0.082$ at time $t = 4.6 \times 10^{-10}$ s.

Assuming that the K_S has a much smaller lifetime than K_L and neglecting CP violating effect, use these data to calculate

(a) the K_L–K_S mass difference and
(b) the decay width of K_S.

[Hint: $\cosh^{-1} x = \ln\left(x + \sqrt{x^2 - 1}\right)$.]

Appendix 21.A: Neutral Kaon Oscillations

The time dependence of the wavefunctions $\Psi_{K^0}(r, t)$ and $\Psi_{\overline{K^0}}(r, t)$, for K^0 and $\overline{K^0}$, respectively, is given by the coupled Schroedinger equations

$$i\hbar \frac{\partial \Psi_{K^0}(r, t)}{\partial t} = \langle K^0|H|K^0\rangle \Psi_{K^0}(r, t) + \langle K^0|H_{WK}|\overline{K^0}\rangle \Psi_{\overline{K^0}}(r, t), \quad (21.14)$$

$$i\hbar \frac{\partial \Psi_{\overline{K^0}}(r, t)}{\partial t} = \langle \overline{K^0}|H|\overline{K^0}\rangle \Psi_{\overline{K^0}}(r, t) + \langle \overline{K^0}|H_{WK}|K^0\rangle \Psi_{K^0}(r, t), \quad (21.15)$$

where the second term on the RHS of (21.14) and (21.15) is due to the $K^0 - \overline{K^0}$ mixing.

Since particles and antiparticles have the same mass, m_K, and decay width, Γ_K, we have

$$\langle K^0|H|K^0\rangle = \langle \overline{K^0}|H|\overline{K^0}\rangle = m_K c^2 - \frac{i}{2}\Gamma_K.$$

Note that for particles moving non-relativistically, their energy is dominated by their rest energy. The decay of unstable particles is encoded by assigning an imaginary part to this rest energy equal to (minus) half the decay width.

The coupled Schroedinger equations (21.14) and (21.15) are solved by diagonalizing the matrix

$$\begin{pmatrix} \langle K^0 | H | K^0 \rangle & \langle \overline{K^0} | H_{\mathrm{WK}} | K^0 \rangle \\ \langle K^0 | H_{\mathrm{WK}} | \overline{K^0} \rangle & \langle \overline{K^0} | H | K^0 \rangle \end{pmatrix}.$$

Up to an overall normalization, the eigenfunctions are

$$\psi_{K_L}(\boldsymbol{r}, t) = \langle K^0 | H_{\mathrm{WK}} | \overline{K^0} \rangle \Psi_{K^0}(\boldsymbol{r}, t) + \langle \overline{K^0} | H | K^0 \rangle \Psi_{\overline{K^0}}(\boldsymbol{r}, t), \qquad (21.16)$$

with eigenvalue

$$m_{K_L} c^2 - \frac{i}{2} \Gamma_L = \langle K^0 | H | K^0 \rangle + \sqrt{\langle \overline{K^0} | H_{\mathrm{WK}} | K^0 \rangle \langle K^0 | H_{\mathrm{WK}} | \overline{K^0} \rangle} \qquad (21.17)$$

and

$$\psi_{K_S}(\boldsymbol{r}, t) = \langle \overline{K^0} | H_{\mathrm{WK}} | K^0 \rangle \Psi_{K^0}(\boldsymbol{r}, t) - \langle K^0 | H | \overline{K^0} \rangle \Psi_{\overline{K^0}}(\boldsymbol{r}, t), \qquad (21.18)$$

with eigenvalue

$$m_{K_S} c^2 - \frac{i}{2} \Gamma_S = \langle K^0 | H | K^0 \rangle - \sqrt{\langle \overline{K^0} | H_{\mathrm{WK}} | K^0 \rangle \langle K^0 | H_{\mathrm{WK}} | \overline{K^0} \rangle}. \qquad (21.19)$$

Defining the CP violating parameter, ϵ as

$$\epsilon = \frac{\langle \overline{K^0} | H_{\mathrm{WK}} | K^0 \rangle - \langle K^0 | H_{\mathrm{WK}} | \overline{K^0} \rangle}{\langle \overline{K^0} | H_{\mathrm{WK}} | K^0 \rangle + \langle K^0 | H_{\mathrm{WK}} | \overline{K^0} \rangle}, \qquad (21.20)$$

we may write the (normalized) eigenfunctions as

$$\Psi_{K_L}(\boldsymbol{r}, t) = \frac{1}{\sqrt{2(1 + |\epsilon|^2)}} \left((1 + \epsilon) \Psi_{K^0}(\boldsymbol{r}, t) + (1 - \epsilon) \Psi_{\overline{K^0}}(\boldsymbol{r}, t) \right) \qquad (21.21)$$

$$\Psi_{K_S}(\boldsymbol{r}, t) = \frac{1}{\sqrt{2(1 + |\epsilon|^2)}} \left((1 - \epsilon) \Psi_{K^0}(\boldsymbol{r}, t) - (1 + \epsilon) \Psi_{\overline{K^0}}(\boldsymbol{r}, t) \right). \qquad (21.22)$$

The time dependences of these eigenfunctions are

$$\Psi_{K_L}(\boldsymbol{r}, t) = e^{-i m_{K_L} c^2 t / \hbar} e^{-\Gamma_L t / 2\hbar} \Psi_{K_L}(\boldsymbol{r}, 0) \qquad (21.23)$$

$$\Psi_{K_S}(\boldsymbol{r}, t) = e^{-i m_{K_S} c^2 t / \hbar} e^{-\Gamma_S t / 2\hbar} \Psi_{K_S}(\boldsymbol{r}, 0). \qquad (21.24)$$

Henceforth, we exploit the fact that ϵ is very small and keep only terms linear in ϵ. Inverting (21.21) and (21.22), we have

$$\Psi_{K^0}(\boldsymbol{r}, t) = \frac{1}{\sqrt{2}} \left((1 + \epsilon) \Psi_{K_L}(\boldsymbol{r}, t) + (1 - \epsilon) \Psi_{K_S}(\boldsymbol{r}, t) \right). \qquad (21.25)$$

Suppose that at time $t = 0$, we prepare a pure K^0 state so that at time $t = 0$ the wavefunction for this state is given by

$$\Psi(\mathbf{r}, 0) = \frac{1}{\sqrt{2}} \left((1 + \epsilon)\Psi_{K_L}(\mathbf{r}, 0) + (1 - \epsilon)\Psi_{K_s}(\mathbf{r}, 0) \right). \tag{21.26}$$

Using the time development of the states K_L and K_S, (21.23) and (21.24), the wavefunction at a later time, t, is

$$\Psi(\mathbf{r}, t) = \frac{1}{\sqrt{2}} \left((1 + \epsilon)\Psi_{K_L}(\mathbf{r}, 0)e^{-im_{K_L}c^2 t/\hbar}e^{-\Gamma_L t/2\hbar} + \right.$$

$$\left. (1 - \epsilon)\Psi_{K_s}(\mathbf{r}, 0)e^{-im_{K_s}c^2 t/\hbar}e^{-\Gamma_{St}/2\hbar} \right). \tag{21.27}$$

The amplitude for this wavefunction to be that of a K^0 state is then

$$A_{K^0}(t) = \int d^3\mathbf{r}\Psi_{K^0}^*(\mathbf{r}, 0)\Psi(\mathbf{r}, t)$$

$$= \frac{1}{2}\left((1 + 2\epsilon)e^{-im_{K_L}c^2 t/\hbar}e^{-\Gamma_L t/2\hbar} + (1 - 2\epsilon)e^{-im_{K_s}c^2 t/\hbar}e^{-\Gamma_{St}/2\hbar} \right), \tag{21.28}$$

where we have used (21.21) and (21.22).

Similarly the amplitude for the wavefunction to be that of a $\overline{K^0}$ is

$$A_{\overline{K^0}}(t) = \frac{1}{2}\left((1 - 2\epsilon)e^{-im_{K_L}c^2 t/\hbar}e^{-\Gamma_L t/2\hbar} - (1 + 2\epsilon)e^{-im_{K_s}c^2 t/\hbar}e^{-\Gamma_{St}/2\hbar} \right). \tag{21.29}$$

The probability, $P(K^0, t)$, for the K^0 state to evolve into K^0 at time t is given by

$$P(K^0, t) = \left|A_{K^0}(t)\right|^2 = \frac{1}{4}\left((1 + 2\Re e\{\epsilon\})e^{-\Gamma_L t/\hbar} + (1 - 2\Re e\{\epsilon\})e^{-\Gamma_{St}/\hbar} \right)$$

$$+ \frac{1}{2}e^{-(\Gamma_L + \Gamma_S)t/2\hbar}\cos\left(\frac{\Delta m_K c^2 t}{\hbar} + 4\Im m\{\epsilon\}\right), \tag{21.30}$$

and the probability, $P(K^0, t)$, for the K^0 state to evolve into $\overline{K^0}$ at time t is given by

$$P(\overline{K^0}, t) = \left|A_{\overline{K^0}}(t)\right|^2 = \frac{1}{4}\left((1 - 2\Re e\{\epsilon\})e^{-\Gamma_L t/\hbar} + (1 + 2\Re e\{\epsilon\})e^{-\Gamma_{St}/\hbar} \right)$$

$$- \frac{1}{2}e^{-(\Gamma_L + \Gamma_S)t/2\hbar}\cos\left(\frac{\Delta m_K c^2 t}{\hbar} + 4\Im m\{\epsilon\}\right). \tag{21.31}$$

In the absence of CP violation ($\epsilon = 0$), we recover (21.8).

Appendix 21.B: The Cabibbo and CKM Matrices

We start with the general case of n generations and denote q_i^+ as the quarks with charge $+\frac{2}{3}$ and q_i^- as the quarks with charge $-\frac{1}{3}$ ($i = 1, \ldots, n$).

The W^+ boson then couples to

$$\sum_{i,j=1}^{n} \bar{q}_i^+ V_{ij} q_j^- ,$$

where V_{ij} is the $n \times n$ generalization of the CKM matrix. If we consider the superposition of the negatively charged quark states

$$\left| q_i'^- \right\rangle \equiv \sum_{j=1}^{n} V_{ij} \left| q_j^- \right\rangle ,$$

with complex conjugate states

$$\left\langle q_i'^- \right| = \sum_{j=1}^{n} \left\langle q_j^- \right| V_{ij}^* ,$$

then W^+ couples simply to

$$\sum_{i=1}^{n} \bar{q}_i^+ q_i^{-\prime} .$$

The quarks states $\left| q_i^- \right\rangle$ are orthonormal states, i.e.

$$\left\langle q_i^- \middle| q_j^- \right\rangle = \delta_{ij}.$$

The superposition states $\left| q_i'^- \right\rangle$ must also be orthonormal:

$$\left\langle q_i'^- \middle| q_j'^- \right\rangle = \sum_{k,l} V_{ik}^* V_{jl} \left\langle q_k^- \middle| q_l^- \right\rangle = \sum_{k,l} V_{ik}^* V_{jl} \delta_{kl}$$

$$= \sum_{k} V_{ik}^* V_{jk}.$$

The orthonormality of the superposition states $|q_i'^-\rangle$ establishes that V_{ij} must be a *"unitary matrix"*

$$\sum_k V_{ik}^* V_{jk} = \delta_{ij}.$$

A general $n \times n$ complex matrix has n^2 complex elements and therefore specified by $2n^2$ real parameters. The requirement that the matrix be unitary reduces this to n^2 parameters, although some of these may still be complex phases.

Cabibbo Matrix

The most general unitary 2×2 matrix with four parameters, θ_C, α, β and γ, may be written as

$$V_C = \begin{pmatrix} \cos\theta_C e^{i\alpha} & \sin\theta_C e^{i(\beta-\gamma)} \\ -\sin\theta_C e^{i\gamma} & \cos\theta_C e^{i(\beta-\alpha)} \end{pmatrix}, \tag{21.32}$$

so that in the case of two generations, the W-boson couples to

$$\begin{pmatrix} \bar{u} & \bar{c} \end{pmatrix} \begin{pmatrix} \cos\theta_C e^{i\alpha} & \sin\theta_C e^{i(\beta-\gamma)} \\ -\sin\theta_C e^{i\gamma} & \cos\theta_C e^{i(\beta-\alpha)} \end{pmatrix} \begin{pmatrix} d \\ s \end{pmatrix}.$$

We may rewrite this as

$$\begin{pmatrix} \bar{u}' & \bar{c}' \end{pmatrix} \begin{pmatrix} \cos\theta_C & \sin\theta_C \\ -\sin\theta_C & \cos\theta_C \end{pmatrix} \begin{pmatrix} d' \\ s' \end{pmatrix},$$

where the phases α, β and γ have been absorbed by making a phase rotation of the quarks u, c and d, relative to the phase of the s-quark:

$$\bar{u}' = e^{i(\beta-\gamma)}\bar{u}, \quad \bar{c}' = e^{i(\beta-\alpha)}\bar{c}, \quad d' = e^{-i(\beta-\alpha-\gamma)}d, \quad s' = s.$$

The s-quark is unchanged because a phase rotation on *all* four quarks by the same amount has no effect on the matrix – only the three relative phases can be used to absorb the phases of the matrix V_C. Since the overall phase of the wavefunction for a particle has no physical significance, the Cabibbo matrix V_C can always be written in terms of a single real parameter, namely the Cabibbo angle, θ_C, and after renaming rename u', d', c' and s' by u, d, c and s, we recover (17.6).

CKM Matrix

In the case of the 3×3 CKM matrix, there are nine parameters for a general unitary matrix. Five of these can be absorbed in the relative complex phases of the six quarks. Of the remaining four, three correspond to angles of rotation, θ_{ij} in the i–j plane in three dimensions, and the remaining parameter is a complex phase. Thus, we see that unlike the 2×2 Cabibbo matrix, the CKM matrix can have complex elements, with phases that cannot be eliminated by a phase rotation of the quarks.

There are many different ways of presenting the CKM matrix. The standard choice is now

$$V_{CKM} = \begin{pmatrix} c_{12}c_{13} & s_{12}c_{13} & s_{13}e^{-i\delta} \\ -s_{12}c_{23} - c_{12}s_{23}s_{13}e^{i\delta} & c_{12}c_{23} - s_{12}s_{23}s_{13}e^{i\delta} & s_{23}c_{13} \\ s_{12}s_{23} - c_{12}s_{23}s_{13}e^{i\delta} & -c_{12}c_{23} - s_{12}c_{23}s_{13}e^{i\delta} & c_{23}c_{13} \end{pmatrix},$$

(21.33)

where $c_{ij} \equiv \cos\theta_{ij}$, $s_{ij} \equiv \sin\theta_{ij}$ $(i < j = 1, \dots, 3)$.

Chapter 22
Beyond the Standard Model (BSM)

22.1 The Status of the Standard Model

The famous Higgs boson discovery in July 2012 by the ATLAS and CMS collaborations at the CERN LHC collider was the culmination of the remarkable success of the Standard Model, which was thitherto missing this last particle. At the same time, this discovery has opened a new chapter in the exploration of physics Beyond the Standard Model (BSM).

One should stress, however, that direct searches of new particles at colliders associated with BSM physics give negative results, so in this respect the SM agrees very well with data. The experimental measurement of a large number of quantities shows remarkable agreement with the predictions of the SM.

This agreement is particularly striking in the Higgs boson sector, where the values of measured Higgs boson couplings with SM particles are consistent with those predicted by the Standard Model to within the experimental uncertainty, as discussed in Chap. 18 and presented in Fig. 18.10.

Despite the impressive overall agreement of the experimental data with the Standard Model predictions, there are several fundamental experimental and theoretical motivations for the existence of BSM physics which is the subject of the discussion below.

22.2 Experimental Motivation for BSM Physics

22.2.1 Neutrino Oscillations and Neutrino Mass

In 1998, the Super-Kamiokande experiment observed oscillations between electron and muon atmospheric neutrinos [167]. This was an indication of the existence of neutrino mass and neutrino flavour mixing. The macroscopic quantum phenomenon

© The Author(s), under exclusive license to Springer Nature Switzerland AG 2021   347
A. Belyaev, D. Ross, *The Basics of Nuclear and Particle Physics*, Undergraduate
Texts in Physics, https://doi.org/10.1007/978-3-030-80116-8_22

of neutrino oscillations was introduced by Bruno Pontecorvo in 1957 [168]. Within the SM framework, the massless neutrino flavour fields (ν_e, ν_μ, ν_τ) are coupled via charged currents (W-bosons) to the charged leptons (e, μ, τ), respectively. What is found experimentally is that these neutrino flavour states which interact with their corresponding charged leptons are *not* mass eigenstates but are linear combinations of the mass eigenstates (ν_1, ν_2, ν_3), and this mixing can be described as

$$\begin{pmatrix} \nu_e \\ \nu_\mu \\ \nu_\tau \end{pmatrix} = U_{PMNS}(\theta_{12}, \theta_{13}, \theta_{23}, \text{phases}) \times \begin{pmatrix} \nu_1 \\ \nu_2 \\ \nu_3 \end{pmatrix}, \tag{22.1}$$

where U_{PMNS} is *"Pontecorvo–Maki–Nakagawa–Sakata"* (PMNS) lepton matrix. This is a 3×3 complex matrix analogous to the CKM matrix for quarks discussed in Chaps. 17 and 21.[1] The observed neutrino flavour oscillations between all neutrino flavours is evidence of non-zero neutrino mass differences, Δm, (with Δm of the order of 10^{-2} eV) between the mass eigenstates. This means that at least two neutrinos are massive with mass of the order of Δm. The mechanism of such an oscillation is the same as that discussed in Chap. 21 for the case of neutral kaon oscillation. The probability for the oscillation of neutrino of flavour α to neutrino with different flavour β is given by

$$P(\nu_\alpha \to \nu_\beta) = -4 \sum_{i>j} \Re e \left\{ U_{\alpha i}^* U_{\beta i} U_{\alpha j} U_{\beta j}^* \right\} \sin^2 \left(\frac{\Delta m_{ij}^2 c^4 L}{4\hbar E} \right)$$

$$+ 2 \sum_{i>j} \Im m \left\{ U_{\alpha i}^* U_{\beta i} U_{\alpha j} U_{\beta j}^* \right\} \sin \left(\frac{\Delta m_{ij}^2 c^4 L}{2\hbar E} \right), \tag{22.2}$$

where $\Delta m_{ij}^2 = m_i^2 - m_j^2$, $U_{\alpha i}$ are the elements of the PMNS matrix (with $\alpha, \beta = e, \mu, \tau$ and $i = 1 \ldots 3$), L is the baseline (i.e. the distance between the point of preparation of a pure ν_α state and the point of observation of the neutrinos) and E is the neutrino energy. We see from (22.2) that the observation of neutrino oscillations implies that at least some of the differences of the squared masses, Δm_{ij}^2, must be non-zero.

The neutrino mass implies the existence of the right-handed neutrino and left-handed antineutrino, which are absent in the SM. This follows from the fact that for

---

[1]In the case where the neutrino and antineutrino are different particles, the PMNS matrix is described by the same number of parameters as the CKM matrix, namely by three angles and one phase. In case where neutrinos and antineutrinos are identical (so-called *"Majorana"* particles), the PMNS matrix has two additional phases, i.e. it is parametrized by three angles and three phases. Current oscillation data are not sufficient to decide whether neutrinos are Majorana particles or not.

a massive particle moving at a velocity below the speed of light, it is always possible to boost into a frame in which the helicity is reversed.

If neutrinos and antineutrinos are different particles (so-called *Dirac neutrinos*), then one can introduce neutrino mass using a minimal modification of the Standard Model simply by adding a Higgs Yukawa coupling to neutrinos, which gives mass to neutrinos described by (18.4). One should note that this minimal SM modification does not introduce any new BSM energy scale, although it does require the addition to the SM of right-handed neutrinos and the corresponding Yukawa interactions. However, if future experiments indicate that the neutrino is a Majorana particle – i.e. identical to their antiparticles – this would be a clear signal of BSM Physics and would require a much more substantial modification of the Standard Model involving the introduction of the new energy scale responsible for neutrino mass generation.

22.2.2   Matter–Antimatter Asymmetry

One of the most striking experimental facts which imply the existence of BSM physics is the value of observed matter–antimatter asymmetry, measured at the cosmological scale. First of all, we know that there is matter–antimatter asymmetry in the Universe from the fact that we exist – and have not annihilated through interaction with antimatter! This asymmetry is usually expressed in terms of the ratio, η_B, of the baryon–antibaryon asymmetry to the photon density, defined as

$$\eta_B = \frac{n_B - n_{\bar{B}}}{n_\gamma}, \tag{22.3}$$

where $n_B, n_{\bar{B}}$ and n_γ are baryon, antibaryon and photon cosmic background number densities. The observed value of η_B is $\simeq 6 \times 10^{-10}$. There are three conditions defined by Andrei Sakharov [163] for the *"baryogenesis"*, the mechanism of the generating such an asymmetry:

1. Interactions which violate the conservation of baryon number,
2. Violation of charge conjugation and CP-symmetries,
3. A deviation from thermal equilibrium – e.g. an adiabatically expanding Universe.

In several well-known studies, it was found that although the SM has several ingredients for baryogenesis, it cannot explain it alone and requires new BSM physics. The SM does *not* contain baryon-number violating interactions. Furthermore, although the SM has CP violation arising from the phase of the CKM matrix, the magnitude of this CP violation is insufficient to reproduce the observed value of η_B. The necessary amount of CP violation can be incorporated for example, in the PMNS neutrino matrix but only in case of Majorana neutrinos, which in turn implies BSM physics. In this case, baryogenesis can be arranged via *"leptogenesis"*, i.e. via mechanism of generation of matter–antimatter asymmetry in the leptonic sector which is then converted into baryon-antibaryon asymmetry through baryon-number violating interactions (the first of Sakharov's conditions). In summary, the observed level of matter–antimatter asymmetry very strongly implies physics beyond the Standard Model.

22.2.3   Dark Matter and Dark Energy

The existence of dark matter (DM) – neutral (non-luminous and non-absorbing) matter – has been now established beyond any reasonable doubt, by several independent cosmological observations (for reviews see e.g. [169]). These include DM evidence from the rotation curves of stars (within galaxies) or entire galaxies (within galaxy clusters), which move faster than expected from the gravitational interactions between visible matter. The other observations are DM gravitational lensing and micro-lensing effects, the anisotropy pattern of the *"cosmic microwave background"* (CMB),[2] the existence of large scale structures and the very important recent observation of the bullet cluster (also known as 1E0657-558), consisting of a cluster of two galaxies which have passed through each other.

Current cosmological data allows an accurate determination of the DM contribution to the total energy of the Universe with a precision of about 1%. In particular it was found from the recent global fit by the Planck collaboration [170] that DM contributes 26% to the total energy of the Universe whilst the contribution from baryonic matter is only 5%. The remaining 69% of the total energy of the Universe comes from *"dark energy"* responsible for a repulsive gravitational effect which balances the attractive gravity of matter. The origin of dark energy is another big puzzle of Cosmology and Particle Physics linked to BSM physics. In principle, it could come from the vacuum expectation of the Higgs boson field, but in this case the contribution to the energy density of the Universe from the original Higgs potential discussed in Chap. 18 would be of the order of 10^{54} GeV/m^3, which is 54 orders of magnitude larger than the observed energy density of the Universe. In order to explain the dark energy of the Universe from the Higgs vacuum energy, one would need to introduce a constant (the famous *"cosmological constant"* postulated by Albert Einstein) tuned to within an incredible accuracy of one part in 10^{54}. In general, the fine-tuning problem is one of the strongest theoretical motivations for BSM physics as we discuss below.

Analyses of the structure formation (such as galaxies or cluster of galaxies) in the Universe show that DM should have been non-relativistic at the onset of its formation in order for such a formation to occur. The additional requirements for

---

[2]The cosmic microwave background is the electromagnetic radiation which originated at an early stage of the evolution of the Universe, and cooled as the Universe expanded, reaching its current temperature of 2.7 K, corresponding to an average wavelength in the microwave region.

DM candidates, which contribute about 84% to the total amount of *matter* in the Universe, are the following:

- They must be stable on cosmological time scale (i.e. the age of the Universe).
- They must be neutral (or interact very weakly with electromagnetic radiation).
- They must be non-baryonic and they must have the observed relic density, meaning that the remaining density (after annihilation in the early Universe) must correspond to the current observed DM density.

Candidates include primordial black holes or new particles, such as weakly interacting massive particles (WIMPs). The scenarios suggested so far in which primordial black holes contribute 100% to the observed DM are quite contrived, but nevertheless this possibility is the subject of detailed ongoing studies.

The best candidate for DM from SM would be neutrinos, but they are too light, would move relativistically (DM should be non-relativistic) and would contribute too little to the DM relic density. Models of BSM physics involving WIMPS with typical mass in the range of $0.1\,\text{GeV}/\text{c}^2$–$10\,\text{TeV}/\text{c}^2$ are very popular candidates for DM. If DM is not too heavy and interacts with Standard Model particles either directly or via some mediators, with a strength larger than that of gravitational interactions, it could be detected in different ways at Earth-based experiments:

(a) From direct production at colliders, resulting in a signature involving an observed SM object, such as jets, photons or leptons that recoil against the missing momentum from the DM particles (which are not visible directly in detectors),
(b) Via the relic density constraint, i.e. the DM candidate must have the observed amount of DM relic density,
(c) From direct DM detection experiments, which are sensitive to elastic scattering of DM particles off nuclei,
(d) From indirect DM detection searches that look for SM particles produced from annihilation of DM into gamma-rays, neutrinos or charged cosmic rays.

22.3 Theoretical Motivations for BSM and Promising Models

22.3.1 Unification of Forces

As discussed in Chap. 20, the electromagnetic and the strong couplings run with the momentum, Q, of the exchanged force carrier. The running of all couplings, α_i, including the weak-interaction coupling, can be described by general formula, analogous to the (20.4) for QCD. To leading order in the couplings, we may write

$$\frac{1}{\alpha_i(Q)} = \frac{1}{\alpha_i(\mu)} - \frac{\beta_{0i}}{4\pi} \ln(Q^2/\mu^2) \, , \tag{22.4}$$

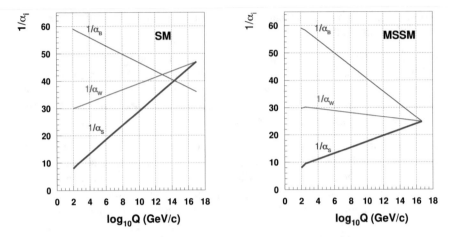

Fig. 22.1 Running of couplings according to (22.4): left – in SM, right – in Minimal Supersymmetric Standard Model (MSSM)

where the index $i = 1, 2, 3$ in α_i and β_{0i} run over electromagnetic, weak and strong interactions, respectively. This formula shows that $1/\alpha_i$ has a linear logarithmic dependence, and therefore plots of $1/\alpha_i$ against $\log(Q)$ are straight lines. Depending on the sign of β_{0i} coefficient (negative for weak and strong interactions and positive for electromagnetic interactions), these lines have negative or positive slopes, as shown in Fig. 22.1(left), where we show $1/\alpha_i$ against Q on a logarithmic scale. The electromagnetic coupling, α, is a linear combination of α_W and the coupling labelled α_B, plotted in Fig. 22.1.[3] It is important to stress, in the context of the $1/\alpha_i$ behaviour, that since the strength of the electromagnetic interactions increases with increasing Q (owing to the positive value of β_{0i}) whereas the strengths of the weak and strong interactions decrease with increasing Q (owing to the negative values of β_{0i}), the strengths of all three interactions become close and even cross each other at very high energies of the order of 10^{14}–10^{16} GeV. This very exciting fact serves as a great hint for the unification of forces within a more general, simple, and elegant framework of some *"Grand Unified Theory"* (GUT). This is a very appealing picture but unfortunately it does not fully work within the SM, since the three lines do not meet in one point as one see from Fig. 22.1(left).

---

[3]In the SM, it is actually α_B which is a fundamental coupling, and in models of Grand Unification it is α_B, rather than α, which is unified with α_S and α_W.

22.3.2 Supersymmetry

The situation changes qualitatively within the paradigm of *"Supersymmetry"* (SUSY), the theory based on a symmetry between bosons and fermions. In the framework of Supersymmetry, each boson has a corresponding fermionic super-partner and vice versa.

The *"Minimal Supersymmetric Standard Model (MSSM)"* – the minimal SUSY extension of the SM – predicts the existence of:

(a) A second Higgs boson accompanied by a charged and another neutral particle, also with spin-zero,

(b) Spin-zero superpartners for quarks and leptons called respectively *"squarks"* and *"sleptons"*,

(c) Fermionic superpartners of neutral SM bosons – four *"neutralinos"*, $\tilde{\chi}^0_{1,2,3,4}$,

(d) Fermionic superpartners of charged SM bosons – two *"charginos"*, $\tilde{\chi}^{\pm}_{1,2}$,

(e) Fermionic superpartners of the gluon – *"gluino"*, $\tilde{g}$, which, like the gluon, has 8 different colour states.

The new particles modify the quantum corrections to screening and anti-screening effects, which in turn affect the running of the couplings. This leads to a dramatic qualitative effect on coupling unification – the couplings unify at momenta of around 10^{16} GeV/c as can be seen in Fig. 22.1(right). Above a certain energy scale, corresponding to the masses of the superpartners, the running of the couplings – especially α_S – changes due to the production and subsequent annihilation of internal virtual superpartner particles. This change leads to slower decrease of α_S and α_W and to faster increase of α_B which happens because the MSSM adds more fermionic than bosonic degrees of freedom. This coupling unification is not a coincidence but a consequence of the boson-fermion symmetry. This unification happens not only in SUSY but also in some other GUT theories. However SUSY has several other very appealing features. For example, SUSY provides an attractive DM candidate which is typically the lightest neutralino, $\tilde{\chi}^0_1$. The symmetry which provides stability for the lightest SUSY particle also provides stability for the proton, even in the presence of baryon-number violating interactions which occur in Grand Unified Theories. The possibility of exact coupling unification is a very appealing aesthetic argument and very good theoretical motivation for BSM physics in general and for SUSY in particular.

22.3.3 Grand Unified Theories (GUTs)

BSM models in which the couplings unify at some unification energy scale, $M_{GUT}c^2$, can be generalized into GUTs. In such theories the strength of *all* the forces of Nature become the same at the unification energy scale, and above it quarks and leptons are treated on the same footing. For this to happen, in addition to interactions between quarks of different colour (QCD) and both quarks and leptons of different flavours (QFD), there are extra force carriers (gauge bosons) with masses of the order of M_{GUT}, which convert a quark into a lepton or vice versa. Interactions with these additional gauge bosons provide the baryon-number violating interactions required by the Sakharov conditions for the generation of matter–antimatter asymmetry.

Below the unification energy scale, the couplings for the strong, weak, and electromagnetic interactions run at different rates as the baryon-number violating interactions become weaker [171]. At an energy scale E these interactions are suppressed by a factor of $\sim E^2 / M_{\text{GUT}}^2 c^4$, in the same way that weak interactions are weak at energies far below the W-mass, as discussed in Chap. 17. The strong, weak and electromagnetic gauge bosons have different coupling to the baryon-number violating gauge bosons, and therefore their corresponding couplings vary differently with momentum as the effect of the interactions involving baryon-number violating gauge bosons diminishes. Nevertheless, there remains some tiny baryon-number violating interactions even at low energies, and these give rise to processes in which a nucleon can decay into a lepton and a meson. However, the lifetime of a proton turns out to be many orders of magnitude longer than the age of the Universe, and for values of M_{GUT} exceeding 10^{16} GeV/c^2 the predicted decay rate is too small to have been detected at searches for proton decay which have been carried out so far, such as the search by the Kamiokande [172] collaboration, which found a lower limit on the proton lifetime of 5.9×10^{33} years.

22.3.4 Quantum Corrections to the Higgs Mass, Hierarchy and Fine-Tuning

There is another very strong theoretical argument in favour of BSM physics. It is related to the quantum corrections to the Higgs boson mass. Representative Feynman diagrams contributing to such corrections for the SM Higgs boson are shown in Fig. 22.2. The feature of quantum corrections to the mass of any spin-zero particles is that they are proportional to the squared value of the momentum cutoff of the particle in the loop, which is the energy scale for the validity of the theory. This scale is of the order of the Planck mass $\Lambda_{UV} \sim M_{PL} c^2 = \sqrt{\hbar c / G} \, c^2 \simeq 1.2 \times 10^{19}$ GeV for the SM. At this energy scale gravity becomes strong and needs to be accounted for, for example in String Theory, which we discuss below. This means that quantum corrections contributing for the Higgs mass are

$$\Delta m_H^2 = K \Lambda_{UV}^2 / c^2 \ , \tag{22.5}$$

where, according to the SM calculations, $K \sim 10^{-2}$. This requires extreme fine-tuning of the theoretical mass parameter m_H^{th} so that together with the quantum correction to the Higgs boson mass Δm_H they reproduce the measured mass of the

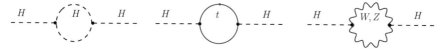

Fig. 22.2 Representative Feynman diagrams contributing to the SM Higgs mass quantum corrections

Fig. 22.3 Representative SUSY diagrams contributing to the Higgs boson mass quantum corrections which supplement the SM corrections in Fig. 22.2. The tilde over the particles inside the loops indicates that they are superpartners

Higgs, m_H:

$$m_H^{th\,2} - \Delta m_H^2 = m_H^2 \equiv (125\,\text{GeV})^2/c^4. \tag{22.6}$$

Since $\Delta m_H^2 \simeq K \Lambda_{UV}^2/c^2 \simeq 10^{36}\,\text{GeV}^2/c^4$, this requires tuning the quantity m_H^{th} to one part in 10^{32}. This problem is known as the *"fine-tuning"* problem and it is related to the *"hierarchy problem"* – namely why is the Higgs boson mass of the order of the electroweak scale and not of the order of the Planck mass, M_{PL}.

These very serious theoretical problems of the Standard Model are solved spectacularly by SUSY: the supersymmetric partners of the SM particles differ by half a unit of spin and therefore obey opposite statistics (Bose ↔ Fermi). As a result, Feynman diagrams involving the production and decay of superpartners shown in Fig. 22.3 generate quantum corrections with the opposite sign from those of the SM and cancel the dependence of Δm_H^2 on Λ_{UV}^2. As a result of this cancellation, the overall Δm_H^2 correction in SUSY is proportional only to the logarithm of the cutoff scale, M_{PL}:

$$\Delta m_H^2 \sim \alpha_W M_{SUSY}^2 \ln(M_{PL}/M_{SUSY}), \tag{22.7}$$

where M_{SUSY} is the mass scale of SUSY particles. From (22.7) one can see that if the M_{SUSY} is not too far from m_H, then the fine-tuning problem is nicely solved by SUSY. Since the SUSY energy scale is now limited by data from the LHC to be of the order of 10^3 GeV or above, the tuning appears at about the 1% level which is much better than the level of fine-tuning required in the absence of SUSY.

The SM fine-tuning problem can be also solved by bringing the cutoff scale down to the TeV level. This can happen in BSM models involving the existence of extra spatial dimensions, which extend over very short distances and therefore cannot be directly observed.

22.3.5   New Strong Interactions and Composite Higgs Models

Another mechanism is realized in theories such as *"Technicolor"*, which postulates the existence of an even stronger copy of QCD with its own *"technigluons"* and *"techniquarks"*. The new interaction is sufficiently strong, even at the electroweak scale, that the product of a techniquark field and its antiparticle field can have a vacuum expectation value, known as a *"condensate"*. This condensate plays the role of the Higgs vacuum expectation value and it is the interaction of this condensate of techniquarks with ordinary matter that generate particle masses. The Higgs particle itself is actually a bound state of a techniquark and its antiparticle, analogous to a meson in QCD. This mass generation via a condensate provides a naturally low scale for the Higgs mass, without fine-tuning. In spite of the qualitatively different nature of the Higgs in such Technicolor models, its properties can be similar to those of the SM Higgs and consistent with the LHC data [173].

There are several other BSM models which involve a composite Higgs, i.e. a Higgs particle which is not a fundamental particle but a bound state of a particle and its antiparticle. In such models, the Higgs mass turns out to be naturally light (compared with the Planck mass) without the need for fine-tuning.

22.3.6   Unification with Gravity and Strings

The one force of Nature that is not included in the SM is gravitation. The application of Quantum Field Theory to gravity leads to inconsistencies in that the calculation of some physical observables involving gravitational interactions generate infinities.

These infinities can be avoided in a development of Quantum Field Theory called *"String Theory"* in which point particles are replaced by very short vibrating strings (length $\sim 10^{-20}$ fm). These strings can accommodate an infinite tower of frequencies (standing waves) corresponding to particles with higher mass and also higher spin. The known particles all correspond to the lowest (fundamental) frequencies of vibration and the higher excitations have energies of order the Planck energy scale ($\sim 10^{19}$ GeV), and so they will almost certainly never be identified. Nevertheless these very high energy scales provide a cutoff to the infinities that arise in the Quantum Field Theory of point particles.

There are two types of strings:

- Open strings which have certain boundary values at their ends,
- Closed strings (i.e. loops) with periodic boundaries.

The lowest frequency state for an open string has spin one, so these strings are identified with the gauge bosons of the SM. The lowest frequency state of a closed string has spin two and is associated with the graviton. String theories *always* contain both open and closed strings, which interact with each other, so that gravitational interactions are automatically incorporated. At energies much smaller than the higher excitation string energies, the interactions between open and closed strings mimic the interactions arising from General Relativity, which may therefore be considered to be the "low-energy" limit of String Theory.

Fermions (quarks and leptons) are included by synthesizing String Theory with Supersymmetry so that the superpartner of an open string has a spin-$\frac{1}{2}$ particle as its lowest frequency vibration. The lowest frequency vibration of the superpartner of the closed string is the spin-$\frac{3}{2}$ "*gravitino*", which is the superpartner of the graviton.

Apart from predicting the existence of higher frequency string vibrations corresponding to particles whose masses are so large that they will almost certainly never be seen, String Theory does not predict the value of any quantity that can currently be measured. Whereas it is a very attractive theory incorporating gravity in a consistent way, and involves some extremely elegant mathematics, it does not contribute in any practical way to Particle Physics.

22.4   Searches and Prospects for BSM Physics

As we have discussed, there are several appealing classes of theories which have the potential to solve the problems of the SM. Note however, that the problem of fine-tuning as applied to dark energy remains unsolved. In these theories, the properties of the Higgs boson (either as a composite state or a fundamental particle) are compatible with those of the Higgs with mass $125\,\text{GeV}/c^2$, discovered at the LHC. There has been a vast effort in searching for signals of various BSM theories both in collider and non-collider experiments. Many of these searches are still ongoing.

No robust signals of any new BSM physics have so far been observed. However, cosmological data on dark matter, neutrino data, matter–antimatter asymmetry as well as the persuasive theoretical arguments discussed above very strongly suggest the existence of BSM physics. Explicit discovery of new particles and/or interactions might be just "around the corner" owing to the improvement in the experimental sensitivity and/or energy in the near future.

Over the years, a number of experimental results have been reported which deviate from the predictions of the SM, but these have turned out to be nothing more than statistical fluctuations of the background, rather than a signal of new physics. Currently, there are two recently announced experimental results which deviate from the prediction of the Standard Model. One of these is the violation of lepton universality, at the 3σ level, in the decay of B-mesons, observed by LHCb [95] (alluded to in Sect. 12.3.1). The second is the measurement of the magnetic moment of the muon [174] at FNAL, which differs at the 4σ level from certain theoretical predictions of the Standard Model. If either (or both) of these results are confirmed, they would constitute the first discovery of the existence of particles and interactions beyond those of the SM.

The scale of new physics – i.e. the mass scale of the models we have discussed – is generically thought to be at the TeV scale or above. However, this limit is not generic and one should also always look at the possibility that new physics is "hiding" in the low energy region but with very small couplings, so that it has not been observed so far. In any case, particle physicists strongly hope that new discoveries will be made in the near future. This hope is based on the strong theoretical motivation for BSM, combined with the rapid experimental developments which are pushing back the energy frontier and sensitivity limits. These include very sensitive direct and indirect DM searches, new neutrino oscillation data which will further enrich our knowledge of neutrino mixing parameters, the success of the LHC, plans for a future 100 TeV hadron collider and future lepton colliders, as well as the success of gravitational wave experiments which are now starting to test alternative DM hypotheses.

Summary

- The Higgs boson discovery completed the verification of the Standard Model. At the same time, this discovery opened a new chapter in the exploration of BSM which is necessary to solve the principal problems of the SM.
- Amongst theoretical motivations for BSM are:

  - The idea of coupling unification. Whereas the SM provides a hint on such a unification, it does not fully provide it,
  - The fine-tuning problem, which is related to the hierarchy between the Higgs mass and the Planck scale,
  - The unification of the SM with gravity.

- Experimental problems of the SM which motivate BSM are:

  - The observation of dark matter at the cosmological scale, whilst the SM does not provide any viable DM candidate,
  - Neutrino oscillations which imply neutrino mass requiring the introduction of right-handed neutrinos,
  - Matter–antimatter asymmetry which requires an amount of CP-violation beyond what we have in the SM,
  - Observation of dark energy.

- There are several very promising theories which include Supersymmetry, Grand Unified Theories, Technicolor, Composite Higgs models and models of extra-dimensions.

  - Supersymmetry solves the hierarchy problem in the fundamental Higgs sector via fermion-boson symmetry and provides DM candidates. It also leads to coupling unification at the energy scale of the order of 10^{16} GeV.
  - In Grand Unified Theories all the forces of Nature unify at some high energy scale. This gives rise to baryon-number violating interactions, leading to a very small but non-zero decay rate for the proton.

- A very attractive alternative to SUSY is Technicolor, in which the electroweak symmetry is broken by strong dynamics in analogy to QCD. In these theories, there is a set of new strong interactions (analogous to QCD) and also extra spin-$\frac{1}{2}$ particles, known as techniquarks. These new strong forces give rise to bound states of a pair of techniquarks and their antiparticles, and the Higgs boson is identified as one of these bound states.
- String Theories in which point particles are replaced by very short strings, which can be open or closed, automatically incorporate gravitational interactions (absent in the SM) since the lowest energy state of a closed string corresponds to a graviton.

• Experimental searches with unprecedented sensitivity and higher energies, aimed a testing these new theories, are currently under way, and have a great potential for the discovery of the underlying theory of Nature.

Answers to Problems

We provide solutions to those problems which have numerical or choice answers.

Chapter 1:

1.1 $17\,\text{s}^{-1}$

1.2 32.5 MeV

1.3 1.6×10^{-4} (1 in 6285)

Chapter 2:

2.1 $50.6°$, $60.3°$

2.2 $34°$

Chapter 3:

3.1 1789.5 MeV; 1189.1 MeV; 756.2 MeV; 155.9 MeV

3.2 11,182 MeV/c^2; 1 a.u. = 931.9 MeV

3.3 58

3.4 $a_c = 0.708$ MeV

Chapter 4:

4.1 0^+; $\frac{5}{2}^+$; 0^+; $\frac{3}{2}^+$; $\frac{1}{2}^+$; (0^+ or 1^+); (1^+ or 2^+)

4.2 $\frac{1}{2}^-$; (0^+, 1^+, 2^+ or 3^+); $\frac{3}{2}^-$

4.3 $-3.82\mu_N$; $-3.82\mu_N$; 0; 0; $-1.27\mu_N$

4.4 (b) $1.21 \pm 0.01 \times 10^6$ MeV/c^{2}fm^2; (c) 2.91×10^6 MeV/c^{2}fm^2

Chapter 5:

5.1 (a) α; (b) α; (c) β^-; (d)β^+; (e) β^-

5.2 3.63×10^3 years

5.3 172 h

5.4 4.1 billion years; 0.28

Chapter 6:

6.2 5.16 MeV; 5.11 MeV

6.3 107.6 s; 6.3 s; 3.2 s

6.4 9.0 MeV

Chapter 7:

7.1 (a) forbidden, $\ell = 2$;
(b) Gamow–Teller transition;
(c) forbidden, $\ell = 1$;
(d) Fermi transition

7.2 971.56 MeV

7.3 (a) 2.5×10^{-3} fm;
(b) 3.29 GeV; 0.73 GeV; 0.22 GeV

Chapter 8:

8.1 (a) E2,M3; (b) M2,E3,M4;
(c) M1,E2; (d) E2,M3,E4,M5

8.2 9.58×10^4 s

8.3 $2.50 \times 10^{-7}\,\text{s}^{-1}$

© The Author(s), under exclusive license to Springer Nature Switzerland AG 2021
A. Belyaev, D. Ross, *The Basics of Nuclear and Particle Physics*, Undergraduate
Texts in Physics, https://doi.org/10.1007/978-3-030-80116-8

8.4 (a) 3.56×10^{-7} eV; (b) 7.35×10^{-7} eV; (c) 1.07 electron-barns

8.5 2.30×10^9 s^{-1}

Chapter 9:

9.1 (a) 4; (b) 181.3 MeV; (c) 5; (d) 198.8 MeV

9.2 $^{237}_{94}$Pu, $^{239}_{94}$Pu, and $^{241}_{94}$Pu

9.3 12.4 kg

9.4 4.9×10^3 kg

Chapter 10:

10.1 0.8 MeV; 3.1×10^{-4} MeV; 2.8 MeV

10.2 0.032

10.3 5.85×10^5 kg

Chapter 11:

11.1 0,1, or 2

11.2 0;1

11.3 -1,0,1

11.4 $\frac{1}{2}^-, +\frac{1}{2}; \frac{1}{2}^-, -\frac{1}{2}; \frac{3}{2}^-, -\frac{3}{2}$

Chapter 12:

12.1 (a) baryon; (b) lepton; (c) meson; (d) meson; (e) meson; (f) baryon

12.2 1.4 fm

12.3 5.85×10^{-24} s

Chapter 13:

13.1 1.4 TeV

13.2 1%

13.3 2.38×10^{-5} cm^2

Chapter 14:

14.1 1.3 m

14.3 62.5 GeV

14.4 (a) 2.1%, 0.71%; (b) 12.6 %, 18.5%

Chapter 15:

15.1 (uuc), $I = 1$, $I_3 = +1$, $S = 0$, $q = +2$

(udc), $I = 1$, $I_3 = 0$, $S = 0$, $q = +1$

(ddc), $I = 1$, $I_3 = -1$, $S = 0$, $q = 0$

(usc), $I = \frac{1}{2}$, $I_3 = +\frac{1}{2}$, $S = -1$, $q = +1$

(dsc), $I = \frac{1}{2}$, $I_3 = -\frac{1}{2}$, $S = -1$, $q = 0$

(ssc), $I = 0$, $I_3 = 0$, $S = -2$, $q = 0$

15.2 (a) 1826 MeV; (b) 2089 MeV

15.3 (a) 938.2 MeV and 493.7 MeV; (b) 1939 MeV

Chapter 17:

17.2 2×10^{-4} fm

17.3 a,c,f,h,i allowed

17.4 0.7 GeV

17.5 1.2×10^7

Chapter 18:

18.2 1.79 GeV/c^2

18.3 (a) 2.2 fb^{-1}; (b) 39 fb^{-1}

Chapter 19:

19.1 a,b

19.3 (a) 2.11 GeV^2/c^2, 0.102, 0.537; (b) 473 GeV^2/c^2, 0.0104, 0.447

Chapter 20:

20.2 (a) 0.0989; (b) 0.0703

Chapter 21:

21.3 (a) 3.8×10^{-6} eV; (b) 8.8×10^{-6} eV

Acronyms and Abbreviations

BNL	Brookhaven National Laboratory
Bq	Becquerel
BR	Branching Ratio
BSM	Beyond the Standard Model
CERN	Centre Européan pour la Recherche Nucléaire
CLIC	Compact Linear Collider
CKM	Cabibbo-Kobayashi-Maskawa
CM	Centre of Mass
CMB	Cosmic Microwave Background
CMS	Compact Muon Solenoid
CP	Charge conjugation and Parity
DESY	Deutsches Elektronen Synchrotron
DIS	Deep Inelastic Scattering
DM	Dark Matter
ECAL	Electromagnetic Calorimeter
EFG	Electric Field Gradient
ESR	Experimental Storage Ring
eV	electron Volt
FCC	Future Circular Collider
FNAL	Fermi National Accelerator Laboratory
fm	fermi (10^{-15} metres)
FWHM	Full Width at Half Maximum
GUT	Grand Unified Theory
HCAL	Hadron Calorimeter
HEP	High Energy Physics
HERA	Hadron Electron Ring Accelerator
HEU	Highly Enriched Uranium
ILC	International Linear Collider
IP	Impact Parameter
IT	Isomer Transition

© The Author(s), under exclusive license to Springer Nature Switzerland AG 2021
A. Belyaev, D. Ross, *The Basics of Nuclear and Particle Physics*, Undergraduate
Texts in Physics, https://doi.org/10.1007/978-3-030-80116-8

ITER	International Thermonuclear Experimental Reactor
JINR	Joint Institute for Nuclear Research
KEK	Kō Enerugī Kasokuki Kenkyū Kikō
LBL	Lawrence Berkeley Laboratory
LEP	Large Electron–Positron Collider
LHC	Large Hadron Collider
LHS	Left-Hand Side
LO	Leading Order
MSSM	Minimal Supersymmetric Standard Model
NLO	Next-to-leading Order
QCD	Quantum Chromodynamics
QED	Quantum Electrodynamics
QFD	Quantum Flavourdynamics
RF	Radio Frequency
RICH	Ring Imaging Cherenkov
RHS	Right-Hand Side
RMS	Root Mean Square
SLAC	Stanford Linear Accelerator Center
SLC	Stanford Linear Collider
SM	Standard Model
SPS	Super Proton Synchrotron
Sp$\bar{\text{p}}$S	Super Proton–Antiproton Synchrotron
SUSY	Supersymmetry
WIMP	Weakly Interacting Massive Particle

Chemical Symbols

Ar	Argon	In	Indium	Pu	Plutonium
Au	Gold	Ir	Iridium	Ra	Radium
B	Boron	Kr	Krypton	Rh	Rhodium
Be	Beryllium	La	Lanthanum	Ru	Ruthenium
Bi	Bismuth	Li	Lithium	Sc	Scandium
Br	Bromine	Lu	Lutetium	Sn	Tin
C	Carbon	Mo	Molybdenum	Sr	Strontium
Ca	Calcium	N	Nitrogen	Ta	Tantalum
Ce	Cerium	Nb	Niobium	Tc	Technetium
Cm	Curium	Ne	Neon	Te	Tellurium
Co	Cobalt	Ni	Nickel	Ti	Titanium
Cu	Copper	No	Nobelium	Tl	Thallium
F	Fluorine	Np	Neptunium	U	Uranium
Fe	Iron	O	Oxygen	V	Vanadium
Fm	Fermium	Os	Osmium	Xe	Xenon
H	Hydrogen	Pa	Protactinium	Y	Yttrium
He	Helium	Pb	Lead	Zn	Zinc
Hf	Hafnium	Po	Polonium	Zr	Zirconium
Hg	Mercury				

© The Author(s), under exclusive license to Springer Nature Switzerland AG 2021
A. Belyaev, D. Ross, *The Basics of Nuclear and Particle Physics*, Undergraduate Texts in Physics, https://doi.org/10.1007/978-3-030-80116-8

Glossary

Accelerator: A machine that accelerates particles (usually protons or electrons or their antiparticles) up to very high energies.

Alpha decay: Radioactivity involving the emission of an α-particle.

Anode: A positively charged electrode.

α-particle: The nucleus of ^4_2He – a tightly bound state of two protons and two neutrons.

Asymptotic freedom: A property of certain Quantum Field Theories, including QCD, in which the effective coupling at a momentum scale Q goes to zero as $Q \to \infty$.

Atomic mass number, A: The total number of nucleons in a nucleus.

Atomic number, Z: The number of protons in a nucleus.

Axial angular frequency: Angular frequency of axial oscillations in a Penning trap.

Azimuthal angle: Angle in the $x - y$ plane in polar coordinates.

Background event: An event originating from a process different from the signal process (the process under investigation) but leading to the same final state.

Barn (b): Unit of scattering cross section equal to $10^{-28}\,\text{m}^2$.

Baryogenesis: The mechanism which is responsible for generation of the baryon–antibaryon asymmetry in the Universe.

Baryons: Hadrons with half-odd-integer spin (fermions).

Becquerel (Bq): Unit of radioactivity corresponding to one decay per second.

Beta decay: Radioactivity involving the emission of an electron (positron) and an antineutrino (neutrino).

β-function: The derivative of the running coupling at momentum scale Q with respect to $\ln Q$.

Bjorken scaling: The independence of structure functions on the square of the momentum transfer, Q^2.

B-meson: A meson containing a b-quark or $\bar{b}$-antiquark.

© The Author(s), under exclusive license to Springer Nature Switzerland AG 2021
A. Belyaev, D. Ross, *The Basics of Nuclear and Particle Physics*, Undergraduate
Texts in Physics, https://doi.org/10.1007/978-3-030-80116-8

Born approximation: Approximation to the (differential) cross section for the scattering of particles (moving non-relativistically).

Boson: Particle with integer intrinsic spin.

Branching ratio: The fraction of decays of an unstable particle into a particular decay channel.

Breeder reactor: A nuclear reactor in which the fissile fuel is manufactured by bombardment of a non-fissile nuclide with thermal neutrons to produce an isotope which decays by β-decay into the fissile nuclide.

Breit–Wigner distribution: The behaviour of a cross section in the vicinity of a resonance peak.

Bremsstrahlung: Radiation of a charged particle due to its acceleration or deceleration.

Bubble chamber: Chamber in which a particle is tracked from the bubbles formed when the supersaturated medium through which it travels is ionized.

Cabibbo angle: Angle, θ_c, which parametrizes the relative strength of the udW^- and usW^- couplings as $g_W \cos\theta_C/\sqrt{2}$ and $g_W \sin\theta_c/\sqrt{2}$ respectively, where g_W is the coupling of the weak interactions. ($\theta_C \approx 13.04°$).

Cabibbo matrix: The mixing matrix of the first two generations of quarks which parametrizes the interactions between either of the up-type quarks (u, c), and either of the down-type quarks (d, s) to the W-boson.

Cabibbo–Kobayashi–Maskawa (CKM) matrix: The matrix which parametrizes the mixing of three quark generations in analogy to the Cabibbo matrix.

Callan–Gross relation: Approximation relation between structure functions F_1 and F_2.

Calorimeter: An electronic device designed to measure the energy of specific particles passing through it.

Cathode: A negatively charged electrode.

Centre of Mass (CM) frame: Frame in which the total momentum of the incoming particles (or a subset of the outgoing particles) is zero.

Centrifugal potential: The effective potential in which a particle moves due to its non-zero orbital angular momentum.

Chain reaction: A reaction which is induced by the absorption of the products of a previous reaction and whose products are then in turn absorbed to give rise to the next "link" in the chain, *etc.*

Channel: A possible set of final-state particles, into which an unstable particle or nucleus can decay.

Charge conjugation: The replacement of particles by their antiparticles, and vice versa.

Charge radius: The expectation value of the radial component for the charged particle distribution.

Charged current: The current associated with $W^\pm$-boson interaction with quarks or leptons. It involves fermions with different charges, such that their charge difference is ± 1.

Charginos: Fermionic supersymmetric partners of the charged SM bosons – $\tilde{\chi}^\pm_{1,2}$.

Charm: Property of hadrons which counts the net number of c-quarks (quarks minus antiquarks) which constitute the hadron. Charm is conserved in strong interactions but not in weak interactions.

Charmonium: A bound state of a c-quark and a $\bar{c}$-antiquark.

Cherenkov radiation: Radiation from charged particles travelling through a medium, emitted when the particle moves faster than the speed of light in that medium.

Chromomagnetic interaction: An interaction between particles which carry charge of the strong force, which is described by Quantum Chromodynamics (QCD), analogous to magnetic interactions in QED.

CNO cycle: A fusion process in stars which begins with the fusion of a proton with a $^{12}_{6}$C nucleus. The net reaction is four protons into a $^{4}_{2}$He nucleus and the carbon nucleus is recycled.

Collider: An accelerator in which two opposing particle beams intersect each other such that the particles from the opposite beams collide.

Compton scattering: The scattering of short wavelength photons off electrons or other particles, in which some of the energy of the photon is imparted to the electron. The scattered photon has a longer wavelength than the incident photon.

Compton wavelength: The shift in wavelength for Compton scattering at a scattering angle of $90°$. For scattering off a particle of mass m, this is equal to h/mc.

Condensate: A vacuum expectation of a certain field or product of fields.

Conduction band: The band of energy levels immediately above the Fermi energy. They are unpopulated at absolute zero temperature. Electrons in this band behave like the free electrons in a conductor.

Confinement: Phenomenon in which quarks or gluons cannot be free particles but only exist bound inside hadrons. They may be exchanged between hadrons.

Cosmic microwave background (CMB): The electromagnetic radiation which originated at an early stage of the evolution of the Universe, and cooled as the Universe expanded, reaching its current temperature of $2.7\,\text{K}$, corresponding to an average wavelength in the microwave region.

Cosmic rays: High energy particles (usually protons) found throughout the Universe.

Cosmological constant: A constant term introduced into Einstein's General Relativity field equations, which generates dark energy density.

Coulomb scattering: The scattering of charged particles due to the electromagnetic interaction between them.

CP: The combined operations of parity inversion and charge conjugation.

Critical mass: The minimum mass of a sample of fissile material required for a chain reaction to be sustained.

Crossing symmetry: The equivalence of the scattering amplitude for a given process and the amplitude for the process in which an outgoing particle is replaced by an incoming antiparticle with negative energy and an incoming particle by an outgoing antiparticle with negative energy.

Curie: The number of decays per second of one gram of $^{226}_{88}$Ra. It is equal to 3.7×10^{10} Bq.

Cyclotron: An accelerator in which particles move in circles under the influence of a magnetic field and accelerated by an oscillating electric field.

Dark energy: An unknown form of energy that contributes about 69% to the total energy of the Universe and which is responsible for negative pressure leading to the accelerated expansion of the Universe.

Dark matter (DM): Neutral (non-luminous and non-absorbing) matter which contributes about 84% to the total amount of matter in the Universe and 26% of the total energy of the Universe.

Daughter nuclide: The nuclide into which a radioactive nuclide decays.

Decay channel: – see **Channel**.

Decay constant: Probability that an unstable particle or radioactive nucleus will decay in unit time.

Decay width: Decay rate of an unstable particle, multiplied by $\hbar$. It is the width of a resonance in which the particle is produced.

Decay rate: – see **Decay constant**.

Deep inelastic scattering (DIS): The high-energy inelastic scattering of a lepton and a nucleon.

Delayed neutron: A secondary neutron in the fission process in which one of the daughter nuclides from the β-decay of the fission fragments is in a sufficiently highly excited state for it to be able to emit a neutron.

Density of states: Number of allowed quantum states per unit energy interval. It is the integral over the phase space of the final-state particles, with fixed total energy.

Depletion zone: Region of a p-n junction semiconductor diode with very few mobile charge carriers.

Differential cross section (with respect to solid angle): The number of particle scattered per unit solid angle at a given scattering angle, per unit incident flux. Differential cross sections can also be defined with respect to other variables such as scattering angle or momentum transferred.

Diode: An electronic device which allows electric charge to pass in one direction only. A semiconductor diode consists of a junction between a p-type semiconductor and an n-type semiconductor.

Dirac equation: Equation for the wavefunction of a relativistic spin-$\frac{1}{2}$ particle.

Dirac fermion: The fermion which is different from its antiparticle.

Dirac neutrinos: Neutrinos which are different from their antiparticles (antineutrinos).

Displaced vertex: – see **Secondary vertex**.

D-meson: A meson containing a c-quark or a $\bar{c}$−antiquark.

Double Beta decay: A β-decay event in which the parent nuclide, with atomic number Z, makes a transition to a daughter nuclide with atomic number $Z + 2$ or $Z - 2$ emitting two electrons or two positrons.

Drift chamber: A wire chamber in which the drift time of the charged ions from the moment they were produced to the moment they reach the detecting wire is determined.

Dynode: An electrode which emits several electrons when a single electron is incident upon it.

Eigenstate: Eigenstate of an operator is a quantum state whose wavefunction is an eigenfunction of that operator.

Eightfold way: A method of classification of hadrons composed of u-, d- and s-quarks and/or antiquarks.

Elastic scattering: A scattering event between two initial particles in which the final state consists only of the two initial particles.

Electric form factor: $F(\theta)$ is the factor by which a scattering amplitude is multiplied in order to account for the finite size of the charge distribution of the target. The differential cross section is the differential cross section for a point-particle multiplied by $|F(\theta)|^2$.

Electric 2^ℓ-pole (multipole) transition: Electromagnetic transition whose amplitude is the matrix element of the electric 2^ℓ-pole (i.e. dipole, quadrupole, *etc.*) operator between initial and final states.

Electric quadrupole moment: For a non-spherically symmetric charge distribution $\rho(r)$, which nevertheless has one axis of symmetry (z-axis), the electric quadrupole moment, Q, is given by
$$Q = \int (3z^2 - r^2)\rho(r)d^3r,$$
where the z-axis is the axis of symmetry of the charge distribution.

Electromagnetic calorimeter (ECAL): Calorimeter which stops electrons and photons and measures their energy.

Electron capture: Inverse β-decay process in which an electron from an inner atomic shell is absorbed by the nucleus and a proton is converted into a neutron.

Electron cooling: A process in which a beam of ions are brought to the same momentum as a result of scattering off electrons.

Electroweak theory: The quantum field theory which unifies the weak and electromagnetic interactions – also known as "Quantum Flavourdynamics".

Enrichment: The process of increasing the concentration of fissile material in a fissionable material.

Exothermic: Releasing more energy is produced than it consumes.

Family: – see **Generation**.

Fermi (fm): Unit of length equal to 10^{-15} m.

Fermi-Dirac statistics: Statistics obeyed by a set of identical particles with half odd-integer spin, such that their total wavefunction is antisymmetric under the interchange of any two particles.

Fermi coupling constant: A constant, G_F which measures the strength of the effective (low-energy) weak interactions. ($G_F = 1.166 \times 10^{-5}\,\text{GeV}^{-2}$).

Fermi energy: The energy of the highest filled energy level in a gas of fermions in their ground state.

Fermi factor: A factor in the Fermi expression for the β-decay rate, which accounts for the Coulomb interaction of the emitted electron or positron with the daughter nucleus.

Fermi golden rule: An approximate formula for a transition rate in terms of the matrix element of the perturbing Hamiltonian between initial and final states, and the density of states.

Fermion: Particle with half odd-integer intrinsic spin.

Fermi transition A β-decay transition in which the total spin of the emitted electron (positron) and antineutrino (neutrino) is zero.

Feynman diagram: A diagrammatic representation of particle scattering or decay from which the amplitude for the scattering or decay process can be deduced.

Feynman graph: – see **Feynman diagram**.

Feynman rules: A set of rules indicating how to deduce the amplitude for a scattering or decay process from a Feynman diagram.

Fine-tuning: The necessity to tune some theory model parameters to a very high degree of accuracy in order to get the correct observable parameter (e.g. particle masses) of the theory.

Fissile: An isotope which can undergo induced fission when bombarded by slow (thermal) neutrons.

Fission: The splitting of a heavy nucleus into two lighter nuclei.

Fissionable: An isotope which can undergo fission when bombarded with neutrons with sufficient energy.

Flavour: Type of quark (u, d, s, c, t, b), or type of lepton (e, μ, τ) together with their neutrinos.

Flux: The number of incident particles per second per unit area.

Forbidden transition: A transition which violates a selection rule for the relationship between the quantum numbers of the initial and final states. In most cases such transitions are not strictly forbidden, but their decay rates are strongly suppressed.

Force carrier: A particle that is exchanged between particles which interact with each other under a given system of forces – also called "gauge bosons".

Fragmentation: The process in which final-state quarks and antiquarks are transformed into observed hadrons.

Fusion: The combining of two light nuclei to form a heavier nucleus whose binding energy is greater than the combined binding energies of the initial nuclei, thereby releasing energy.

Full width at half maximum (FWHM): The difference between two (adjacent) arguments of a function, either side of a maximum at which the value of the function is one-half of its peak value.

Fusion reactivity: The product of the fusion cross section and the relative velocity of the fusion nuclei.

g-factor: A dimensionless quantity relating a nuclear (atomic) magnetic moment to the nuclear (Bohr) magneton and the nuclear (atomic) total angular momentum.

Gamma decay: Radioactivity in which a nucleus in an excited state decays to a lower state emitting a very short wavelength photon, with energy more than 100 keV.

Gamow–Teller transition: A β-decay transition in which the total spin of the emitted electron (positron) and ant-neutrino (neutrino) is one.

Gauge boson: – see **Force carrier**.

Geiger counter: Tube containing low-pressure gas for detecting radioactivity.

Generation: A set of elementary particles consisting of a charged lepton, and a neutrino, a quark with electric charge $+\frac{2}{3}$, and a quark with electric charge $-\frac{1}{3}$ (each quark having three possible colours).

Girdler sulphide process: A method of increasing the concentration of deuterium in water by bringing hydrogen sulphide in contact with water at two different temperatures.

Gluon: The force carrier (gauge boson) of the strong interactions.

Gluino: Supersymmetric partner of a gluon which is a fermion and, like the gluon, has 8 different colours.

Grand Unified Theory (GUT): The theory which predicts unification of forces within a more general and simple framework.

Gravitino: The spin-$\frac{3}{2}$ superpartner of the graviton.

Graviton: The spin-2 force carrier of the gravitational interactions.

Hadron: A strongly interacting particle.

Hadron Calorimeter: A calorimeter placed outside the electromagnetic calorimeter, which absorbs hadrons and measures their energy.

Half-life: Time taken for a radioactive nuclide to decay to a quantity equal to one-half of its original quantity.

Helicity: The component of the spin, $\mathbf{s}$ of a particle in the direction of its momentum $\mathbf{p}$, normalized by $s\hbar$.

Hierarchy problem: The lack of explanation as to why the Higgs boson mass is of the order of the electroweak scale and not of the order of the Planck mass, M_{PL}.

Higgs boson: The massive scalar particle appearing as a result of the Higgs mechanism. It can be viewed as an excitation of the Higgs field.

Higgs field: The scalar field which is the core of the mechanism of particle mass generation in the Standard Model. In the absence of a Higgs particle, it has constant value everywhere in space called the "vacuum expectation value".

Higgs mechanism: The mechanism of particle mass generation in the Standard Model. In this mechanism, the Higgs field acquires non-zero vacuum expectation value, $\langle\phi\rangle$, and the interaction of the vacuum expectation value component of the Higgs field with other particles (including Higgs bosons themselves) generates particle masses.

Higgs particle: – see **Higgs boson**.

Higgs potential: The function $V(\phi)$ of the Higgs field, ϕ, proportional to the energy density. The vacuum state is defined by the minimum of this function. The value of ϕ for which $V(\phi)$ is minimum is then its vacuum expectation value, $\langle\phi\rangle$.

High Energy Physics: Particle Physics

Hindrance factor: A factor which suppresses the α-decay rate from parent nuclides with an odd number of protons or neutrons (or both) owing to the reduced probability of forming an α-particle (quasi-bound state) inside the parent nucleus.

Hypercharge: Twice the difference between the electric charge of a particle and its third component of isospin.

Hyperfine energy shift: The shift in energy of atomic states due to the interaction of electron and nuclear magnetic moments.

Hyperon: A baryon containing one or more s-quarks, but *no b-* or *t*-quarks.

Impact parameter: The impact parameter of a projectile scattering off a target is the perpendicular distance between the initial velocity of the projectile and the target.

Inelastic scattering: A scattering event between two initial particles in which the final state contains particles which are different from the two initial particles (it may or may not also contain the initial particles).

Internal conversion: The emission of an inner shell atomic electron when a nucleus makes a transition from one excited state to another state.

Internal conversion coefficient: The ratio of the number of internal conversion electrons emitted to the number of γ-rays emitted.

Intrinsic parity: The parity, η, of a particle. It is the eigenvalue of the parity operator $\mathcal{P}$. The particle wavefunction $\Psi(r)$ obeys the relation $\Psi(-r) = \eta\Psi(r)$.

Ion: An atom with one or more electrons missing (positive ion) or one or more extra electrons (negative ion).

Islands of isomer: Regions of the Periodic Table where isomers are quite common.

Isobar: Nuclides with the same atomic mass number but different numbers of protons (neutrons).

Isochronous: Occurring at the same time.

Isomer: A metastable excited state of a nucleus.

Isomeric shift: The shift in a nuclear energy level due to the Coulomb interactions between the electrons with the nucleus.

Isomer transition: The electromagnetic decay of a nuclear isomer.

Isospace: An (abstract) internal space used to describe isospin. Isospin transformations are described by rotations in isospace.

Isospin: A quantum number, I, assigned to a multiplet of (2I+1) particles which are almost identical except for their electric charge, which differs by one unit between members of the multiplet. The third component of the isospin varies in integer steps from -I to I.

Isotone: Nuclei with the same number of neutrons but different numbers of protons.

Isotope: Nuclei with the same atomic number but different atomic mass number.

Isotopic spin: – see **Isospin**.

Isotopic spin vector: A vector in isospace.

***jj*-Coupling:** Coupling scheme in which the spin–orbit interaction is stronger than the interaction between the fermions (electrons or nucleons), so that the energy level of each spin-$\frac{1}{2}$ fermion with orbital angular momentum ℓ splits into two states with total angular momentum $j = \ell + \frac{1}{2}$ and $j = \ell - \frac{1}{2}$.

K-shell: The innermost electron shell in an atom.

Källén function: The function, $\lambda(x, y, z)$, used to calculate magnitude of momentum for relativistically moving particles:

$$\lambda(x, y, z) = x^2 + y^2 + z^2 - 2xy - 2xz - 2yz.$$

Kaon: Lightest K-mesons with strangeness ± 1, and electric charge zero or ± 1.

Klystron: The high power microwave vacuum tube used as an amplifier or oscillator, particularly in linear accelerators.

Kurie plot: The graph of the square root of the quotient of the number of β-particles emitted with a given kinetic energy and the Fermi function, plotted against β-particle kinetic energy.

L-shell: The second innermost electron shell in an atom.

Lab frame: Frame in which a projectile particle is incident on a stationary target.

Lattice QCD: A version of QCD in which the continua of space and time are replaced by a discrete set of lattice sites, thereby reducing the field theory to multi-variable Quantum Mechanics, enabling various low-momentum physical quantities to be calculated by numerical methods.

Lawson criterion: The criterion required for a plasma of fusion reactants to be exothermic.

Leading order (LO): The first term in a perturbative expansion in powers of the coupling constant.

Leptogenesis: The mechanism which is responsible for generation of the lepton–antilepton asymmetry in the Universe.

Lepton: A particle which does not participate in the strong (nuclear) interaction. A lepton may have electric charge.

Leptonic decay: Decay of a meson into leptons (and antileptons) only.

Lepton flavour: The different leptons. There are three flavours of charged lepton e, μ, τ and each has its own neutrino.

Lifetime: The reciprocal of the decay rate of an unstable particle or nuclide. It is the time taken for a sample of a radioactive nuclide to decay to a quantity equal to $1/e$ times its initial quantity.

Linac: A linear accelerator.

Liquid Drop Model: The model of a nucleus as a liquid drop composed of nucleons, from which the semi-empirical mass formula is derived.

Luminosity: Rate of particle collisions per unit area (usually quoted in cm^{-2} per second).

Magic numbers: Number of protons or neutrons (or both) contained in nuclides which are particularly stable and tightly bound.

Magnetron angular frequency: Orbital angular frequency of a Penning trap.

Magnetic 2^ℓ–pole (multipole) transition: Electromagnetic transition whose amplitude is the matrix element of the magnetic 2^ℓ-pole (i.e. dipole, quadrupole, *etc.*) operator between initial and final states.

Majorana particle: A fermion which is its own antiparticle (unlike a Dirac fermion).

Mandelstam variables: s is a relativistically invariant quantity equal to the square of total energy of the incident particles in the CM frame. The invariant quantity t is a measure of the momentum transferred. For scattering in which the incoming and outgoing particle have the same mass, $-t$ is the square of the momentum transfer for elastic scattering in the CM frame.

Manhattan project: The development of the atomic bomb between 1942 and 1946 in Los Alamos, New Mexico.

Mass defect: The difference between the total mass of the nucleons in a nucleus and the mass of the nucleus itself (equal to minus the binding energy divided by c^2).

Mass eigenstate: A particle with well-defined mass and lifetime, whose wavefunction is an eigenfunction of the free-particle Hamiltonian.

Mass spectrometer: A device for measuring the ratio of mass to electric charge of ions.

Matrix Element: A property of a quantum operator, $\mathcal{O}$, and two quantum states $|i\rangle$ and $|j\rangle$ with wavefunctions $\Psi_i(\boldsymbol{r}_1, \cdots)$ and $\Psi_j(\boldsymbol{r}_1, \cdots)$ respectively, given by

$$\langle j|\mathcal{O}|i\rangle = \int \Psi_j^*(\boldsymbol{r}_1, \cdots)\, \mathcal{O}\Psi_i(\boldsymbol{r}_1, \cdots)\, d^3\boldsymbol{r}_1 \cdots .$$

Mean lifetime: – see **Lifetime**.

Mesons: Hadrons with integer spin (bosons).

Microwaves: Electromagnetic radiation with frequencies between 10^9 Hz and 10^{11} Hz.

Minimal Supersymmetric Standard Model (MSSM): The minimal Supersymmetric extension of the Standard Model predicting partners of SM particles (super-partners) that have the same quantum numbers with the exception of intrinsic spin which differs by $\frac{1}{2}$.

Mirror nuclei: – see **Mirror nuclides**.

Mirror nuclides: A pair of nuclides in which the number of protons in one nuclide is equal to the number of neutrons in the other and vice versa.

Modified cyclotron angular frequency: Angular frequency of radial oscillations in a Penning trap.

Mössbauer effect: The recoil-free absorption of γ-rays from a nucleus embedded in a crystal.

Mössbauer spectroscopy: The measurement of very small shifts in the energy of γ-rays exploiting the Mössbauer effect, but where the absorber (or source) moves with a small constant velocity so that the energy of the examined γ-ray is Doppler shifted.

Multipolarity: The angular momentum quantum number, ℓ, carried away by a photon in a nuclear transition $\ell = 1$ is called "dipole transition", $\ell = 2$ is called "quadrupole transition" *etc.*

Muon calorimeter: Calorimeter placed on the periphery of a detector, which stops muons and measures their energy.

Neutral weak current: The current associated with the interaction of the Z-boson with quarks or leptons.

Neutralinos: Fermionic supersymmetric partners of the neutral SM bosons.

Neutrino: A very low-mass particle with spin-$\frac{1}{2}$, which interacts through the weak interactions and is emitted along with a positron in β-decay processes. For β-decay processes with emission of electrons, an antineutrino is emitted.

Neutron multiplication factor: The ratio of the number of neutrons produced in a given stage of a chain reaction to the number produced in the previous stage.

Next-to-leading order (NLO): The second term in a perturbative expansion in powers of the coupling constant.

Noble gas: An inert (chemically and optically inactive) gas whose atoms contain complete shells of electrons.

n-type semiconductor: A semiconductor doped with electron donor impurities, giving rise to a surplus of electrons in the valence band.

Nuclear isomer: A metastable excited state of a nucleus.

Nuclear magneton: The magnetic moment of a proton with one unit of angular momentum, defined by

$$\mu_N = \frac{e}{2m_p}$$

Nucleon: A proton or neutron.

Nuclide A nucleus with a given atomic number and a given atomic mass number.

Null hypothesis assumption: Hypothesis that a quantity to be measured is zero, i.e. that there are no signal events.

Off mass-shell: Particle whose energy and momentum do not obey the relativistic energy–momentum relation.

Parity: The behaviour of the wavefunction of a system under the parity transformation $r \rightarrow -r$. Both elementary and composite particles also possess an intrinsic parity.

Parent nuclide: The initial nuclide in a radioactive process.

Partial width: The width of a particle resonance multiplied by the branching ratio for decay into a particular channel.

Parton: A constituent of a hadron – a quark, antiquark, or gluon.

Parton distribution function: The probability, $f_i^h(x, Q)$, of finding a parton of type i inside a hadron h with a fraction x of the momentum of the hadron, in the infinite momentum frame (in which the hadron is moving with a velocity close to the speed of light). The function depends on the momentum, Q, transferred by the probe used to measure the distribution function.

Penning trap: A cylindrical device which uses a uniform magnetic field and a quadrupole electric field to confine ions both radially and axially.

Pentaquark: A baryon composed of four quarks and an antiquark.

Perturbation theory: A technique for the calculation of a physical quantity or quantum amplitude as an expansion in powers of a small coupling which parametrizes the interaction Hamiltonian.

Perturbative calculation: – see **Perturbation theory.**

Phase space: A space of points in $3n$-dimensional momentum space, each point representing an allowed quantum state of an ensemble of n particles confined within a given volume. The integral over phase space for a fixed total energy is the density of states at that energy.

Photoelectric effect: The emission of electrons from certain metals when light of sufficiently short wavelength in incident upon it.

Photocathode: A negatively charged electrode made from a photoelectric metal.

Photomultiplier tube: A device for amplifying a very weak light signal, as low as a single photon, into a measurable electric pulse.

Pion: A π-meson with charge zero or ± 1. These are the lightest hadrons.

Pion decay constant: A dimensionful constant which parametrizes the coupling of the pion to its constituent quark–antiquark pair. It is related to the quark-antiquark wavefunction inside the pion.

Plasma: A gas of ions and electrons (usually at high temperature).

Plasma accelerator: A compact ultra-high energy accelerator, which uses very high electric fields associated with charged particle waves in a plasma.

Pontecorvo–Maki–Nakagawa–Sakata (PMNS) matrix: The matrix which parametrizes the mixing of three neutrino generations in analogy to the CKM matrix for quarks.

p-p cycle: Chain of fusion reactions in stars in which four protons fuse, through several stages, into a nucleus of $^4_2$He.

Primary Vertex: The point of collision of two colliding beams.

Prompt neutron: A neutron emitted at the same time as an initial fission reaction.

Propagator: The quantum amplitude for the propagation of a particle with a given energy and momentum.

p-type semiconductor: A semiconductor doped with electron acceptor impurities whose, giving rise to a deficit of electrons in the valence band.

p-value: The probability $p(x)$ that the measurement of a given quantity will return a value greater than x.

p-wave: The wavefunction of a particle which has one unit of orbital angular momentum ($\ell = 1$).

Quantum Chromodynamics (QCD): The relativistic quantum field theory of the strong interactions between gluons and quarks.

Quantum Electrodynamics (QED): The relativistic quantum field theory of Electromagnetism.

Quantum Flavourdynamics (QFD): – see **Electroweak theory**.

Quantum operator: An operator, $\mathcal{A}$, acting on a wavefunction, usually representing a physically measurable quantity, A. It is a function of the coordinates, r_i, of the wavefunction and their canonical momentum operators, $-i\hbar\partial_i$. For a system in an eigenstate of the quantity A, (i.e. it has well-defined value for that quantity), the wavefunction is an eigenfunction of the operator.

Quantum tunnelling: A mechanism in Quantum Physics by which a particle can tunnel through a potential barrier, whose peak is greater than the kinetic energy of the particle so that in classical physics the particle cannot traverse the barrier.

Quark: Elementary strongly interacting particle with spin $\frac{1}{2}$ and electric charge either $+\frac{2}{3}$ or $-\frac{1}{3}$, from which all hadrons are built.

Quark confinement: The mechanism under which it is impossible to isolate a free quark. It is equivalent to the statement that all physical particles are colourless.

Quasi-bound state: A temporary bound state of a daughter nucleus and an α-particle formed during the process of α-decay.

Q-value: The total energy released in a nuclear process.

Radiation length: The product of the density of a medium and the distance through which a charged particle must move for its energy to be attenuated by a factor of e.

Radio frequency (RF): Frequency of electromagnetic waves below about 10^9 Hz.

Radiocarbon dating: A method of dating fossils by comparing the abundance of the isotope $^{14}_{6}C$ in a sample with the relative abundance in the atmosphere.

Radiometric dating: Method of dating rock samples by measuring the ratio of the concentrations of a parent radioactive isotope to that of its daughter isotope.

Readout: Electronic device which converts a signal from the detector into a digital input.

Reduced mass: Effective mass of a two-body interacting system.

Relativistically invariant: A quantity which does *not* change under Lorentz transformations and therefore has the same value in any reference frame.

Resonance: Peak in cross section at a given centre-of-mass energy of the incident particles or of a subset of the final-state particles, due to the formation of an unstable intermediate particle.

Rest energy: Relativistic energy of a particle at rest $(= Mc^2$, for a particle of mass, $M)$.

Reverse bias: Connection of a semiconductor diode to electrodes for which the positive electrode is connected to the n-type semiconductor and the negative electrode the p-type semiconductor.

Running coupling: Effective coupling as a function of transferred momentum.

Scintillation counter: A device for detecting radioactivity using a crystal of a scintillating material which emits photons when radioactive particles are incident upon it.

Scission point: The point of separation of two fission products.

Sea quarks: Quark–antiquark pairs which exist inside a hadron at any one instant, apart from the valence quarks. Sea quarks have no effect on the properties of the hadron.

Secondary vertex: The point of decay of a relatively long-lived particle created at the primary vertex.

Secular equilibrium: The situation in a radioactive decay chain, with a long-lived first parent, in which the number of daughter nuclei remains unchanged since they are being produced as fast as they decay.

Semi-empirical mass formula: A formula for calculating the binding energy (and hence the mass) of a nucleus from the Liquid Drop Model.

Semi-leptonic decay: Decay of a meson into one or more lighter mesons plus leptons.

Shape factor: A factor which amends the linear Kurie plot by accounting for the non-zero orbital angular momentum carried off by the emitted leptons in a forbidden β-decay transition.

Signal significance: Deviation from the null hypothesis (no signal) assumption, often expressed in terms of the number of standard deviations for a Gaussian distribution.

Signal process: A process which leads to events which are characteristic of the particular physical phenomenon under investigation.

Silicon tracker: A detector placed close to the interaction point of two beams, which allows the identification of charged particle tracks.

Slepton: Supersymmetric partner of charged lepton with the same quantum numbers except its spin which is zero.

Solar neutrinos: Neutrinos produced in the fusion processes in the sun.

Solid angle: The area element of a sphere of unit radius centred at the apex (the point at which the solid angle is measured).

Spherical harmonics: The angular part of the wavefunction for a particle moving in a spherically symmetric potential. It is classified by the orbital angular momentum quantum number, ℓ, and its third component (the magnetic quantum number) m.

Spherical polar coordinates: Coordinates of a vector v in terms of its magnitude v, its polar angle, θ, between the vector and the z-axis, and the azimuthal angle, ϕ, in the $x - y$ plane.

Spin–orbit interaction The interaction between the magnetic field due to the orbital angular momentum of a charged particle and its magnetic moment due to its spin. It leads to shifts the energy levels of the particle, which are classified by their total angular momentum quantum number, j.

Squark: Supersymmetric partner of a quark with the same quantum numbers except its spin, which is zero.

Standard deviation: A measure of the error in the predicted number, n of events. It is equal to the square root of the mean square deviation. For a random process, the standard deviation is $\sqrt{n}$.

Standard Model: A model of fundamental particles and force carriers for the strong, weak, and electromagnetic interactions.

Steradian (sr): The unit of solid angle. It is the solid angle subtended at the centre of a sphere of unit radius by a spherical sector of unit area.

Stopping power: The rate of loss of energy of a particle in a detector.

Strangeness: Property of hadrons which counts the net number of $\bar{s}$-antiquarks (antiquarks minus quarks) that constitute the hadron. Strangeness is conserved in strong interactions but not in weak interactions.

String Theory: A theory in which particles are considered to be short vibrating strings. This theory automatically incorporates gravity.

Strong interactions: The strong, short-range interaction between nucleons inside a nucleus or partons inside a hadron.

Strong isospin: A symmetry between u- and $d-$quarks obeyed by the strong interactions. Other flavours of quarks do *not* transform under strong isospin.

Structure function: A function containing information about the distribution of partons in a hadron, probed in deep inelastic scattering.

s-wave: The wavefunction of a particle which has zero orbital angular momentum ($\ell = 0$).

SU(2): A symmetry group whose basic building block has two components. It is used in a mathematical description of rotations and can also be applied to isospin transformations, which can be considered to be rotations in isospace.

SU(3): A symmetry group whose basic building block has three components. It can be used in a mathematical description of the transformation between particles built out of u-, d- and s-quarks. It is also applied to the three quark colours in QCD.

Supersymmetry (SUSY): A theory which is based on a symmetry between bosons and fermions. In the framework of SUSY each boson has a corresponding fermionic superpartner and each fermion has a bosonic superpartner.

Symmetry group: A set of transformations under which certain interactions are invariant. The strict mathematical definition of a group imposes certain other requirements on the set of transformations.

Synchrocyclotron: A cyclotron in which the frequency of the alternating electric field is varied in order to account for relativistic effects.

Synchrotron: A cyclotron in which the applied magnetic field is varied in order to account for relativistic effects.

Technicolor: The theory which postulates the existence of new strong interactions and new particles (technigluons and techniquarks) at the TeV energy scale or above. The new interaction is sufficiently strong, even at the electroweak scale, that the product of a techniquark field and its antiparticle field can have a vacuum expectation value, known as a "condensate", which plays the role of the vacuum expectation value of the Higgs boson field.

Technigluon: The strongly interacting spin-one boson appearing in Technicolor theory – the technicolor analogy of a gluon in QCD.

Techniquark: The strongly interacting spin-$\frac{1}{2}$ fermion appearing in Technicolor theory – the technicolor analogy of a quark in QCD.

Tetraquark: A meson composed of two quarks and two antiquarks.

Thermal neutron: A slow neutron whose kinetic energy is of the order of thermal fluctuations at room temperature.

Threshold energy: Minimum energy of incoming particles in the CM frame for a particular inelastic scattering process to occur.

Tokamak: A fusion reactor containing a magnetically confined toroidal shaped plasma.

Townsend avalanche: An avalanche of electrons caused by an initial electron accelerated through a high voltage and colliding with atoms in a low-pressure gas.

Tracker: A solid-state detector which enables a path of a particle to be tracked.

Transition radiation: Radiation which occurs when a charged particle crosses the boundary between two media with different refractive indices.

Transmutation: A nuclear reaction in which nuclides are converted into different nuclides by exchanging protons and/or neutrons.

Transuranic element: An element with atomic number exceeding 92. Such elements do not occur naturally.

Transverse momentum: The component of the momentum of a final-state particle, perpendicular to the direction of the incident scattering particles. Although not fully Lorentz invariant, it *is* independent of whether the scattering is analysed in the Lab or CM frame – or any frame between.

Trigger: A modification to readout software such that data from a scattering event is only recorded if certain criteria on the final state are satisfied.

Unitary matrix: An $n \times n$ matrix, M, whose complex matrix elements, M_{ij} obey the relation $\sum_k M_{ki} M_{kj}^* = \delta_{ij}$. In matrix notation $M^\dagger M = I$.

Valence band: The band of energy levels in a semiconductor just below the Fermi energy. When an electron in this band is excited into the conduction band it leaves a positively charged "hole" in the vacated energy level.

Vacuum expectation value: The constant value which is possessed by a quantum field in the state with no particles (vacuum state).

Vacuum state: The state consisting of no particles.

Valence quark: One of the principal quarks or antiquarks which constitute a hadron, determining its properties.

Vector meson: A meson with spin one.

Vector meson dominance: The direct coupling of a vector meson to an off-shell photon.

Virtual particles: Particles that are exchanged between other particles during an interaction, and therefore exist only for a short time so that they are off mass-shell.

Wakefield: The electric field associated with a wave of charged particles in a plasma.

W-boson: The charged force carrier of the weak interactions.

Weak interactions: Interactions between particles responsible for β-decay.

Weak isospin: An isospin symmetry which differs from the isospin under which the strong interactions are invariant. Weak isospin is approximately obeyed by weak interactions. All fundamental fermions (quarks and leptons) are members of $I = \frac{1}{2}$ doublets of weak isospin.

Weinberg angle: The angle θ_W, whose cosine is equal to the ratio of the W-boson to Z-boson mass, M_W/M_Z. The strengths of the couplings of the Z-boson and W-boson are related to the Weinberg angle.

Wire chamber: A chamber filled with a low-pressure gas, containing a network of wires held at high potential.

WKB approximation: A method for approximating the wavefunction of a particle moving in a space-dependent potential, by assuming that the de Broglie wavelength is a slowly varying function of position.

Yukawa potential: A potential, which contains an exponentially decreasing factor as well as a $1/r$ fall-off. Such a potential occurs when the interaction is mediated by a massive particle.

Yukawa coupling: Dimensionless coupling constant characteristic of a spin-zero boson interaction with fermions.

Z-**boson:** The neutral force carrier (gauge boson) of the weak interactions.

Zeeman effect: The splitting of atomic energy levels when a magnetic field is applied. The energy levels are proportional to the magnetic quantum number m, which denotes the component of the total angular momentum of the electrons in the direction of the applied magnetic field.

References

1. E. Goldstein, Uber eine noch nicht untersuchte Strahlungsform an der Kathode inducirter Entladungen, Sitzungsberichie der Königlichen Akademie der Wissenschaften zu Berlin **39**, 691 (1886)
2. J.J. Thomson, Cathode rays, Phil. Mag. Ser.5 **44**, 293 (1897). https://doi.org/10.1080/14786449708621070
3. C.D. Anderson, S.H. Neddermeyer, Cloud Chamber Observations of Cosmic Rays at 4300 Meters Elevation and Near Sea-Level, Phys. Rev. **50**, 263 (1936). https://doi.org/10.1103/PhysRev.50.263
4. C.M.G. Lattes, G.P.S. Occhialini, C.F. Powell, Observations on the Tracks of Slow Mesons in Photographic Emulsions. 1, Nature **160**, 453 (1947). https://doi.org/10.1038/160453a0. [,99(1947)]
5. G.D. Rochester, C.C. Butler, Evidence for the Existence of New Unstable Elementary Particles, Nature **160**, 855 (1947). https://doi.org/10.1038/160855a0
6. G. Arnison, et al., Experimental Observation of Isolated Large Transverse Energy Electrons with Associated Missing Energy at $\sqrt{s} = 540$-GeV, Phys. Lett. **122B**, 103 (1983). https://doi.org/10.1016/0370-2693(83)91177-2. [,611(1983)]
7. M. Banner, et al., Observation of single isolated electrons of high transverse momentum in events with missing transverse energy at the cern pp collider, Physics Letters B **122**(5), 476 (1983). https://doi.org/10.1016/0370-2693(83)91605-2. http://www.sciencedirect.com/science/article/pii/0370269383916052
8. G. Arnison, et al., Experimental Observation of Lepton Pairs of Invariant Mass Around 95-GeV/c^2 at the CERN SPS Collider, Phys. Lett. **126B**, 398 (1983). https://doi.org/10.1016/0370-2693(83)90188-0. [,7.55(1983)]
9. P. Bagnaia, et al., Evidence for $z^0 \to e^+ e^-$ at the cern pp collider, Physics Letters B **129**(1), 130 (1983). https://doi.org/10.1016/0370-2693(83)90744-X. http://www.sciencedirect.com/science/article/pii/037026938390744X
10. G. Aad, et al., Observation of a new particle in the search for the Standard Model Higgs boson with the ATLAS detector at the LHC, Phys. Lett. **B716**, 1 (2012). https://doi.org/10.1016/j.physletb.2012.08.020
11. S. Chatrchyan, et al., Observation of a New Boson at a Mass of 125 GeV with the CMS Experiment at the LHC, Phys. Lett. **B716**, 30 (2012). https://doi.org/10.1016/j.physletb.2012.08.021

© The Author(s), under exclusive license to Springer Nature Switzerland AG 2021
A. Belyaev, D. Ross, *The Basics of Nuclear and Particle Physics*, Undergraduate Texts in Physics, https://doi.org/10.1007/978-3-030-80116-8

12. J.J. Thomson, On the Structure of the Atom: an Investigation of the Stability and Periods of Oscillation of a number of Corpuscles arranged at equal intervals around the Circumference of a Circle; with Application of the Results to the Theory of Atomic Structure, Phil. Mag. Ser.6 **7**, 237 (1904). https://doi.org/10.1080/14786440409463107

13. H. Geiger, E. Marsden, On the Scattering of α-particles by Matter, Proc. Roy. Soc. **A81**, 174 (1908). https://doi.org/10.1098/rspa.1908.0067

14. H. Geiger, E. Marsden, The Scattering of the α-particles by Matter, Proc. Roy. Soc. **A83**, 492 (1910). https://doi.org/10.1098/rspa.1910.0038

15. H. Geiger, E. Marsden, On a Diffuse Relection of the α-particles, Proc. Roy. Soc. **A82**, 495 (1909). https://doi.org/10.1098/rspa.1909.0054

16. H. Nagoka, Kinetics of a System of Particles illustrating the Line and the Band Spectrum and the Phenomena of Radioactivity, Phil. Mag. Ser.6 **7**, 445 (1904). https://doi.org/10.1080/14786440409463141

17. E. Rutherford, The Scattering of α and β Particles by Matter and the Structure of the Atom, Phil. Mag. Ser.6 **21**, 669 (1911). https://doi.org/10.1080/14786440508637080

18. H. Geiger, E. Marsden, The Laws of Deflexion of α-particles Through Large Angles, Phil. Mag. Ser.6 **25**, 604 (1913). https://doi.org/10.1080/14786440408634197

19. T. Bayram, et al., New Parameters for Nuclear Charge Radius Formulas, Acta. Physica Polonica B **44**, 1791 (2013). https://doi.org/:10.5506/APhysPolB.44.1791

20. N. Mott, H. Massey, *The Theory of Atomic Collisions* (Oxford at the Clarendon Press, 1965)

21. J. Bellicard, et al., Scattering of 750 Mev Electrons by Calcium Isotopes, Phys. Rev. Lett. **19**, 527 (1967). https://doi.org/10.1103/PhysRevLett.19.527

22. R. Woods, D. Saxon, Diffuse Surface Optical Model for Nucleon-Nuclei Scattering, Phys. Rev. **95**, 577 (1954). https://doi.org/10.1103/PhysRev.95.577

23. M. Born, Quantenmechanik der stossvorgänger, Zeitschrift für Physik **38**, 803 (1926). https://doi.org/10.1007/BF01397184

24. E. Fermi, *Nuclear Physics* (University of Chicago Press, 1950)

25. J. Rapaport, An Optical Model; Analysis of Neutron Scattering, Phys. Rep. **87**, 25 (1982). https://doi.org/10.1016/0370-1573(82)90105-3

26. W.R. Plass, T. Dickel, C. Scheidenberger, Multiple-reflection time-of-flight mass spectrometry, International Journal of Mass Spectrometry **349-350**, 134 (2013). https://doi.org/10.1016/j.ijms.2013.06.005. http://www.sciencedirect.com/science/article/pii/S138738061300239X. 100 years of Mass Spectrometry

27. G. Gamow, Mass Defect Curve and Nuclear Constitution, Proc. Roy. Soc. A **126 (803)**, 632 (1930)

28. C. von Weizsacker, Zur Theorie der Kernmassen, Zeitschrift fur Physik **96 (7-8)**, 431 (1935). https://doi.org/10.1007/BF01337700

29. H. Bethe, R. Bacher, Stationary States of Nuclei, Rev. Mod. Phys. **8**, 82 (1936)

30. G. Royer, C. Gautier, On the Coefficients and Terms of the Liquid Drop Model and Mass Formula, Phys. Rev. **C73**, 067302 (2006). https://doi.org/10.1103/RevModPhys.8.82

31. G. Audi, et al., The AME2003 Atomic Mass Evaluation (II). Tables, Graphs and References, Nuclear Physics **A729**, 337 (2003). https://doi.org/10.1016/j.nuclphysa.2003.11.003

32. D. Ivanenko, E. Gapon, Zur Bestimmung der Isotopenzahl, Die Naturwissenschaften **20(43)**, 792–793 (1932). https://doi.org/10.1007/BF01494007

33. E. Feenberg, E. Wigner, On the structure of the nuclei between helium and oxygen, Phys. Rev. **51**, 95 (1937). https://doi.org/10.1103/PhysRev.51.95. https://link.aps.org/doi/10.1103/PhysRev.51.95

34. M.G. Mayer, On Closed Shells in Nuclei, Phys. Rev. **74**, 235 (1948). https://doi.org/10.1103/PhysRev.74.235

35. M.G. Mayer, On closed shells in nuclei. 2, Phys. Rev. **75**, 1969 (1949). https://doi.org/10.1103/PhysRev.75.1969

36. O. Haxel, J.H.D. Jensen, H.E. Suess, On the "Magic Numbers" in Nuclear Structure, Phys. Rev. **75(11)**, 1766 (1949). https://doi.org/10.1103/PhysRev.75.1766.2

37. H. Becquerel, Sur les Radiations Invisibles Emises par les Corps Phosphorescents, Comptes Rendus **122**, 501 (1896)
38. J.J. Thomson, On the Charge of Electricity by the Ions Produced by Rongen Rays, Phil. Mag. Ser.5 **46**, 528 (1898). https://doi.org/10.1080/14786449808621229
39. H. Geiger, W. Muller, Das Elektronenzählrohr, Physikalische Zeitschrift **29**, 839 (1928)
40. S. Townsend, J, The theory of ionization of gases by collision, Nature **85**, 400 (1911). https://doi.org/10.1038/085400b0
41. S. Curran, *Luminescence and the Scintillation Counter* (Academic Press, 1953)
42. P.V. Heerden, A new apparatus in nuclear physics for the investigation of β- and γ-rays. part i, Physica **16**(6), 505 (1950). https://doi.org/10.1016/0031-8914(50)90007-3
43. P.V. Heerden, J. Milatz, A new apparatus in nuclear physics for the investigation of β- and γ-rays. part II, Physica **16**(6), 517 (1950). https://doi.org/10.1016/0031-8914(50)90008-5
44. M. Curie, P. Curie, G. Bemont, Sur une nouvelle substance fortement radio-active, contenu dans la pechblende, Comptes Rendus **127**, 1215–1217 (1898)
45. W. Libby, Atmospheric helium-3 and radiocarbon from cosmic radiation, Phys. Rev. **69**, 671 (1946). https://doi.org/10.1103/physrev.69.671.2
46. G. Gamow, On the quantum theory of the atomic nucleus, Zeitschrift für Physik **51**, 204 (1928)
47. H. Geiger, J. Nuttall, The Ranges of the Alpha Particles from Various Radioactive Substances and a Relation between Range and Period of Transformation, Phil. Mag. Ser. 6 **22**, 613 (1911). https://doi.org/10.1080/14786441008637156
48. G. Royer, Alpha Emission and Spontaneous Fission through Quasi-molecular Shapes, J. Phys. G. **26**, 1149 (2000). https://doi.org/10.1088/0954-3899/26/8/305
49. H. Viola, G. Seaborg, Nuclear Systematics of the Heavy Elements II. Lifetimes for alpha, beta and spontaneous fission decay, J. Inorg. Nucl. Chem. **28**, 741 (1966). https://doi.org/10.1016/0022-1902(66)80412-8
50. T. Don, Z. Ren, New calculations of alpha-decay half-lives by the Viola-Seaborg formula, Eur. Phys. J. A **26**, 69 (1966). https://doi.org/10.1140/epja/i2005-10142-y
51. G. Neary, The Beta-ray Spectrum of Radium E, Proc. Roy. Soc. A **175**, 71 (1940). https://doi.org/10.1098/rspa.1940.0044
52. C. Cowan, et al., Detection of the Free Neutrino: A Confirmation, Science **124**, 103 (1956). https://doi.org/10.1126/science.124.3212.103
53. E. Fermi, Tentativo di una teoria dei raggi β, Il Nuovo Cimento **11**, 1 (1934). https://doi.org/10.1007/BF02959820
54. F. Kurie, On the Use of the Kurie Plot, Phys. Rev. **73**, 1207 (1948). https://doi.org/10.1103/PhysRev.73.1207
55. M. Aker, et al., Improved Upper Limit on the Neutrino Mass from a Direct Kinematic Method by KATRIN, Phys. Rev. Lett. **123**, 221802 (2019). https://doi.org/10.1103/PhysRevLett.123.221802
56. E. Kearns, et al., Detecting Massive Neutrinos, Scientific American **281**, 64 (1999). https://doi.org/10.1103/PhysRevL
57. G. Wick, Sugli Elementi Radioattivi di F. Joliot el. Curie, Rend. Accademia dei Lincei **19**, 319 (1934)
58. L. Alvarez, Nuclear K capture, Phys. Rev. D **52**, 134 (1937). https://doi.org/10.1103/PhysRev.52.134
59. T. Lee, C. Yang, Question of Parity Conservation in Weak Interactions, Phys. Rev. **104**, 254 (1956). https://doi.org/10.1103/PhysRev.104.254
60. C. Wu, et al., Experimental Test of Parity Conservation in Beta-decays, Phys. Rev. **105**, 1413 (1957). https://doi.org/10.1103/PhysRev.105.1413
61. R. Arnold, et al., Measurement of the Double-beta Half-life and Search for the Neutrinoless Double Beta Decay of Calcium-48 with the NEMO-3 Detector , Phys. Rev. **D93**, 112008 (2016). https://doi.org/10.1103/PhysRevD.93.112008
62. P. Dirac, The Quantum Theory of Emission and Absorption of Radiation, Proc. Roy. Soc. A **114**, 243–265 (1898). https://doi.org/10.1098/rspa.1927.0039

63. V.F. Weisskopf, Radiative transition probabilities in nuclei, Phys. Rev. **83**, 1073 (1951). https://doi.org/10.1103/PhysRev.83.1073. https://link.aps.org/doi/10.1103/PhysRev.83.1073

64. R. Weiner, Nuclear isomeric shift on spectral lines, Nuov. Cim. **iV**, 1587 (1956). https://doi.org/10.1007/BF02746390

65. S. DeBenedetti, G. Lang, R. Ingalls, Electric quadrupole splitting and the nuclear volume effect in the ions of fe^{57}, Phys. Rev. Lett. **6**, 60 (1961). https://doi.org/10.1103/PhysRevLett.6.60. https://link.aps.org/doi/10.1103/PhysRevLett.6.60

66. R. Mossbauer, Kernresonanzabsorption von gammastrahlung in ir191, Zeitschrift fur Naturforschung A **14**, 211 (1958). https://doi.org/10.1515/zna-1959-0303

67. R.V. Pound, G.A. Rebka, Apparent weight of photons, Phys. Rev. Lett. **4**, 337 (1960). https://doi.org/10.1103/PhysRevLett.4.337. https://link.aps.org/doi/10.1103/PhysRevLett.4.337

68. R.V. Pound, J.L. Snider, Effect of gravity on nuclear resonance, Phys. Rev. Lett. **13**, 539 (1964). https://doi.org/10.1103/PhysRevLett.13.539. https://link.aps.org/doi/10.1103/PhysRevLett.13.539

69. J. Cockroft, E. Walton, Experiments with high energy positiv ions ii.- the disintegration of elements by high energy protons, Proc. Roy. Soc. Lond. **A137**, 229 (1932). https://doi.org/10.1098/rspa1932.0133

70. E. Fermi, Possible production of elements of atomic number higher than 92, Nature **133**, 898 (1934). https://doi.org/10.1038/133898a0

71. I. Noddack, Über das element 93, Angewandte Chemie **47**, 653 (1934). https://doi.org/10.1002/ange.19340473707

72. O. Hahn, F. Strassman, Über den nachweis und das verhalten der bei der bestrahlung des urans mittels neutronen entstehenden erdalkalimetalle, Naturwissenschaft **27**, 11 (1939). https://doi.org/10.1007/BF01488241

73. N. Bohr, J.A. Wheeler, The mechanism of nuclear fission, Phys. Rev. **56**, 426 (1939). https://doi.org/10.1103/PhysRev.56.426. https://link.aps.org/doi/10.1103/PhysRev.56.426

74. B. Carlson, Ellipsoidal distribution of charge and mass, J.Math. Phys. **2**, 441 (1961). https://doi.org/10.1063/1.1703729

75. G.N. Flerov, K.A. Petrzhak, Spontaneous fission of uranium, Proc. USSR Acad. Sci. **28**, 500 (1940)

76. W. Myers, W. Swiatecki, Nuclear masses and deformations, Nucl. Phys. **81**, 1 (1966). https://doi.org/10.1016/0029-5582(66)90639-0

77. H.L. Anderson, E. Fermi, L. Szilard, Neutron production and absorption in uranium, Phys. Rev. **56**, 284 (1939). https://doi.org/10.1103/PhysRev.56.284. https://link.aps.org/doi/10.1103/PhysRev.56.284

78. C. Chyba, C. Milne, Simple calculation of the critical mass for highly enriched uranium and plutonium-239, American Journal of Physics **82**, 977 (2014). https://doi.org/10.1103/PhysRev.56.426

79. C. Chyba, C. Milne, Simple calculation of the critical mass for highly enriched uranium and plutonium-239, American Journal of Physics **82**, 977 (2014). https://doi.org/10.1119/1.4885379

80. C. Rubbia, A high gain energy amplifier operated with fast neutrons, AIP. Conf. Proc. **346**, 44 (1994)

81. M. Oliphant, P. Harteck, E. Rutherford, Transmutation effects observed with heavy hydrogen, Proc. Roy. Soc. A **144**, 692 (1933). https://doi.org/10.1098/rspa.1934.0077

82. D. Keefe, Inertial confinement fusion, Ann. Rev. Nucl. Part. Sci. **32**, 391 (1982)

83. A. Eddington, Thw internal constitution of the stars, Nature **106**, 14 (1920)

84. H.A. Bethe, Energy production in stars, Phys. Rev. **55**, 103 (1939). https://doi.org/10.1103/PhysRev.55.103. https://link.aps.org/doi/10.1103/PhysRev.55.103

85. C. von Weiszäcker, Über elementumwandiunger in innem der sterne i, Physikalische Zeitschrift **38**, 176 (1937)

86. C. von Weiszäcker, Über elementumwandiunger in innem der sterne ii, Physikalische Zeitschrift **39**, 633 (1938)

87. W. Gerlach, O. Stern, Das magnetische moment des silberatoms, Z. Physik **9**, 352 (1922). https://doi.org/10.1007/BF01326984

88. W. Gerlach, O. Stern, Experimental test of the applicability of the quantum their to the magnetic field, Z. Phys. **9**, 349 (1922). https://doi.org/10.1007/BF01326983

89. R.P. Feynman, Space-time approach to quantum electrodynamics, Phys. Rev. **76**, 769 (1949). https://doi.org/10.1103/PhysRev.76.769. https://link.aps.org/doi/10.1103/PhysRev.76.769

90. J. Schwinger, Quantum electrodynamics 1. a covariant formulation, Phys. Rev. **74**, 1439 (1948). https://doi.org/10.1103/PhysRev.74.1439. https://link.aps.org/doi/10.1103/PhysRev.74.1439

91. S. Tomonaga, On a relativistically invariant formulation of the quantum theory of wave fields, Progress of Theorertical Physics **1**, 27 (1946). https://doi.org/10.1143/PTP.1.27

92. S.L. Glashow, Partial-symmetries of weak interactions, Nuclear Physics **22**(4), 579 (1961). https://doi.org/10.1016/0029-5582(61)90469-2. http://www.sciencedirect.com/science/article/pii/0029558261904692

93. S. Weinberg, A model of leptons, Phys. Rev. Lett. **19**, 1264 (1967). https://doi.org/10.1103/PhysRevLett.19.1264. https://link.aps.org/doi/10.1103/PhysRevLett.19.1264

94. A. Salam, J. Ward, Weak and electromagnetic interactions, Il Nuovo Cimento pp. 568–577 (1959). https://doi.org/10.1007/BF02726525

95. R. Aaij, et al., Test of lepton universality in beauty-quark decays (arXiv:2103.11769. LHCB-PAPER-2021-004) (2021). https://cds.cern.ch/record/2758740

96. G. Occhialini, C. Powell, Nuclear disintegrations produced by slow charged particles of low mass, Nature **159**, 186 (1947). https://doi.org/10.1038/159186a0

97. B.P. Abbott, et al., Observation of gravitational waves from a binary black hole merger, Phys. Rev. Lett. **116**, 061102 (2016). https://doi.org/10.1103/PhysRevLett.116.061102. https://link.aps.org/doi/10.1103/PhysRevLett.116.061102

98. R. Barate, Others, Study of muon-pair production at centre-of-mass energies from 20 to 136 gev with the aleph detector, Physics Letters B **399**(3), 329 (1997). https://doi.org/10.1016/S0370-2693(97)00353-5. http://www.sciencedirect.com/science/article/pii/S0370269397003535

99. C.D. Anderson, The positive electron, Phys. Rev. **43**, 491 (1933). https://doi.org/10.1103/PhysRev.43.491. https://link.aps.org/doi/10.1103/PhysRev.43.491

100. S.H. Neddermeyer, C.D. Anderson, Note on the nature of cosmic-ray particles, Phys. Rev. **51**, 884 (1937). https://doi.org/10.1103/PhysRev.51.884. https://link.aps.org/doi/10.1103/PhysRev.51.884

101. S. Mandelstam, Determination of the pion-nucleon scattering amplitude from dispersion relations and unitarity. general theory, Phys. Rev. **112**, 1344 (1958). https://doi.org/10.1103/PhysRev.112.1344. https://link.aps.org/doi/10.1103/PhysRev.112.1344

102. D.A. Glaser, Some Effects of Ionizing Radiation on the Formation of Bubbles in Liquids, Phys. Rev. **87**, 665 (1952). https://doi.org/10.1103/PhysRev.87.665

103. W. Abdallah, A. Hammad, A. Kasem, S. Khalil, Long-lived $b - l$ symmetric ssm particles at the lhc, Phys. Rev. D **98**, 095019 (2018). https://doi.org/10.1103/PhysRevD.98.095019. https://link.aps.org/doi/10.1103/PhysRevD.98.095019

104. F. Cavallari, Performance of calorimeters at the lhc, J. PHys. Conf. Ser. **293**, 012001 (2011). https://doi.org/10.1103/PhysRev.51.884

105. H. Yukawa, On the interaction of elementary particles, Proc. Phys. Math. Soc. Jap. **17**, 48 (1935)

106. M. Gell-Mann, A schematic model of baryons and mesons, Physics Letters **8**(3), 214 (1964). https://doi.org/10.1016/S0031-9163(64)92001-3. http://www.sciencedirect.com/science/article/pii/S0031916364920013

107. G. Zweig, An su(3) model for strong interaction symmetry and its breakiing, Developments of the Quark Theory of Hadrons **1**, 22 (1964)

108. V.E. Barnes, et al., Observation of a hyperon with strangeness minus three, Phys. Rev. Lett. **12**, 204 (1964). https://doi.org/10.1103/PhysRevLett.12.204. https://link.aps.org/doi/10.1103/PhysRevLett.12.204

109. M. Gell-Mann, Symmetries of baryons and mesons, Phys. Rev. **125**(3), 1067 (1961). https://doi.org/10.1103/PhysRev.125.1067

110. Y. Ne'eman, Derivation of strong interactions from a gauge invariance, Nuclear Physics **26**(2), 222 (1961). https://doi.org/10.1016/0029-5582(61)90134-1. http://www.sciencedirect.com/science/article/pii/0029558261901341

111. S.L. Glashow, J. Iliopoulos, L. Maiani, Weak interactions with lepton-hadron symmetry, Phys. Rev. D **2**, 1285 (1970). https://doi.org/10.1103/PhysRevD.2.1285. https://link.aps.org/doi/10.1103/PhysRevD.2.1285

112. J.E. Augustin, et al., Discovery of a narrow resonance in e^+e^- annihilation, Phys. Rev. Lett. **33**, 1406 (1974). https://doi.org/10.1103/PhysRevLett.33.1406. https://link.aps.org/doi/10.1103/PhysRevLett.33.1406

113. J.J. Aubert, et al., Experimental observation of a heavy particle j, Phys. Rev. Lett. **33**, 1404 (1974). https://doi.org/10.1103/PhysRevLett.33.1404. https://link.aps.org/doi/10.1103/PhysRevLett.33.1404

114. S.W. Herb, et al., Observation of a dimuon resonance at 9.5 gev in 400-gev proton-nucleus collisions, Phys. Rev. Lett. **39**, 252 (1977). https://doi.org/10.1103/PhysRevLett.39.252. https://link.aps.org/doi/10.1103/PhysRevLett.39.252

115. F. Abe, et al., Observation of top quark production in $\bar{p}p$ collisions with the collider detector at fermilab, Phys. Rev. Lett. **74**, 2626 (1995). https://doi.org/10.1103/PhysRevLett.74.2626. https://link.aps.org/doi/10.1103/PhysRevLett.74.2626

116. R. Aaij, et al., Observation of $j/\psi p$ resonances consistent with pentaquark states in $\Lambda_b^0 \to j/\psi K^- p$ decays, Phys. Rev. Lett. **115**, 072001 (2015). https://doi.org/10.1103/PhysRevLett.115.072001. https://link.aps.org/doi/10.1103/PhysRevLett.115.072001

117. Aaij, et al., Observation of a narrow pentaquark state, $P_c(4312)^+$, and of the two-peak structure of the $P_c(4450)^+$, Phys. Rev. Lett. **122**, 222001 (2019). https://doi.org/10.1103/PhysRevLett.122.222001. https://link.aps.org/doi/10.1103/PhysRevLett.122.222001

118. T. Xiao, S. Dobbs, A. Tomaradze, K.K. Seth, Observation of the charged hadron $z_c^\pm(3900)$ and evidence for the neutral $z_c^0(3900)$ in $e^+e^- \to \pi\pi j/\psi$ at s=4170mev, Physics Letters B **727**(4), 366 (2013). https://doi.org/10.1016/j.physletb.2013.10.041. http://www.sciencedirect.com/science/article/pii/S0370269313008484

119. M. Ablikim, M.N. Achasov, X.C. Ai, O. Albayrak, D.J. Ambrose, et al., Observation of a charged charmoniumlike structure in $e^+e^- \to \pi^+\pi^- j/\psi$ at $\sqrt{s}$=4.26 GeV, Phys. Rev. Lett. **110**, 252001 (2013). https://doi.org/10.1103/PhysRevLett.110.252001. https://link.aps.org/doi/10.1103/PhysRevLett.110.252001

120. Z.Q. Liu, et al., Study of $e^+e^- \to \pi^+\pi^- j/\psi$ and observation of a charged charmoniumlike state at belle, Phys. Rev. Lett. **110**, 252002 (2013). https://doi.org/10.1103/PhysRevLett.110.252002. https://link.aps.org/doi/10.1103/PhysRevLett.110.252002

121. R. Aaij, et al. Observation of structure in the j/ψ-pair mass spectrum (2020), arXiv:2006.16957

122. R. Aaij, et al. A model-independent study of resonant structure in $b^+ \to d^+d^-k^+$ decays (2020), arXiv:2009.00025

123. N. Cabibbo, Unitary symmetry and leptonic decays, Phys. Rev. Lett. **10**, 531 (1963). https://doi.org/10.1103/PhysRevLett.10.531. https://link.aps.org/doi/10.1103/PhysRevLett.10.531

124. M. Kobayashi, T. Maskawa, CP-Violation in the Renormalizable Theory of Weak Interaction, Progress of Theoretical Physics **49**(2), 652 (1973). https://doi.org/10.1143/PTP.49.652

125. Electroweak measurements in electron–positron collisions at w-boson-pair energies at LEP, Physics Reports **532**(4), 119 (2013). https://doi.org/10.1016/j.physrep.2013.07.004

126. P.W. Higgs, Broken Symmetries and the Masses of Gauge Bosons, Phys. Rev. Lett. **13**, 508 (1964). https://doi.org/10.1103/PhysRevLett.13.508

127. F. Englert, R. Brout, Broken Symmetry and the Mass of Gauge Vector Mesons, Phys. Rev. Lett. **13**, 321 (1964). https://doi.org/10.1103/PhysRevLett.13.321

128. G.S. Guralnik, C.R. Hagen, T.W.B. Kibble, Global Conservation Laws and Massless Particles, Phys. Rev. Lett. **13**, 585 (1964). https://doi.org/10.1103/PhysRevLett.13.585

129. G. 't Hooft, Renormalizable Lagrangians for Massive Yang-Mills Fields, Nucl. Phys. B **35**, 167 (1971). https://doi.org/10.1016/0550-3213(71)90139-8

130. D. de Florian, et al., Handbook of LHC Higgs Cross Sections: 4. Deciphering the Nature of the Higgs Sector (2016). https://doi.org/10.2172/1345634, https://doi.org/10.23731/CYRM-2017-002

131. A.M. Sirunyan, et al. Evidence for Higgs boson decay to a pair of muons (2020), arXiv:2009.04363

132. R. Miskimen, Neutral pion decay, Annual Review of Nuclear and Particle Science **61**, 1 (2011). DOI https://doi.org/10.1146/annurev-nucl-102010-130426

133. P.A. Zyla, et al., Review of Particle Physics, PTEP **2020**(8), 083C01 (2020). https://doi.org/10.1093/ptep/ptaa104

134. K. Olive, et al., Review of Particle Physics, Chin. Phys. C **38**, 090001 (2014). https://doi.org/10.1088/1674-1137/38/9/090001

135. J. Sakurai, Theory of strong interactions, Annals of Physics **11**(1), 1 (1960). https://doi.org/10.1016/0003-4916(60)90126-3. http://www.sciencedirect.com/science/article/pii/0003491660901263

136. C.G. Callan, D.J. Gross, High-energy electroproduction and the constitution of the electric current, Phys. Rev. Lett. **22**, 156 (1969). https://doi.org/10.1103/PhysRevLett.22.156. https://link.aps.org/doi/10.1103/PhysRevLett.22.156

137. J.D. Bjorken, E.A. Paschos, Inelastic electron-proton and γ-proton scattering and the structure of the nucleon, Phys. Rev. **185**, 1975 (1969). https://doi.org/10.1103/PhysRev.185.1975. https://link.aps.org/doi/10.1103/PhysRev.185.1975

138. H. Fritzsch, M. Gell-Mann, H. Leutwyler, Advantages of the color octet gluon picture, Physics Letters B **47**(4), 365 (1973). https://doi.org/10.1016/0370-2693(73)90625-4. http://www.sciencedirect.com/science/article/pii/0370269373906254

139. H.D. Politzer, Reliable perturbative results for strong interactions?, Phys. Rev. Lett. **30**, 1346 (1973). https://doi.org/10.1103/PhysRevLett.30.1346. https://link.aps.org/doi/10.1103/PhysRevLett.30.1346

140. D.J. Gross, F. Wilczek, Ultraviolet behavior of non-abelian gauge theories, Phys. Rev. Lett. **30**, 1343 (1973). https://doi.org/10.1103/PhysRevLett.30.1343. https://link.aps.org/doi/10.1103/PhysRevLett.30.1343

141. V. Kher, A.K. Rai, Spectroscopy and decay properties of charmonium, Chinese Physics C **42**(8), 083101 (2018). https://doi.org/10.1088/1674-1137/42/8/083101

142. J. Ellis, M.K. Gaillard, G.G. Ross, Search for gluons in e^+e^- annihilation, Nuclear Physics B **111**(2), 253 (1976). https://doi.org/10.1016/0550-3213(76)90542-3. http://www.sciencedirect.com/science/article/pii/0550321376905423

143. R. Brandelik, et al., Evidence for planar events in e^+e^- annihilation at high energies, Physics Letters B **86**(2), 243 (1979). https://doi.org/10.1016/0370-2693(79)90830-X. http://www.sciencedirect.com/science/article/pii/037026937990830X

144. D. Buskulic, et al., Studies of qcd in $e^+e^- \rightarrow$ hadrons at e_{cm} = 130 and 136 gev, Zeit. Phys. C **73**, 409 (1997). https://doi.org/10.1007/s002880050330

145. Y. Dokshitser, Calculation of structure functions of deep-inelastic scattering - and e +e - annihilation by perturbation theory in quantum chromodynamics, Sov. Phys. JETP **73**, 641 (1977)

146. V. Gribov, L. Lipatov, Deep inelastic e p scattering in perturbation theory, Sov. J. Nucl. Phys. **15**, 438 (1972)

147. G. Altarelli, G. Parisi, Asymptotic freedom in parton language, Nuclear Physics B **126**(2), 298 (1977). https://doi.org/10.1016/0550-3213(77)90384-4. http://www.sciencedirect.com/science/article/pii/0550321377903844

148. A. Martin, et al., Parton distributions for the LHC, Eur. Phys. J. **C63**, 189–285 (2009). https://doi.org/10.1140/epjc/s10052-009-1072-5

149. R. Ellis, J. Sexton, QCD Radiative Corrections to Parton Parton Scattering, Nucl. Phys. B **269**, 445 (1986). https://doi.org/10.1016/0550-3213(86)90232-4

150. G. Latino, QCD physics at the Tevatron, Frascati Phys. Ser. **34**, 217 (2004)

151. S. Glazov, Measurement of dis cross section at hera, Braz. J. Phys. (2007). https://doi.org/10. 1590/S0103-97332007000500030

152. K.G. Wilson, Confinement of quarks, Phys. Rev. D **10**, 2445 (1974). https://doi.org/10.1103/ PhysRevD.10.2445. https://link.aps.org/doi/10.1103/PhysRevD.10.2445

153. M. Creutz, Confinement and lattice gauge theory, Physica Scripta **23**(5B), 973 (1981). https:// doi.org/10.1088/0031-8949/23/5b/011

154. F.A. Browder, Thomas E. Harris, J. Kumar (eds.). *Proceedings, 5th International Workshop on Charm Physics (Charm 2012): Honolulu, Hawaii, USA, May 14-17, 2012* (2012)

155. Z. Bai, N.H. Christ, , C.T. Sachrajda, $K_L - K_S$ mass difference, EPJ Web of Conferences **175**, 13017 (2018). https://doi.org/10.1051/epjconf/201817513016

156. M. Gell-Mann, A. Pais, Behavior of neutral particles under charge conjugation, Phys. Rev. **97**, 1387 (1955). https://doi.org/10.1103/PhysRev.97.1387. https://link.aps.org/doi/10.1103/ PhysRev.97.1387

157. M. Bardon, K. Lande, L. Lederman, W. Chinowsky, Long-lived neutral k mesons, Annals of Physics **5**(2), 156 (1958). https://doi.org/10.1016/0003-4916(58)90048-4. http://www. sciencedirect.com/science/article/pii/0003491658900484

158. H.D. Schulz, Observation of b0b0 oscillations, Nuclear Physics A **478**, 283 (1988). https:// doi.org/10.1016/0375-9474(88)90858-5. http://www.sciencedirect.com/science/article/pii/ 0375947488908585

159. B. Aubert, et al., Evidence for $D^0 - \overline{d}^0$ mixing, Phys. Rev. Lett. **98**, 211802 (2007). https:// doi.org/10.1103/PhysRevLett.98.211802. https://link.aps.org/doi/10.1103/PhysRevLett.98. 211802

160. M. Starič, et al., Evidence for $D^0 - \overline{d}^0$ mixing, Phys. Rev. Lett. **98**, 211803 (2007). https:// doi.org/10.1103/PhysRevLett.98.211803. https://link.aps.org/doi/10.1103/PhysRevLett.98. 211803

161. J.H. Christenson, J.W. Cronin, V.L. Fitch, R. Turlay, Evidence for the 2π decay of the k_2^0 meson, Phys. Rev. Lett. **13**, 138 (1964). https://doi.org/10.1103/PhysRevLett.13.138. https:// link.aps.org/doi/10.1103/PhysRevLett.13.138

162. H. Burkhardt, et al., First Evidence for Direct CP Violation, Phys. Lett. B **206**, 169 (1988). https://doi.org/10.1016/0370-2693(88)91282-8

163. A.D. Sakharov, Violation of cp invariance, c asymmetry, and baryon asymmetry of the universe, ZhETF Pis'ma **5**, 32–35 (1967). https://doi.org/10.1070/ PU1991v034n05ABEH002497

164. B. Aubert, et al., Observation of *CP* violation in the b^0 meson system, Phys. Rev. Lett. **87**, 091801 (2001). https://doi.org/10.1103/PhysRevLett.87.091801. https://link.aps.org/doi/10. 1103/PhysRevLett.87.091801

165. R. Aaij, et al., Observation of cp violation in charm decays, Phys. Rev. Lett. **122**, 211803 (2019). https://doi.org/10.1103/PhysRevLett.122.211803. https://link.aps.org/doi/10.1103/ PhysRevLett.122.211803

166. S.W. Herb, D.C. Hom, L.M. Lederman, J.C. Sens, H.D. Snyder, J.K. Yoh, J.A. Appel, B.C. Brown, C.N. Brown, W.R. Innes, K. Ueno, T. Yamanouchi, A.S. Ito, H. Jöstlein, D.M. Kaplan, R.D. Kephart, Observation of a dimuon resonance at 9.5 gev in 400-gev proton-nucleus collisions, Phys. Rev. Lett. **39**, 252 (1977). https://doi.org/10.1103/PhysRevLett.39. 252. https://link.aps.org/doi/10.1103/PhysRevLett.39.252

167. Y. Fukuda, et al., Evidence for oscillation of atmospheric neutrinos, Phys. Rev. Lett. **81**, 1562 (1998). https://doi.org/10.1103/PhysRevLett.81.1562

168. B. Pontecorvo, Mesonium and anti-mesonium, Sov. Phys. JETP **6**, 429 (1957)

169. G. Bertone, D. Hooper, J. Silk, Particle dark matter: Evidence, candidates and constraints, Phys. Rept. **405**, 279 (2005). https://doi.org/10.1016/j.physrep.2004.08.031

170. N. Aghanim, et al., Planck 2018 results. V. CMB power spectra and likelihoods, Astron. Astrophys. **641**, A5 (2020). https://doi.org/10.1051/0004-6361/201936386

171. H. Georgi, H.R. Quinn, S. Weinberg, Hierarchy of interactions in unified gauge theories, Phys. Rev. Lett. **33**, 451 (1974). https://doi.org/10.1103/PhysRevLett.33.451. https://link.aps.org/ doi/10.1103/PhysRevLett.33.451

172. M. Miura, Search for nucleon decay in super-kamiokande, Nuclear and Particle Physics Proceedings **273-275**, 516 (2016). https://doi.org/10.1016/j.nuclphysbps.2015.09.076. https://www.sciencedirect.com/science/article/pii/S2405601415005659. 37th International Conference on High Energy Physics (ICHEP)

173. A. Belyaev, M.S. Brown, R. Foadi, M.T. Frandsen, The Technicolor Higgs in the Light of LHC Data, Phys. Rev. **D90**, 035012 (2014). https://doi.org/10.1103/PhysRevD.90.035012

174. B. Abi, et al., Measurement of the positive muon anomalous magnetic moment to 0.46 ppm, Phys. Rev. Lett. **126**, 141801 (2021). https://doi.org/10.1103/PhysRevLett.126.141801. https://link.aps.org/doi/10.1103/PhysRevLett.126.141801

Index

© The Author(s), under exclusive license to Springer Nature Switzerland AG 2021
A. Belyaev, D. Ross, *The Basics of Nuclear and Particle Physics*, Undergraduate
Texts in Physics, https://doi.org/10.1007/978-3-030-80116-8

Printed in the United States
by Baker & Taylor Publisher Services